Monographs on Theoretical and Applied Genetics 15

Monographs on Theoretical and Applied Genetics

Volume 1 **Meiotic Configurations**
A Source of Information for Estimating Genetic Parameters
By J. Sybenga (1975)

Volume 2 **Pollination Mechanisms, Reproduction and Plant Breeding**
By R. Frankel and E. Galun (1977)

Volume 3 **Incompatibility in Angiosperms**
By D. de Nettancourt (1977)

Volume 4 **Gene Interactions in Development**
By L. I. Korochkin (1981)

Volume 5 **The Molecular Theory of Radiation Biology**
By K. H. Chadwick and H. P. Leenhouts (1981)

Volume 6 **Heterosis**
Reappraisal of Theory and Practice
Editor: R. Frankel (1983)

Volume 7 **Induced Mutations in Plant Breeding**
By W. Gottschalk and G. Wolff (1983)

Volume 8 **Protoplast Fusion**
Genetic Engineering in Higher Plants
By Y. Y. Gleba and K. M. Sytnik (1984)

Volume 9 **Petunia**
Editor: K. C. Sink (1984)

Volume 10 **Male Sterility in Higher Plants**
By M. L. H. Kaul (1988)

Volume 11 **Tree Breeding: Principles and Strategies**
By G. Namkoong, H. C. Kang, and J. S. Brouard (1988)

Volume 12 **The Wheat Rusts – Breeding for Resistance**
By D. R. Knott (1989)

Volume 13 **Population Dynamics and the Tribolium Model**
Genetics and Demography
By R. F. Costantino and R. A. Desharnais (1991)

Volume 14 **Genetic Improvement of Tomato**
Editor: G. Kalloo (1991)

Volume 15 **Evolution and Function of Heterostyly**
Editor: S. C. H. Barrett (1992)

Spencer C. H. Barrett (Ed.)

Evolution and Function of Heterostyly

With 63 Figures

Springer-Verlag Berlin Heidelberg GmbH

Professor SPENCER C. H. BARRETT
University of Toronto
Department of Botany
Toronto, Ontario
Canada M5S 3B2

DOI 10.1007/978-3-642-86656-2

Library of Congress Cataloging-in-Publication Data. Evolution and function of heterostyly / Spencer C. H. Barrett, ed. p. cm. – (Monographs on theoretical and applied genetics; 15) Includes bibliographical references and index. 1. Heterostylism. 2. Plants – Evolution. I. Series. QK659.E86 1992 582'.038 – dc29 91-29490

Originally published by Springer-Verlag Berlin Heidelberg New York in 1992.
MyCopy version of the original edition 1992

Reproduction of the figures: Gustav Dreher GmbH, Stuttgart
Typesetting: International Typesetters Inc., Makati, Philippines
31/3145-5 4 3 2 1 0 – Printed on acid-free paper
www.springer.com/mycopy

Preface

"I do not think anything in my scientific life has given me so much satisfaction as making out the meaning of the structure of heterostyled flowers"

CHARLES DARWIN (1876)[1]

The subject of heterostyly has fascinated biologists since Darwin's early observations and experiments on heterostylous species conducted at Down House in Kent during the middle of the last century. Darwin's studies were summarized in his classic book *The Different Forms of Flowers on Plants of the Same Species* published in 1877. Despite the widespread attention that heterostyly has received as a model system for genetic and evolutionary studies, there has been no comprehensive monograph on the subject since the publication of Darwin's book. The present volume is an attempt to remedy this deficiency by providing an up-to-date review of current research on heterostylous breeding systems from an evolutionary perspective.

Heterostylous plants offer excellent opportunities to investigate evolutionary patterns and processes at a variety of levels of biological organization. This is because of the relatively straightforward link between genes, development, morphology and fitness in comparison with many other reproductive adaptations. In this regard the polymorphisms provide unsurpassed opportunities for integrated evolutionary studies involving experimental field and laboratory techniques for the study of plant adaptation. The diversity of approaches used to study heterostyly are reflected in the topics covered in this volume written by leading authorities in the field of plant reproductive biology. Chapters on the early history, structure, development, genetics, evolution, adaptive significance and population biology of heterostylous plants are represented. In addition, the importance of pollen-pistil interactions, gametophytic competition and the optimal allocation of resources to female and male sex function is also covered along with a review of the ways that heterostyly has been evolutionarily modified into other breeding systems. Some chapters synthesize a large and often scattered literature; others provide new theoretical insights and experimental data not previously published.

This project would not have got off the ground if it were not for the enthusiasm of Marc Nicholls. On behalf of the authors I thank him for his early interest and initiative. Finally, I thank the authors for their patience and assistance in preparing this volume, Brenda Casper and Robert Ornduff for their generosity of spirit and Suzanne Barrett, William Cole and the staff at Springer-Verlag for their editorial help.

SPENCER C.H. BARRETT

[1] from Charles Darwin's autobiographical recollections; see "The Autobiography of Charles Darwin and Selected Letters" edited by F. Darwin (1958) Dover Publications Inc., New York

Contents

Chapter 8 The Application of Sex Allocation Theory to Heterostylous Plants
B. B. CASPER

Chapter 9 Pollen Competition in Heterostylous Plants
M. A. MCKENNA

Chapter 10 Evolutionary Modifications of Tristylous Breeding Systems
S. G. WELLER

Chapter 1

Heterostylous Genetic Polymorphisms: Model Systems for Evolutionary Analysis

S.C.H. Barrett[1]

1 Introduction

Heterostyly is a genetic polymorphism in which plant populations are composed of two (distyly) or three (tristyly) morphs that differ reciprocally in the heights of stigmas and anthers in flowers (Fig. 1). The style-stamen polymorphism is usually accompanied by a sporophytically controlled, diallelic self-incompatibility system that prevents self- and intramorph fertilizations, and a suite of ancillary morphological polymorphisms, particularly of the stigmas and pollen of floral morphs. Heterostyly is reported from approximately 25 angiosperm families and has usually been viewed as a floral device that promotes outcrossing, hence reducing the harmful effects of close inbreeding in plant populations.

Since the pioneering work of Darwin and Hildebrand in the last century (see Chap. 2), evolutionary biologists have been intrigued by the complex sexual arrangements of reproductive organs in heterostylous plants. "In their manner of fertilisation" Darwin wrote of *Lythrum* species (Darwin 1865), "these plants offer a more remarkable case than can be found in any other plant or animal." How heterostyly originated, what selective forces maintain the polymorphism, and why it often becomes evolutionarily modified into other breeding systems are questions often posed by workers investigating heterostylous groups. The attention heterostyly has received during this century, considering its infrequent occurrence, resides in several outstanding features that has made it a model system for addressing a variety of questions in evolutionary biology.

First, heterostyly is a simply inherited polymorphism in which the floral morphs are easily identified under field conditions. Population studies using ecological genetic approaches (Ford 1964) therefore offer attractive opportunities for investigations of the natural selection, maintenance, and breakdown of heterostyly (Crosby 1949; Bodmer 1960; Weller 1976a; Barrett 1985a). Second, experimental field studies of the pollination biology of heterostylous plants have enabled analysis of the function and adaptive significance of the polymorphism (Ganders 1979; Barrett 1990). Studies of this type are facilitated by the limited number of mating groups in heterostylous populations and the conspicuous size differences of pollen produced by the floral morphs. These features enable measurements of a variety of reproductive processes associated with pollen transport that are more difficult to investigate in monomorphic species (Barrett and Wolfe 1986). Finally, because of the more

[1]Department of Botany, University of Toronto, Toronto, Ontario, Canada M5S 3B2

Monographs on Theoretical and Applied Genetics 15
Evolution and Function of Heterostyly (ed. by S.C.H. Barrett)

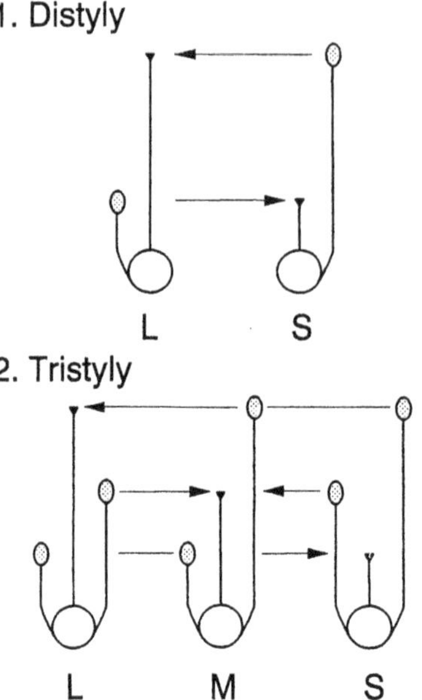

Fig. 1. The heterostylous genetic polymorphisms distyly and tristyly. Legitimate (compatible) pollinations are indicated by the *arrows*, other pollen-pistil combinations are termed illegitimate and usually result in reduced or no seed set. L, M, and S refer to the long-, mid- and short-styled morphs, respectively. Distyly is controlled by a single locus with two alleles. The L morph is usually of genotype *ss* and the S morph *Ss*. In tristyly, the most common mode of inheritance involves two diallelic loci (*S* and *M*), with *S* epistatic to *M*. See Chapter 5 for further details of the inheritance of heterostyly

obvious links between genes, development, morphology, and fitness, in comparison with most other reproductive adaptations, the polymorphism provides opportunities for integrated studies in genetics, development, and population biology. Thus, heterostylous plants provide a rich source of material for evolutionary biologists and represent one of the classic research paradigms for neo-Darwinian approaches to the study of evolution and adaptation.

In this chapter the main themes covered in the book are briefly introduced in the order in which they appear. I begin by examining the nature of heterostyly, its morphological and developmental characteristics, and how it is inherited. This is followed by a consideration of models for the evolution and selection of heterostyly and discussion of the functional basis and reproductive consequences of the floral polymorphisms. The evolutionary breakdown of heterostyly is then reviewed and the chapter concludes by outlining research avenues likely to prove profitable in the future. Throughout, an attempt is made to highlight contrasting viewpoints, cover literature not dealt with in other chapters, and raise unanswered questions to assist forthcoming work on heterostyly.

2 Nature and Occurrence

Research workers differ in opinion regarding the types of variation considered essential for defining a given species as heterostylous. For Darwin and most subsequent workers, particularly those with a genetical perspective, the term heterostyly has usually been reserved for plants with both a reciprocal arrangement of stigma and anther heights (hereafter reciprocal herkogamy) and a diallelic incompatibility system (as illustrated in Fig. 1). Following this view, the litmus test for proof of the occurrence of "true" heterostyly has been the demonstration, by controlled pollinations, of the presence of an intramorph incompatibility system in a species with reciprocal herkogamy. Early on, however, Hildebrand (1866) used the term heterostyly in a strictly morphological sense, and because of recent discoveries, discussed below, there seem to be good grounds for using this approach.

Although the majority of heterostylous plants possess reciprocal herkogamy, diallelic incompatibility, and various ancillary floral polymorphisms, research over the last few decades (reviewed in Barrett and Richards 1990) has revealed a significant number of cases where plants with style length polymorphisms exhibit various combinations of heterostylous and "non-heterostylous characters". The latter include strong self-compatibility, multiallelic incompatibility, monomorphic stamen heights, and an absence of ancillary polymorphisms (Table 1). In some cases taxa with unusual character combinations are related to heterostylous taxa (e.g., *Linum grandiflorum*); in other cases (e.g., *Epacris impressa*) they are not. Because of the spectrum of variation associated with plants displaying style length polymorphisms, it would seem to make more sense to reserve the term heterostyly for species that are polymorphic for a reciprocal arrangement of stigma and anther heights at the population level. At the same time, however, it should be recognized that reciprocal herkogamy can vary greatly in expression (J.H. Richards, D.G. Lloyd, and S.C.H. Barrett, unpubl. data), be associated with various compatibility and incompatibility systems, and need not be accompanied by a suite of ancillary floral polymorphisms.

The number of families containing heterostylous species has grown with increased botanical exploration, particularly of tropical regions. On the other hand, many species originally reported as heterostylous have on closer examination proven to be otherwise (e.g., *Mirabilis, Phlox, Veronica* see Barrett and Richards 1990). Taxonomists working with herbarium specimens have often confused interpopulation discontinuities in floral organ size or developmental variability with heterostyly. Figure 2 illustrates the taxonomic distribution of heterostyly among Dahlgren's superorders of angiosperms (Dahlgren 1980). In Chapter 6, Lloyd and Webb estimate that the polymorphisms are likely to have evolved on at least 23 separate occasions and possibly more if heterostyly has arisen more than once in a family (e.g., Rubiaceae, Anderson 1973).

Two additional families (Ericaceae and Polemoniaceae) may also belong on the list of families containing heterostylous taxa. Recently, R.J. Marquis (unpubl. data) has documented style and stamen length variation in populations of the long-lived shrub *Kalmiopsis leachiana* (Ericaceae) from S. Oregon. Some populations apparently contain two floral morphs differing in style length and stamen height, while others are composed of a single floral morph. Within dimorphic populations style

Table 1. Character combinations in plants with style length polymorphisms

Taxon	Stamen position	Incompatibility	Ancillary polymorphisms	Reference
1. Stylar dimorphism				
Primula vulgaris	Dimorphic	DSI	+++	Darwin (1877)
Linum grandiflorum	Monomorphic*	DSI	++	Darwin (1877)
Villarsia parnassiifolia	Monomorphic	DSI + MSI?	++	Ornduff (1986)
Amsinckia grandiflora	Dimorphic	SC	++	Ornduff (1976)
Quinchamalium chilense	Monomorphic	SC	++	Riveros et al. (1987)
Anchusa officinalis	Monomorphic	MSI	+	Philipp and Schou (1981)
Epacris impressa	Monomorphic	MSI	–	O'Brien and Calder (1989)
Chlorogalum angustifolium	Monomorphic	SC	–	Jernstedt (1982)
2. Stylar trimorphism				
Lythrum salicaria	Trimorphic	DSI	+++	Darwin (1865)
Eichhornia paniculata	Trimorphic	SC	++	Barrett (1985b)
Narcissus triandrus	Trimorphic	MSI	–	S.C.H. Barrett D.G. Lloyd and J. Arroyo (unpubl.data)

DSI = diallelic self-incompatibility, MSI - multiallelic self-incompatibility, SC = self-compatible. Ancillary polymorphisms: +++ well developed, ++ moderately developed, + weakly developed, – absent. *But see chapter 3, Table 1 and Chapter 6, page 166.

length variation is more pronounced than anther height variation and pollen from the two floral morphs is uniform in size. Flowers of *K. leachiana* are atypical for a heterostylous species in being bowl-shaped, although this condition does occur in distylous *Fagopyrum* and *Turnera* (see Chap. 6).

The second putative case of a new heterostylous family involves *Gilia nyensis* (Polemoniaceae). In a floristic treatment of the Polemoniaceae for various western states of the USA, Cronquist reported heterostylous populations of this species from Nye County, Nevada (Cronquist et al. 1984). Recently, D. Wilken (pers. commun.) has investigated these populations and found that they contain long- and short-styled morphs in approximately equal proportions. Studies of the pollen and stigmas of the two morphs failed to reveal any significant dimorphisms.

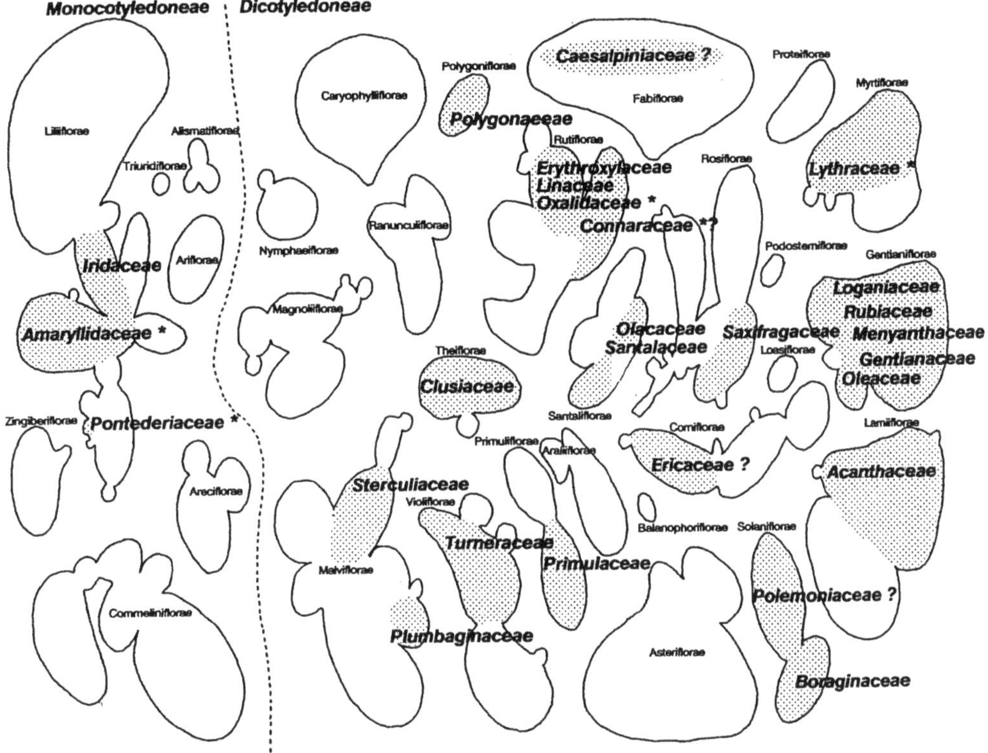

Fig. 2. Taxonomic distribution of heterostyly among the superorders of angiosperms according to Dahlgren's classification of the angiosperms (Dahlgren 1980). Families are positioned within superorders according to Dahlgren's placement of orders and families. Families with tristylous species are indicated by an *asterisk*, those in which the presence of heterostyly needs confirmation are indicated by a *question mark*

3 Structure and Development

The defining features of heterostyly involve morphological polymorphisms. Yet in comparison with the wealth of genetic and ecological work on heterostyly there is a paucity of detailed structural and developmental data for most heterostylous groups. Information on the reciprocal arrangement of stamens and styles and the array of ancillary polymorphisms in heterostylous taxa is comprehensively reviewed by Dulberger in Chapter 3. She points out that population biologists have devoted considerable attention to investigating the adaptive significance of reciprocal herkogamy, with less effort devoted to determining the function of structural differences between pollen grains and stigmas of the floral morphs. Considerable scope would appear to exist for manipulative experiments that investigate the role of stigma and pollen polymorphisms in affecting the capture, hydration, and germination of compatible and incompatible pollen in different heterostylous groups. Dulberger develops earlier views (Mather and de Winton 1941; Dulberger 1975a,b) that the polymorphic properties of pistils and pollen most likely participate directly in the incompatibility mechanism of heterostylous plants, with style length differences involved in the synthesis of incompatibility specifities. While not denying some role for morphological polymorphisms in promoting cross-pollination, she emphasizes their primary function as mechanisms to prevent inbreeding through incompatibility. This unified view of heterostyly is based on the assumption that there is functional integration of morphological, developmental, and biochemical components of the entire syndrome.

A somewhat different perspective on the morphological components of heterostyly is presented by Lloyd and Webb in Chapters 6 and 7, which deal with the evolution and selection of heterostyly, respectively. They follow Yeo (1975) and Ganders (1979) in rejecting a single primary function for the different components of the heterostylous syndrome and instead consider that the function of the various morphological polymorphisms should be considered separately. Lloyd and Webb view the morphological components of heterostyly principally as adaptations that influence different aspects of the pollination process, rather than as a mechanism to avoid inbreeding. Some traits (e.g., reciprocal herkogamy) actively promote cross-pollination, while other polymorphisms (e.g., of pollen and stigmas) function to reduce levels of self-pollination and self-interference.

Attempts to determine the relationships between the structure and function of different polymorphic traits would be aided by developmental studies of heterostylous plants. Despite some early investigations on this topic (Stirling 1932, 1933, 1936; Schaeppi 1935; Bräm 1943) there has been little modern work, using SEM techniques, that has attempted to define the developmental processes responsible for structural differences between the floral morphs, particularly in distylous plants. In Chapter 4, Richards and Barrett review the existing information, primarily from tristylous families, and outline a variety of growth models to account for differences in style and stamen length in heterostylous plants. They suggest that comparative developmental studies of heterostylous groups with inter- and intra-specific floral variation would be valuable to determine whether recurrent associations among heterostylous characters (e.g., long styles, large stigmas, and long stigmatic papillae) are manifestations of common developmental processes. This type of study would be particularly interesting in taxa (e.g., *Palicourea, Amsinckia, Fauria*) where the normal associations are either lacking or reversed in certain species or populations. Anomalies of this type are reviewed by Dulberger in Chapter 3.

Comparative developmental approaches in heterostylous groups displaying variation in breeding systems (e.g., tristyly and distyly in *Lythrum* and *Oxalis*, distyly and dioecy in *Cordia* and *Nymphoides*, and heterostyly and homostyly in most groups) would help to determine the kinds of developmental modifications that underlie shifts in mating patterns. Since changes of this type are frequently associated with the evolution of reproductive isolation (Barrett 1989a), developmental data may shed light on controversies surrounding the morphological and genetic basis of speciation in plants (Gottlieb 1984; Coyne and Lande 1985).

4 Genetics and Molecular Biology

Early experimental studies on heterostyly were largely genetical in nature and concerned with determining the inheritance of the polymorphism. Many leading geneticists (e.g., W. Bateson, R.A. Fisher, J.B.S. Haldane, A. Ernst, A.B. Stout, K. Mather), working in the early to mid-periods of this century, were attracted to working on distyly and tristyly as a model system for studies on inheritance, linkage, recombination, epistasis, supergenes, and polymorphic equilibria. These investigations made a significant contribution to the overall growth of Mendelian and population genetics, and because of this work, heterostyly is frequently presented in textbooks on genetics and evolution as one of the classic examples of a balanced genetic polymorphism involving supergenes.

The inheritance of heterostyly has now been determined for 13 genera in 11 families; this work is reviewed by Lewis and Jones in Chapter 5. The most striking feature of the data on inheritance is the uniformity of one diallelic locus *S, s* in distyly and two loci *S, s* and *M, m* in tristyly, and the dominance of the short-styled morph in both systems. Interestingly, models of the evolution of heterostyly by Lloyd and Webb in Chapter 7 provide an explanation for this common pattern of inheritance of the long- and short-styled morphs in most heterostylous plants. Only three exceptions to the dominance of the short-styled morph have been reported. These occur in *Limonium, Hypericum* and *Oxalis* (see Chap. 5).

Recently, Bennett et al. (1986) have proposed a three-locus model for the control of style length in tristylous *Oxalis rosea* (and see Leach 1983). In this species some short-styled plants are dominant to non-short-styled plants (mid- or long-styled plants) while others are recessive. Breeding experiments demonstrated that the short-styled morph is governed by two gene pairs (*A, a* and *S, s*). In plants segregating for *A, a* on an *SS* background, short styles are recessive, while in plants segregating for *S, s* on an *aa* background, short styles are dominant. Genotypes for the three style morphs under the three-locus model are given in Table 2. The model raises the possibility that in other species of *Oxalis*, with the conventional two diallelic locus control of tristyly [e.g., *O. valdiviensis*, Fisher and Martin (1948) and members of the section *Ionoxalis*, Weller (1976b)] a third locus is present but fixed in the homozygous recessive condition (*aa*). Whether a third locus occurs in other tristylous families is

Table 2. Genotypes for the long-, mid-, and short-styled morphs of *Oxalis rosea* according to the three-locus model of Bennett et al. (1986)

Long-styled morph	Mid-styled morph		Short-styled morph
AASSmm	*AASSMM*	*AaSSMm*	*aaSSMM*
AASsmm	*AASSMm*	*AsSsMM*	*aaSSMm*
AAssmm	*AASsMM*	*AaSsMm*	*aaSSmm*
AaSSmm	*AASsMm*	*AassMM*	*aaSsMM*
AaSsmm	*AAssMM*	*AassMm*	*aaSsMm*
Aassmm	*AAssMm*	*aassMM*	*aaSsmm*
aassmm	*AaSSMM*	*aassMm*	

not known, but the data for *O. rosea* are of some interest in view of D. Charlesworth's suggestion that a third gene may have been involved in the evolution of tristyly (Charlesworth 1979).

Despite progress on the inheritance of style length there are still major gaps in our understanding of the genetical architecture of heterostyly. While generalizations concerning supergene control of heterostyly are frequently made, the evidence comes largely from A. Ernst's extensive work on *Primula* reviewed in Chapter 5. The supergene model may be applicable to most distylous plants; however the number and organization of loci controlling characters of the syndrome are likely to vary among different groups. Whether supergenes are involved in the control of tristyly remains a contentious issue (Chap. 4 and 5). Studies of the genetics of semi-homostylous variants in tristylous taxa would be valuable in addressing this problem. While semi-homostylous variants in *Eichhornia* spp. (S.C.H. Barrett unpubl. data) and *Decodon verticillatus* (C.G. Eckert and S.C.H. Barrett unpubl. data) do not behave as though they have arisen by recombination, those in *Pemphis acidula* (D. Lewis, pers. commun.) appear to do so. Studies of the genetics of homostyle and semi-homostyle formation in heterostylous plants are of general significance to studies of mating-system evolution because floral modifications influencing selfing rate are often under major gene control. This enables tests of theoretical models which frequently assume this mode of inheritance for mating-system modification (Wells 1979; Holsinger et al. 1984; Lande and Schemske 1985).

Modifier genes nonallelic to major genes governing the morphological and physiological features of heterostyly appear to be widespread in heterostylous species. In self-compatible species genes of this type may be of evolutionary significance because of their influence on mating systems. Wide variation in stamen and style length in *Amsinckia* and *Turnera* species directly influences the outcrossing rate of populations (Ganders et al. 1985; Barrett and Shore 1987; S. Belaoussoff and J.S. Shore, unpubl. data). In *Turnera ulmifolia* quantitative genetic studies have demonstrated that this variation is polygenically controlled (Shore and Barrett 1990). The genetic basis of floral variation in *Amsinckia* is unknown, but would certainly be worth investigating since it appears likely that selection on this variation is responsible for the multiple origins of selfing in the genus. Variation in the strengh of self-incompatibility is also common in heterostyly species but its genetic basis has rarely been investigated (but see Beale 1939; DeWinton and Haldane 1933; Mather 1950; Barrett and Anderson 1985; Shore and Barrett 1986) despite its obvious significance for the mating system of populations. Leaky or cryptic systems of self-incompatibility may enable the adjustment of outcrossing levels, depending on the supply of outcross and self-pollen delivered to plants by pollinators.

Recent advances in molecular biology offer exciting new opportunities for understanding more about the genes that control heterostyly. As yet no work on the molecular genetics of heterostyly has been undertaken, but it seems probable that molecular techniques will soon be employed in addressing questions concerned with the number, location, organization, and regulation of genes controlling floral organogenesis and incompatibility. One of the goals of these studies should be the elucidation of molecular mechanisms underlying the contrasting development processes (see Chap. 4) that give rise to the different phenotypes of the floral morphs. The availability of a range of recombinant genotypes (see Chap. 5) and other variants with altered floral traits may be useful in this regard. Elsewhere, molecular studies of floral mutants in *Arabidopsis* and *Antirrhinum* have provided novel insights into the genetic control of flower development in these species (Schwarz-Sommer et al. 1990). If approaches used in these taxa can be successfully transferred to heterostylous plants, they are likely to resolve many unanswered questions concerning the genetic architecture and development of heterostyly.

For molecular studies of heterostyly to be successful, it is necessary that the genes governing the floral polymorphisms are identified and their DNA sequences obtained by molecular cloning techniques. This has been achieved for several *S* alleles in both gametophytic and sporophytic systems of homomorphic incompatibility (reviewed by Haring et al. 1990). Sequences of *S*-glycoproteins in *Nicotiana* and *S* locus-specific glycoproteins in *Brassica* show little similarity, supporting the early suggestion by Bateman (1952) of independent origins for the two systems of homomorphic incompatibility. Comparisons of DNA sequences from different families, particularly those containing taxa with both diallelic and multiallelic incompatibility (e.g., Boraginaceae) may provide the most convincing evidence concerning the evolutionary relationships, if any, of heterostyly to other systems of incompatibility. Studies of this type would be based on the assumption that there are genes specifically for diallelic incompatibility in heterostylous plants analogous in function to those found in families with homomorphic incompatibility. However, if, as Lloyd and Webb suggest in Chapter 6, diallelic incompatibility arises separately in each floral morph, through selection on pollen performance in a particular style length, the search for specific incompatibility genes may be a fruitless exercise since a common molecular mechanism may not exist.

Once sequence data become available for genes controlling heterostylous characters, attempts can be made to characterize how, when, and where the genes are expressed in development. Of particular interest in this regard will be to determine whether the same genes, organized and expressed in a similar way, are responsible for floral differentiation in different heterostylous groups. A supergene model involving several tightly linked genes is most often used to explain the close association between heterostylous characters. However, other genetic models should also be considered, particularly those that involve regulatory genes. Such genes could act to generate the polymorphism by switching development along alternate pathways, or by controlling hormonal gradients responsible for differential organ growth in a morph- or tissue-specific manner, analogous to sex-limited gene expression. With such models, physical linkage of genes controlling the syndrome of floral traits is unnecessary. Models of this type might be particularly useful for explaining the unique properties of tristylous polymorphisms. Although at present these issues fall within the realm of speculation, the rapid progress in development of molecular techniques makes it likely that in the future it may be relatively straightforward to locate and characterize the genes controlling heterostyly. When this is done, we may find that different genetic systems control similar phenotypic expressions in unrelated heterostylous families.

5 Origin and Evolution

Heterostyly has originated on more than 20 separate occasions among angiosperm families (Fig. 2), yet understanding its evolutionary development remains one of the most difficult problems in plant breeding-system evolution. Perhaps this is because the course of evolution in heterostylous groups, "may have been complex and circuitous" (Fisher 1958). A major stumbling block has been our inability to identify, among close relatives of heterostylous taxa, patterns of floral variation which clearly represent stages in the assembly of the polymorphisms. Apart from the Plumbaginaceae, where Baker's studies suggest the build-up of distyly in several steps, beginning with diallelic incompatibility (Baker 1966), in most families the polymorphisms appear to arise de novo, without obvious clues as to the intermediate stages involved. In this respect heterostyly differs from other polymorphic sexual systems, such as dioecy, where variation patterns among related taxa have enabled inferences on the evolutionary pathways involved in the separation of sexes (Bawa 1980).

Our ability to identify stages involved in the evolution of heterostyly, depends on the particular model that is being evaluated. Most modern workers (for exceptions see Anderson 1973; Richards 1986, p. 254) have favored the view that diallelic incompatibility precedes the evolution of reciprocal herkogamy in heterostylous plants (Baker 1966; Yeo 1975; D. Charlesworth and B. Charlesworth 1979; Ganders 1979; Lewis 1982; and Chap. 5). However, Lloyd and Webb in Chapters 6 and 7 revive Darwin's original idea (Darwin 1877) that reciprocal herkogamy developed first, followed by the evolution of an intramorph incompatibility system[2]. Theoretical models of the evolution of distyly, by D. Charlesworth and B. Charlesworth (1979) and Lloyd and Webb (Chap. 7), differ in the ancestral conditions invoked and in the sequence in which heterostylous characters are assembled (Fig. 3). The models therefore make different predictions about the evolutionary build-up of heterostyly and the types of variation patterns likely to be found in the immediate ancestors of heterostylous plants. In this respect both models differ from Mather and DeWinton's (1941) suggestion that the morphological and physiological components of heterostyly arise simultaneously.

The phylogenetic status of self-compatible heterostylous taxa is of particular interest in evaluating models for the evolution and function of heterostyly. In recent years, more cases have been reported in which the morphological components of the polymorphism are accompanied by high levels of self-fertility. Several genera, most notably *Amsinckia* (Ray and Chisaki 1957), *Cryptantha* (Casper 1985), *Eichhornia* (Barrett 1988a), *Melochia* (Martin 1967), *Nivenia* (Goldblatt and Bernhardt 1990), *Quinchamalium* (Riveros et al. 1987), and the monotypic *Decodon* (C.G. Eckert and S.C.H. Barrett unpubl. data) contain species that are highly self-compatible. It has usually been assumed that this condition is derived through relaxation and eventual

[2] Some disagreement exists over the interpretation of Darwin's views on this point. Yeo (1975, p. 149) states "He did, however, suggest (Darwin 1877, p. 262) that the parent-species of heterostyled plants were "in some degree self-sterile"...... In other words, Darwin thought incompatibility came first and heterostyly followed". However, while Darwin may have considered that the ancestors of heterostylous plants exhibited some degree of self-sterility, it is quite clear from the discussion on pp. 260–268 (Darwin 1877) that he believed the evolution of the intramorph incompatibility system *followed* the establishment of reciprocal herkogamy.

1. Selfing Avoidance Model

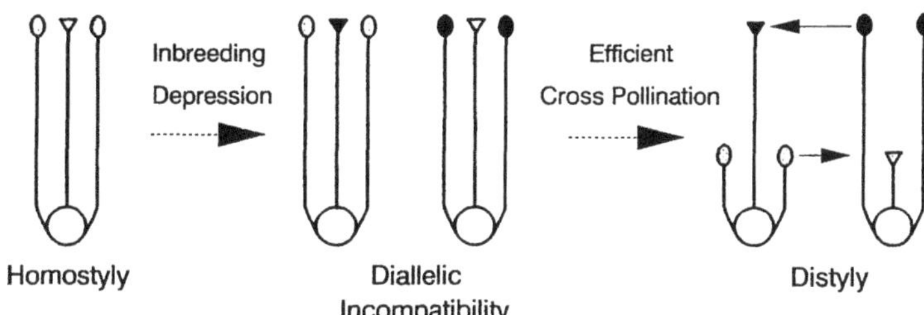

2. Pollen Transfer Model

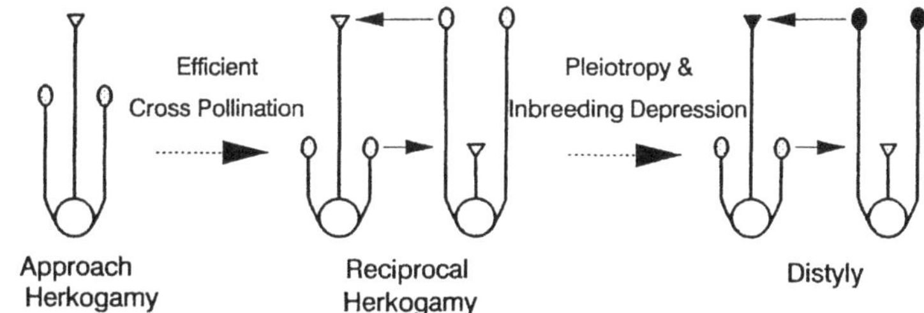

Fig. 3. The two major models of the evolution of distyly. The anti-selfing model follows D. Charlesworth and B. Charlesworth (1979) and the pollen transfer model is developed by Lloyd and Webb in Chapters 6 and 7. The models differ in the ancestral phenotypes invoked, the sequence of establishment of reciprocal herkogamy and diallelic incompatibility, and the emphasis placed on different selective forces. Only major stages in the models are shown for simplicity. Phenotypes with *uniform* pollen and stigmas are self-compatible, those with *shaded* pollen or stigmas are self-incompatible

loss of diallelic incompatibility (Ganders 1979). Genes modifying the strength of incompatibility commonly occur in heterostylous species (see above), and in some of the genera listed above related taxa possess normally functioning diallelic incompatibility systems. However, Lloyd and Webb's model of the evolution of heterostyly (Chap. 7) predicts the occurrence of self-compatible heterostylous populations as an ancestral condition in heterostylous groups (Fig. 3). This contrast with the Charlesworths' model (D. Charlesworth and B. Charlesworth 1979) where self-compatible heterostyly is always likely to be a derived condition. Sound phylogenetic data on the relationships between self-compatible and self-incompatible heterostylous taxa would therefore be useful in evaluating the validity of these models.

Another fertile area of relevance to models of the evolution of heterostyly concerns the apparent association between heterostylous polymorphisms and multi-allelic self-incompatibility in *Anchusa* (Dulberger 1970; Schou and Philipp 1984), *Narcissus* (Fernandes 1935; Dulberger 1964; Bateman 1954a; Lloyd et al. 1990) and possibly *Villarsia* (Ornduff 1988). This association is unexpected under a model in which diallelic incompatibility evolves first and reciprocal herkogamy functions to reduce illegitimate pollen transfer between the small number of mating groups associated with this type of incompatibility. However, this association can be accommodated under Lloyd and Webb's model since the selective forces they invoke to explain the evolution of reciprocal herkogamy are independent of whether the ancestral condition is self-incompatible or self-compatible.

The most complex problem concerned with the origins of heterostyly is the evolution of tristyly. Briggs and Walters (1984) recently remarked that "speculation about the evolution of the tristylous condition... is likely to be the hobby of the man who plays three-dimensional chess!". Notwithstanding this pessimistic view, some limited progress has been made on the topic. D. Charlesworth (1979) has developed a quantitative model for the evolution of tristyly and several features of her model have subsequently proven to be correct for particular tristylous groups (Bennett et al. 1986; Kohn and Barrett 1992). In addition, Richards and Barrett (Chap. 4) have investigated the developmental basis of tristyly and attempted to integrate this information with existing knowledge about the inheritance of the polymorphism. Their studies question the commonly held assumption that two separate anther whorls are a necessary precondition for the evolution of tristyly. Before more progress on this topic can be made, however, the question of whether supergenes exist in tristylous species and whether the *S* and *M* loci represent duplicated loci must be answered (Richards 1986; Olmstead 1989). In addition, basic data on the morphology, genetics, and incompatibility relationships from taxa in two additional families[3] (Amaryllidaceae and Connaraceae) recently reported as tristylous (see Chap. 6) are required to determine which features they share with the more widely known tristylous families reviewed by Weller in Chapter 10.

Patterns of floral variation in *Narcissus* (Amaryllidaceae) are particularly intriguing for models of the evolution of tristyly because species possessing stylar monomorphism, dimorphism and trimorphism occur (Fig. 4). Recent work on floral variation in *Narcissus* populations in southern Spain (S.C.H. Barrett, D.G. Lloyd and J. Arroyo unpubl. data) indicates that species with monomorphic populations (e.g., *N. bulbocodium*) exhibit approach herkogamy with two stamen levels; dimorphic populations (e.g., *N. assoanus*) are polymorphic for stigma height with two stamen levels at similar heights; and trimorphic populations (e.g., *N. triandrus*) exhibit reciprocal herkogamy, with three distinct stigma and stamen heights. The functional significance of this variation, and whether it represents stages in the evolutionary build-up of tristyly, are not known. Once again, sound phylogenetic information on species relationships, as well as field studies of the reproductive biology of populations, are required to assess this possibility.

6 Function and Adaptive Significance

Models of the evolution of heterostyly not only differ in the sequence in which morphological and physiological components of the polymorphisms are thought to arise, but also in the emphasis placed on different selective forces (Fig. 3). Most modern workers interpret heterostyly as an outbreeding mechanism. Following this

[3] Note added in proof, M.S. Zavada and T.K. Lowrey (unpubl. ms.) have recently reported the possible occurrence of tristyly in the south African shrub *Dais continifolia* L. (Thymelaeceae). The three floral morphs differ in style length, stigmatic papillae size and pollen exine sculpturing.

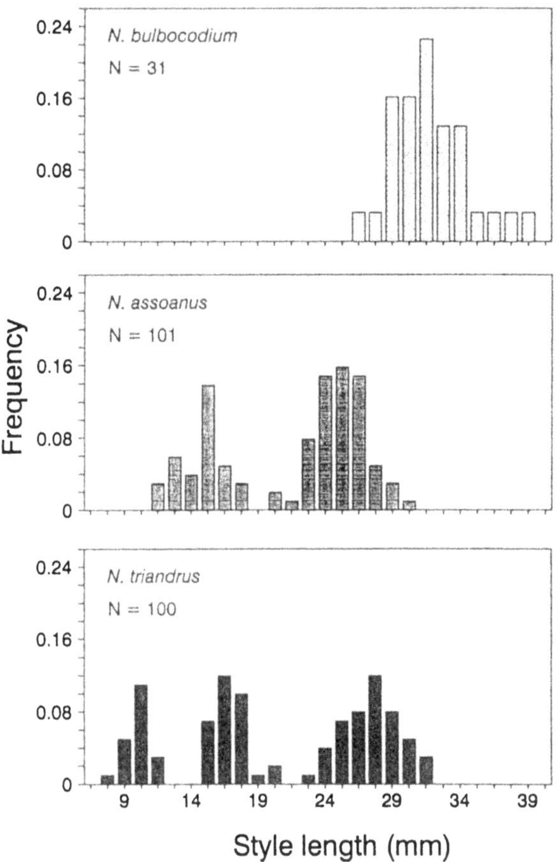

Fig. 4. Patterns of style length variation in *Narcissus* species from southern Spain. The distributions are based on a random sample of flowers collected from a single population of each species. *Narcissus bulbocodium* is monomorphic, *N. assoanus* is dimorphic and *N. triandrus* is trimorphic for style length. (Unpubl. data of S.C.H. Barrett, D.G. Lloyd and J. Arroyo)

view, diallelic incompatibility evolved first as a selfing avoidance mechanism, with inbreeding depression the primary selective force. According to this hypothesis, reciprocal herkogamy then arose secondarily to promote efficient pollen transfer between incompatibility types, hence reducing pollen wastage (see Chap. 7).

Although diallelic incompatibility prevents self-fertilization it also prevents intramorph mating; and, because of the small number of mating groups in heterostylous populations, the extent to which outbreeding is promoted is severely restricted in comparison with multiallelic incompatibility systems. This observation led Darwin (1877), and more recently Lloyd and Webb (in Chaps. 6 and 7), to question whether heterostyly is likely to have evolved primarily as an outbreeding mechanism. Instead, they suggest that reciprocal herkogamy evolved prior to incompatibility to promote efficient pollen transfer between individuals. Incompatibility may then develop secondarily as a pleiotropic byproduct of selection for increased pollen competitive ability on the style type to which pollen is most frequently transferred. Lloyd and Webb (Chap. 7) also show that diallelic incompatibility can be selected as an anti-selfing device, arising when most interplant pollen transfer is already intermorph as a result of reciprocal herkogamy. Thus the cost of diallelic incompatibility, in terms of lost ability to fertilize plants of the same incompatibility type, is reduced.

What evidence exists for Darwin's hypothesis that the function of reciprocal herkogamy is to actively promote efficient cross-pollination between anthers and stigmas of equivalent height in the floral morphs? In Chapter 7, Lloyd and Webb review a variety of experimental studies on the pollination biology of heterostylous populations that have attempted to investigate the adaptive significance of reciprocal herkogamy. Through the development of a novel procedure for analyzing pollen deposition patterns on naturally pollinated stigmas they demonstrate, in contrast to the conclusions of previous workers, these data from earlier studies actually provide impressive support for the Darwinian hypothesis that heterostyly functions to promote legitimate pollination.

A recent study by Kohn and Barrett (1992), using genetic markers and the experimental manipulation of garden arrays of tristylous self-compatible *Eichhornia paniculata*, investigated the two primary hypotheses (anti-selfing and improved cross-pollination) concerned with the function of reciprocal herkogamy. Outcrossing rates, levels of intermorph mating and morph-specific male and female reproductive success were compared in replicate trimorphic and monomorphic populations. They found that both outcrossing rates and seed set were higher in all three morphs in trimorphic than in monomorphic populations (Fig. 5). A large proportion (95%) of outcrossed matings in trimorphic populations were due to intermorph mating. Floral polymorphism therefore increased both outcrossing rate and seed set but the magnitude of the differences varied significantly among the floral morphs. If the ancestral condition in heterostylous groups resembled the long-styled morph, as suggested by Lloyd and Webb in Chapters 6 and 7, then the large increase in seed set in this morph when in trimorphic arrays (see Fig. 5) suggests that the selective basis for the evolution of floral polymorphism may have been increased pollen transfer, rather than higher levels of outcrossing.

Although much has been written concerning the adaptive significance of heterostylous characters (reviewed in Chaps. 3, 6, and 7), little work has been conducted in natural populations where both plants and pollinators have been experimentally manipulated. Yet heterostylous plants offer excellent opportunities for this type of work which is critical for testing adaptive hypotheses (see Chap. 7). By altering the floral morphology (e.g., emasculated versus intact flowers) and morph structure of populations (monomorphic versus polymorphic, homostylous versus heterostylous morphs) hypotheses that invoke fitness differences based on floral morphology can be tested experimentally. In these experiments, use of a variety of techniques previously employed in field studies of heterostylous populations should enable quantitative data to be collected on each of the elementary stages in the pollination and mating process. These include data on the foraging behaviour of pollinators (Weller 1981; Husband and Barrett 1991a), pollen removal from flowers (Wolfe and Barrett 1989; L.D. Harder and S.C.H. Barrett, unpubl. data), pollen deposition on pollinators (Olesen 1979; Wolfe and Barrett 1989) and intact and

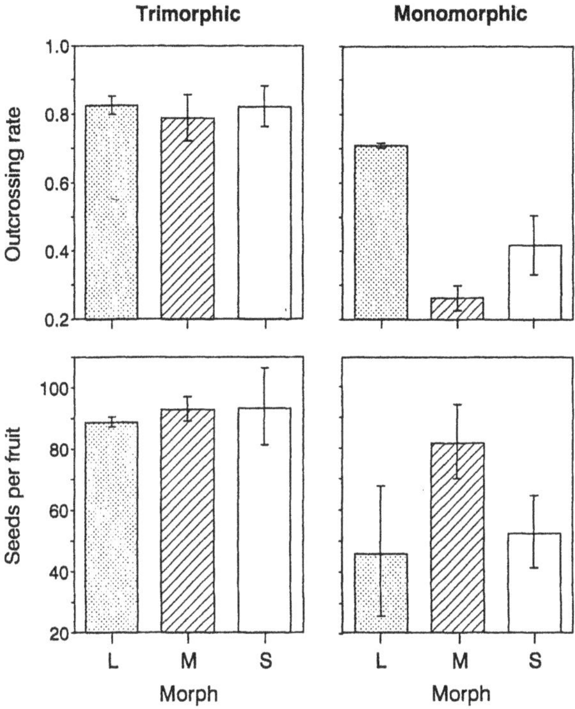

Fig. 5. Mean morph-specific outcrossing rates and seed set in experimental trimorphic and monomorphic populations of tristylous, self-compatible, *Eichhornia paniculata*. Bars represent two standard errors based on three replicate populations of each treatment. (After Kohn and Barrett 1992)

emasculated flowers (Ganders 1974; Barrett and Glover 1985), pollen carryover (Feinsinger and Busby 1987), pollen tube growth (M.B. Cruzan and S.C.H. Barrett, unpubl. data), and mating system parameters including rates of outcrossing (Glover and Barrett 1986), disassortative mating (Ganders 1975a; Barrett et al. 1987), correlated mating (Morgan and Barrett 1990), and male reproductive success (Kohn and Barrett 1992).

In Chapters 6 and 7, Lloyd and Webb introduce an additional selective force that could account for the evolution of floral polymorphisms associated with the heterostylous syndrome. They suggest that various ancillary characters associated with pollen and stigmas may encourage cross-fertilization by reducing the likelihood of mutual interference between the reproductive functions of stamens and carpels within flowers. The recognition that reproductive success occurs through both female and male function in hermaphrodite flowers, and that conflicts may arise between the two, is associated with a growing awareness that various facets of plant reproduction can be viewed from the perspectives of sexual selection and sex allocation theory (Charnov 1982).

In Chapter 8, Casper examines the application of sex allocation theory to heterostylous plants by focusing attention on how heterostyly may function in ways that are unrelated to its role as an incompatibility system. The idea that fitness through female and male function might differ between the morphs in heterostylous species is reviewed. In addition, the issues of the optimal allocation to pollen and seed production in the floral morphs and the selective mechanisms by which morph ratios may be controlled are addressed. Sex allocation models of heterostyly have not been subjected to empirical tests, but Casper discusses a range of studies that address a major assumption of the models; that plants have the ability to control offspring morph ratios through genes unlinked to the heterostyly loci.

Unfortunately, although morph-specific differences in reproductive traits are ubiquitous in heterostylous populations (reviewed in Chaps. 3 and 8), their adaptive significance and influence on fitness gain through female and, particularly, male function are largely unknown. The major limitation to assessing the role of sexual selection in heterostylous plants is a lack of data on male reproductive success. Most of the evidence that the floral morphs might specialize in functional gender comes from comparisons of maternal reproductive characters (reviewed in Chap. 8). A recent attempt to measure the functional gender of floral morphs in tristylous *Eichhnornia paniculata* using genetic markers demonstrated large differences in male reproductive success, particularly between the long- and short-styled morphs (Kohn and Barrett 1992). Short-styled plants consistently sired more mid-styled offspring than long-styled plants. Whether this type of mating asymmetry commonly occurs in tristylous populations is not known, but similar kinds of gender differences are a prerequisite for models concerned with evolution of dioecism from heterostyly (see below).

7 Pollen-Pistil Interactions

Studies of pollen-pistil interactions in heterostylous plants (reviewed in Chaps. 3 and 6) demonstrate that the general properties of diallelic incompatibility systems are fundamentally distinct from those found in multiallelic homomorphic systems. As Dulberger points out in Chapter 3, a variety of different inhibition sites for incompatible pollen tubes, including the stigma, style and ovary, are evident in heterostylous plants. A characteristic feature of most heterostylous taxa in which pollen tube growth has been investigated is the difference in sites of inhibition between the floral morphs (Gibbs 1986).

Recent studies of pollen tube growth following legitimate and illegitimate pollinations in distylous *Primula* spp. (Wedderburn and Richards 1990) and tristylous *Pontederia* spp. (Anderson and Barrett 1986; Scribailo and Barrett 1991) clearly demonstrate the absence of a unitary rejection response, as occurs in homomorphic systems. In *P. sagittata* inhibition sites are particularly complex and depend on the specific pollen size-style length combination that is employed (Fig. 6). The occurrence of variable inhibition sites in heterostylous plants is used as evidence by Lloyd and Webb (Chap. 6) to argue that incompatibility reactions in heterostylous plants may have evolved separately in each floral morph and, as a result, need not share a common molecular basis involving matching recognition factors in the pollen and style. Observations on pollen-pistil interactions in heterostylous plants, com-

A.

Pollen	Style L	Style M	Style S
l	**13.7**	7.2	1.7
m	7.9	**7.6**	1.7
s	1.1	1.0	**1.3**

B.

Pollen	Style L	Style M	Style S
l	**81.2**	72.6	7.4
m	11.1	**91.8**	27.5
s	0.7	16.4	**78.4**

Fig. 6 A,B. Pollen-pistil interactions in self-incompatible, tristylous *Pontederia sagittata*. **A** Mean distance (mm) at which pollen tubes terminate growth in the style following controlled pollinations. **B** Mean percentage seed set. Values in *bold type* are the legitimate combinations. [After **A** Scribailo and Barrett (1991) and **B** Glover and Barrett (1983)]

bined with taxonomic information on the distribution of heteromorphic and homomorphic incompatibility, provide evidence against the view that diallelic incompatibility evolved by degeneration of a multiallelic system until only two alleles remained (Crowe 1964; Muenchow 1982; Wyatt 1983).

Another facet of pollen-pistil interactions, reviewed by McKenna in Chapter 9, concerns the potential for microgametophytic competition to occur in heterostylous plants as a consequence of differences between the floral morphs in style length and pollen size. Gametophytic selection is also discussed by Casper in Chapter 8 in her evaluation of mechanisms that may contribute to the regulation of floral morph ratios in heterostylous populations.

Several approaches have been used in studies of pollen competition in heterostylous plants. These include: (1) Mixed pollinations and progeny tests to determine whether pollen carrying alternate alleles at loci governing style length differs in competitive ability (Tseng 1938; Baker 1975; Barrett et al. 1989). (2) Marker gene and pollen tube growth studies to determine whether differences in the competitive ability of self versus outcrossed or legitimate versus illegitimate pollen occur in self-compatible species (Weller and Ornduff 1977; Glover and Barrett 1986; Casper et al. 1988; M.B. Cruzan and S.C.H. Barrett, unpubl. data). (3) Use of the style length polymorphism as an experimental system for determining whether the intensity of pollen competition (data of McKenna in Chap. 9) or pollen precedence (Graham and Barrett 1990) varies with style length. (4) Studies of the inhibitory effects of prior application of incompatible pollen on compatible pollen tube growth and seed set (Shore and Barrett 1984; Barrett and Glover 1985; Nicholls 1987; Murray 1990; Scribailo and Barrett 1992). This latter phenomenon, known as "stigma or stylar clogging", can be viewed as one aspect of pollen-stigma interference discussed above. Considerable scope for future studies on pollen competition exist in heterostylous plants. Although functional work on heterostyly has focused primarily on the pollination process, it is clear that a variety of post-pollination mechanisms can potentially influence the mating system and morph structure of heterostylous populations.

Studies of pollen-pistil interactions in self-compatible heterostylous taxa are particularly important because of the question of whether self-compatibility is an ancestral or derived condition in heterostylous groups (see Sect. 5 above). Progeny data from *Amsinckia* spp. (Weller and Ornduff 1977; Casper et al. 1988) and *Eichhornia paniculata* (Glover and Barrett 1986 and Fig. 7) indicate significant differences in the siring ability of legitimate and illegitimate pollen when applied in mixtures. This type of phenomenon has been referred to as cryptic incompatibility. It is not clear, however, whether the mechanisms responsible for discrimination between the different pollen types in these species simply represent a weak expression of diallelic incompatibility or whether other aspects of pollen-pistil interactions are involved. Do these species exhibit a rudimentary and primitive form of diallelic incompatibility of the type anticipated in Lloyd and Webb's model of the evolution of heterostyly? Or are these species secondarily self-compatible, through the action of modifier genes that have weakened, but not completely abolished, a pre-existing diallelic incompatibility system? Studies of additional taxa are needed to determine whether differences in the siring ability of legitimate and illegitimate pollen tubes commonly occur in self-compatible heterostylous plants.

8 Maintenance and Breakdown

Mating types in self-incompatible plants are maintained in populations by frequency-dependent selection. In taxa with homomorphic incompatibility the number and frequency of mating types can only be determined by extensive crossing programs. Data are thus limited to a handful of species (Emerson 1939; Bateman 1954b;

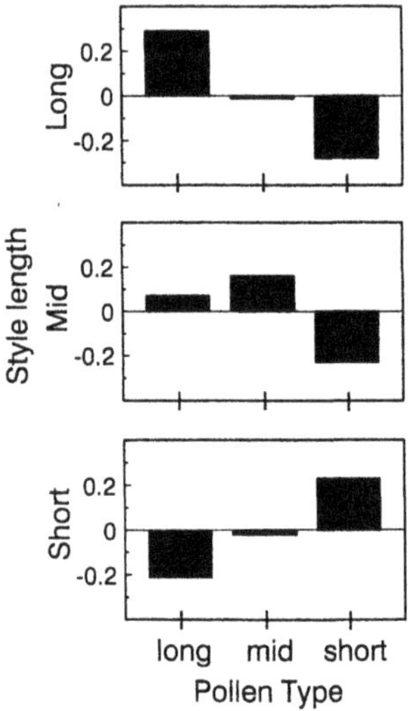

Fig. 7. Fertilization success of pollen from long-, mid-, and short-level anthers when applied in mixtures to stigmas of the long-, mid-, and short-styled morphs of *Eichhornia paniculata*. Genetic markers were used to determine the fertilization success of different pollen types in the mixtures. A value of O indicates random fertilization. (Unpubl. data of M.B. Cruzan and S.C.H. Barrett)

Sampson 1967; Campbell and Lawrence 1981; Karron et al. 1990). In contrast, considerable information on population structure is available for both distylous and tristylous species, because of the small number of mating types and the ease with which they can be identified (reviewed in Chap. 8). Equilibrium morph ratios of 1:1 should be reached in one generation in self-incompatible distylous populations because of the simple mode of inheritance and disassortative mating. In tristylous species, however, the more complex pattern of inheritance results in a slower approach to equilibrium, and morph frequencies at any given time are more strongly influenced by the genotypes of individuals initiating populations and the effects of finite population size (Morgan and Barrett 1988). In large tristylous populations at equilibrium, however, a 1:1:1 ratio of floral morphs will prevail, provided that all morphs have equal fitness, and some degree of disassortative mating occurs (Heuch 1979).

Population surveys of distylous species frequently report equality of the floral morphs (isoplethy) although exceptions are known (e.g., *Primula vulgaris*, Crosby 1949, but see Richards and Ibrahim 1982; *Lythrum* spp., Ornduff 1978; *Hedyotis nigricans*, Levin 1974; *H. caerulea*, Ornduff 1980). Morph frequencies in populations of self-compatible species (e.g., *Amsinckia* spp., Ganders 1975a; *Quinchamalium chilense*, Riveros et al. 1987) or those with extensive clonal growth (e.g., *Oxalis* spp., Mulcahy 1964; *Nymphoides indica*, Barrett 1980; *Menyanthes trifoliata*, F.R. Ganders, unpubl. data) frequently display unequal morph ratios (anisoplethy) because of morph-specific differences in selfing rate or founder effects.

Large scale surveys of floral morph frequencies in tristylous species indicate a variety of complex patterns. Figure 8 illustrates data from four species (*Lythrum salicaria, Decodon verticillatus, Pontederia cordata* and *Eichhornia paniculata*). The striking feature of the data is that despite similar modes of inheritance of tristyly (Chap. 5 and unpubl. data of S.C.H. Barrett and C.G. Eckert) morph frequencies in each species differ in unique ways that reflect their contrasting life history, population ecology, and mating systems. Theoretical models and computer simulation studies of the population dynamics of tristyly in these species have provided insights into the relative importance of stochastic and deterministic forces in explaining morph frequencies (Heuch 1980; Barrett et al. 1983; Morgan and Barrett 1988; Barrett et al. 1989; Eckert and Barrett, 1992).

Equilibrium floral morph frequencies in heteromorphic species with putative multiallelic incompatibility systems (e.g., *Anchusa, Narcissus*) are likely to be quite different from species with diallelic incompatibility. This is because the floral polymorphisms are apparently unlinked to loci controlling incompatibility. In *Villarsia parnassiifolia*, Ornduff (1988) has suggested that incompatibility alleles may be linked to loci controlling morphological polymorphisms in the long-styled morph, but not in the short-styled morph. Where morphological and physiological characters are unlinked, morph frequencies will depend on the relative fitness of the floral phenotypes as female and male parents. This is likely to be largely determined by the pollination biology of populations. In *Anchusa officinalis*, the long-styled morph predominated in all eleven populations sampled by Philipp and Schou (1981). A similar pattern was evident in 10 out of 11 dimorphic and trimorphic populations of *Narcissus* spp. surveyed in southern Spain (S.C.H. Barrett, D.G. Lloyd, and J. Arroyo unpubl. data). Interestingly, the long-styled morph also predominates in the only two

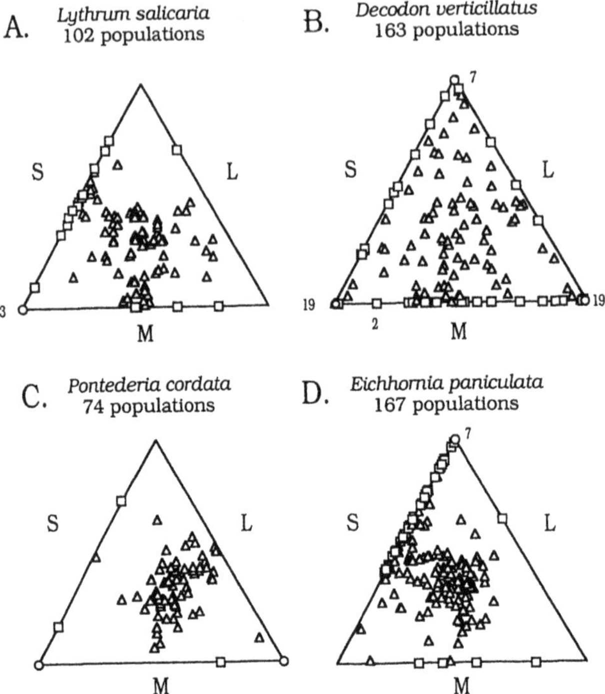

Fig. 8A-D. Floral morph frequences in populations of four tristylous species. **A** *Lythrum salicaria*, **B** *Decodon verticillatus*, **C** *Pontederia cordata*, **D** *Eichhornia paniculata*. *Triangles* represent populations with three floral morphs, *squares* are populations with two floral morphs, and *circles* are populations containing a single floral morph. Isoplethy is equidistant from all axes and the distance of a symbol from each axis is proportional to the frequency of the morph in a population, e.g., *triangles* close to the *S* axis have a low frequency of the short-styled morph. [*Lythrum salicaria* and *Decodon verticillatus*, Eckert and Barrett (1992); *Pontederia cordata*, Barrett et al. (1983); *Eichhornia paniculata*, Barrett et al. (1989) and (unpubl. data)]

well-documented nonheterostylous species with stigma height polymorphisms (*Chlorogalum angustifolium*, Jernstedt 1982; *Epacris impressa*, O'Brien and Calder 1989).

These data suggest that the long-styled phenotype commonly has a selective advantage over the short-styled phenotype, a finding consistent with the overall prevalence of approach herkogamy and rarity of reverse herkogamy in angiosperm families (Webb and Lloyd 1986). Field studies of the pollination biology and mating systems of taxa with stigma height polymorphisms are of special interest because of Lloyd and Webb's suggestion (Chaps. 6 and 7) that this type of variation may represent an early stage in the evolution of heterostyly. Critical testing of their model requires a demonstration that significant levels of disassortative mating occur in populations polymorphic for stigma height.

Surveys of morph frequencies in heterostylous populations have provided valuable clues on the evolutionary processes responsible for the breakdown of heterostyly. In Chapter 10, Weller reviews some of this work in his discussion of the ways that tristyly has become evolutionarily modified in the Lythraceae, Oxalidaceae and Pontederiaceae. Two contrasting evolutionary pathways are most commonly found in these families; the evolution of distyly, by loss of one of the style morphs (Lythraceae and Oxalidaceae), and the evolution of predominant self-fertilization via semi-homostyly. Other modifications include the evolution of facultative apomixis in *Oxalis dillenii* ssp. *filipes* (Lovett Doust et al. 1981), and the possibility of gender specialization in *Sarcotheca celebica* (Lack and Kevan 1987) and *Decodon verticillatus* (C.G. Eckert and S.C.H. Barrett unpubl. data). In addition, Lemmens (1989) has recently reported tristyly, distyly, homostyly, and dioecy in various genera of tropical Connaraceae and suggested that evolutionary trends in this family are similar to those documented in the Lythraceae and Oxalidaceae. Population studies of these genera and other putatively heterostylous taxa of tropical origin (reviewed in Barrett and Richards 1990) are needed to confirm the true nature of their breeding systems.

A variety of evolutionary modifications have also been documented in distylous groups (reviewed in Ganders 1979; Richards 1986; Barrett 1988b). These include the shift from outcrossing to different degrees of selfing through the evolution of homostyly (e.g., *Amsinckia*, Ganders et al. 1985; *Psychotria*, Hamilton 1990; *Turnera ulmifolia*, Fig. 9), the development of apomixis in *Limonium* (Baker 1966) and *Erythroxylum* (Berry et al. 1991), and the evolution of various forms of gender specialization including gynodioecy (*Armeria*, Baker 1966; *Nymphoides*, Vasudevan Nair 1975) and dioecy (primarily in the Rubiaceae, Menyanthaceae and Boraginaceae; Baker 1958; Ornduff 1966; Opler et al. 1975; Lloyd 1979; Beach and Bawa 1980; Muenchow and Grebus 1989; Charlesworth 1989). These shifts in breeding behavior provide a rich, but to date largely untapped, source of experimental material for studies aimed at determining the ecological and genetic basis of mating-system evolution in plants.

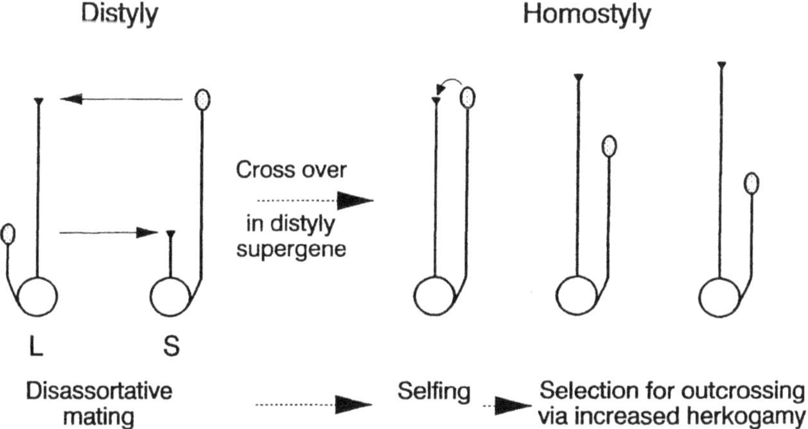

Fig. 9. Model of the evolutionary breakdown of distyly to homostyly in the *Turnera ulmifolia* complex after Barrett and Shore (1987). Homostylous populations vary in stigma-anther separation; the variation is polygenically controlled (Shore and Barrett 1990) and influences the outcrossing rate of populations (S. Belaoussoff and J.S. Shore, unpubl. data)

Most progress on the evolution of mating systems in heterostylous groups has been made by investigating the selective forces responsible for the dissolution of heterostyly to homostyly (reviewed in Richards 1986; Barrett 1989b). The frequent breakdown of floral polymorphism to monomorphism in distylous and tristylous groups represents a model system for studies of the evolution of self-fertilization in plants. This is because the direction of evolutionary change is usually unambiguous, genetic modifications are often simply inherited (Chap. 5), and alterations in floral morphology that influence breeding behavior are usually of large phenotypic effect and therefore readily detected under field conditions. These features have aided both theoretical modeling and experimental studies of natural and artificial populations (B. Charlesworth and D. Charlesworth 1979; Piper et al. 1986; J.R. Kohn and S.C.H. Barrett unpubl. data). To disentangle the selective forces responsible for the evolution of self-fertilization, however, data on selfing rates, inbreeding depression, male and female fertility, patterns of sex allocation, and the genetic basis of mating system modification are required (D. Charlesworth and B. Charlesworth 1981, 1987; Lande and Schemske 1985; Barrett and Eckert 1990). In addition, an understanding of the ecological and demographic circumstances under which selfing variants are favored is also needed (Lloyd 1980; Holsinger 1991).

Some of this information is available for the best-studied heterostylous groups (e.g., *Primula*, Piper et al. 1984, 1986; Curtis and Curtis 1985; Richards 1986; Boyd et al. 1990; *Eichhornia*, Barrett 1979; Barrett et al. 1989; Morgan and Barrett 1989; Barrett and Husband 1990; Husband and Barrett 1991a,b; *Turnera*, Shore and Barrett 1985, 1990; Barrett and Shore 1987; and *Amsinckia*, Ray and Chisaki 1957; Ganders 1975a,b; Ganders et al. 1985), and several lines of evidence indicate that homostyles can experience a selective advantage under conditions where pollinator service is unreliable. However, where fitness components have been compared in heterostylous and homostylous morphs, considerable variation in both space and time has been evident, complicating simple interpretation based on reproductive assurance alone. Long-term studies of the demographic genetics of mixed populations of short-lived taxa are most likely to provide convincing evidence of the selective mechanisms responsible for the evolution of self-fertilization in heterostylous taxa.

9 Conclusions

A century of research on heterostyly has passed since the publication in 1877 of Charles Darwin's book *The Different Forms of Flowers on Plants of the Same Species*, which summarized his extensive investigations of distylous and tristylous plants. The subsequent period can be divided into two phases, differing in research emphasis. The first, lasting until the 1960s, largely involved genetic and biosystematic studies by European workers on a small number of herbaceous taxa (e.g., *Primula, Lythrum, Oxalis, Linum,* and taxa of Plumbaginaceae). At the time, this work was part of mainstream genetics and aided the development of theories concerned with the regulation of recombination and evolution of genetic systems (Darlington 1939; Darlington and Mather 1949; Grant 1975).

Since then, the growth of population biology and increased opportunities for tropical research have stimulated a diversification of approaches involving a broader range of heterostylous taxa. Over the past two decades field investigations of the reproductive ecology and genetics of populations, in concert with the development of theoretical models, have enabled some testing of specific hypotheses concerned with the evolution and breakdown of heterostyly. In addition, structural and developmental studies and work on pollen-pistil interactions have provided the necessary morphological and physiological context for investigations concerned with the adaptive significance of floral polymorphisms.

Although considerable headway has been made during the past century in understanding various aspects of the evolution and breakdown of heterostyly, most of the work has focused at or below the species level. As a result, most interpretations have been conducted in an ahistorical context with little evidence, aside from character correlations, on the number of times characters may have evolved, or the direction and temporal sequence of character transformation. In order that hypotheses on evolutionary pathways in heterostylous groups can be tested more rigorously, phylogenetic analyses using cladistic techniques must be undertaken (Donoghue 1989). Studies of this type are likely to benefit from recent advances in molecular systematics which have provided an array of molecular characters that can be used in phylogenetic reconstruction (Felsenstein 1988; Palmer et al. 1988). Molecular data are of particular value because the characters used to assess relationships are likely to be independent of changes in morphological traits associated with breeding-system evolution. High levels of homoplasy in morphological characters can complicate attempts at phylogenetic reconstruction, particularly those involving character syndromes associated with floral evolution (Eckenwalder and Barrett 1986).

Phylogenetic analysis of heterostylous families in which specific evolutionary hypotheses have been proposed concerning breeding-system variation would be particularly worthwhile. These include: Plumbaginaceae — is the sequence of character build-up for the heterostylous syndrome proposed by Baker (1966) supported? Rubiaceae — has distyly originated more than once in the family, as suggested by Anderson (1973), and how often has it broken down to dioecy? Boraginaceae — what are the phylogenetic relationships between distylous taxa with and without diallelic incompatibility and between these taxa and those with multiallelic incompatibility? Lythraceae and Oxalidaceae — how often has tristyly broken down to distyly in each family, and are similar pathways involved? Finally, phylogenetic analysis would be especially valuable for determining the character syndromes of the immediate ancestors of heterostylous genera. Such information is important for understanding why and how heterostyly has evolved in particular families and provides the necessary historical background for microevolutionary studies of the selection process.

Acknowledgments. I thank Juan Arroyo, Svenja Belaoussoff, Paul Berry, Mitch Cruzan, Jim Eckenwalder, Chris Eckert, Fred Ganders, Sean Graham, Brian Husband, Josh Kohn, David Lloyd, Bob Marquis, Bob Ornduff, Robin Scribailo, Joel Shore, and Dieter Wilken for valuable discussion, providing manuscripts and permission to cite unpublished data, Bill Cole for producing the figures and the Natural Sciences and Engineering Research Council of Canada for supporting my research on heterostylous plants.

References

Anderson JM, Barrett SCH (1986) Pollen tube growth in tristylous *Pontederia cordata* L. (Pontederiaceae). Can J Bot 64:2602–2607

Anderson WR (1973) A morphological hypothesis for the origin of heterostyly in the Rubiaceae. Taxon 22:537–542

Baker HG (1958) Studies in the reproductive biology of West African Rubiaceae. J West Afr Sci Assoc Fr 4:9–24

Baker HG (1966) The evolution, functioning and breakdown of heteromorphic incompatibility systems I. The Plumbaginaceae. Evolution 18:507–512

Baker HG (1975) Sporophyte-gametophyte interactions in *Linum* and other genera with heteromorphic self-incompatibility. In: Mulcahy DL (ed) Gamete competition in plants and animals. North Holland, Amsterdam, pp 191–200

Barrett SCH (1979) The evolutionary breakdown of tristyly in *Eichhornia crassipes* (Mart.) Solms (water hyacinth). Evolution 33:499–510

Barrett SCH (1980) Dimorphic incompatibility and gender in *Nymphoides indica* (Menyanthaceae). Can J Bot 58:1938–1942

Barrett SCH (1985a) Ecological genetics of breakdown in tristyly. In: Haeck J, Woldendorp JW (eds) Structure and functioning of plant populations II. Phenotypic and genotypic variation in plant populations. North Holland, Amsterdam, pp 267–275

Barrett SCH (1985b) Floral trimorphism and monomorphism in continental and island populations of *Eichhornia paniculata* (Spreng.) Solms (Pontederiaceae). Biol J Linn Soc 25:41–60

Barrett SCH (1988a) Evolution of breeding systems in *Eichhornia* (Pontederiaceae): A review. Ann M Bot Gard 75:741–760

Barrett SCH (1988b) The evolution, maintenance and loss of self-incompatibility systems. In: Lovett Doust J, Lovett Doust L (eds) Plant reproductive ecology patterns and strategies. Oxford Univ Press, Oxford, pp 98–124

Barrett SCH (1989a) Mating system evolution and speciation in heterostylous plants. In: Otte D, and Endler J (eds) Speciation and its consequences. Sinauer, Sunderland, MA, pp 257–283

Barrett SCH (1989b) Evolutionary breakdown of heterostyly. In: Bock JH, and Linhart YB (eds) The evolutionary ecology of plants. Westview Press, Colorado, pp 151–170

Barrett SCH (1990) The evolution and adaptive significance of heterostyly. Trends Ecol Evol 5:144–148

Barrett SCH, Anderson JM (1985) Variation in expression of trimorphic incompatibility in *Pontederia cordata* L. (Pontederiaceae). Theor Appl Genet 70:355–362

Barrett SCH, Eckert CG (1990) Variation and evolution of mating systems in seed plants. In: Kawano S (ed) Biological approaches and evolutionary trends in plants. Academic Press, Lond New York, pp 229–254

Barrett SCH, Glover DE (1985) On the Darwinian hypothesis of the adaptive significance of tristyly. Evolution 39:766–774

Barrett SCH, Husband BC (1990) Variation in outcrossing rates in *Eichhornia paniculata*: The role of demographic and reproductive factors. Plant Spec Biol 5:41–56

Barrett SCH, Richards JH (1990) Heterostyly in tropical plants. Mem N Y Bot Gard 55:35–61

Barrett SCH, Shore JS (1987) Variation and evolution of breeding systems in the *Turnera ulmifolia* L. complex (Turneraceae). Evolution 41:340–354

Barrett SCH, Wolfe LM (1986) Pollen heteromorphism as a tool in studies of the pollination process in *Pontederia cordata*. In: Mulachy DL, Mulcahy GB, Ottaviano E (eds) Biotechnology and ecology of pollen. Springer Berlin Heidelberg, New York, pp 435–442

Barrett SCH, Price SD, Shore JS (1983) Male fertility and anisoplethic population structure in tristylous *Pontederia cordata* (Pontederiaceae). Evolution 37:745–759

Barrett SCH, Brown AHD, Shore JS (1987) Disassortative mating in tristylous *Eichhornia paniculata* (Spreng.) Solms (Pontederiaceae). Heredity 58:49–55

Barrett SCH, Morgan MT, Husband BC (1989) The dissolution of a complex genetic polymorphism: the evolution of self-fertilization in tristylous *Eichhornia paniculata* (Pontederiaceae). Evolution 43:1398–1416

Bateman AJ (1952) Self-incompatibility systems in angiosperms I Theory. Heredity 6:285–310
Bateman AJ (1954a) The genetics of *Narcissus* I-sterility. Daffodil Tulip Year Book 1954:23–29
Bateman AJ (1954b) Self-incompatibility systems in angiosperms II *Iberis amara*. Heredity 8:305–332
Bawa KS (1980) Evolution of dioecy in flowering plants. Annu Rev Ecol Syst 11:15–40
Beach JH, Bawa KS (1980) Role of pollinators in the evolution of dioecy from distyly. Evolution 34:1138–1143
Beale GH (1939) Further studies of pollen-tube competition in *Primula sinensis*. Ann Eugenics 9:259–268
Bennett JH, Leach CR, Goodwins IR (1986) The inheritance of style length in *Oxalis rosea*. Heredity 56:393–396
Berry PE, Tobe H, Gómez JA (1991) Agamospermy and the loss of distyly in *Erythroxylum undulatum* (Erythroxylaceae) from northern Venezuela. Am J Bot 78:595–600
Bodmer WF (1960) Genetics of homostyly in populations of *Primula vulgaris*. Philos Trans R Soc Lond Ser B 242:517–549
Boyd M, Silvertown J, Tucker C (1990) Population ecology of heterostyle and homostyle *Primula vulgaris*: growth, survival, and reproduction in field populations. J Ecol 78:799–813
Bräm M (1943) Untersuchungen zur Phänalyse und Entwicklungsgeschichte der Blüten von *Primula pulerulenta, P. cockburniana* und ihrer F_1-Bastarde. Arch Julius Klaus-Stift Vererbungsforsch Sozialanthropol Rassenhyg 18:235–259
Briggs D, Walters SM (1984) Plant variation and evolution, 2nd edn. Cambridge Univ Press, Cambridge
Campbell JM, Lawrence MJ (1981) The population genetics of the self-incompatibility polymorphism in Papaver rhoeas I. The number of S-alleles in families from three localities. Heredity 46:69–79
Casper BB (1985) Self-compatibility in distylous *Cryptantha flava* (Boraginaceae). New Phytol 99:149–154
Casper BB, Sayigh LS, Lee SS (1988) Demonstration of cryptic incompatibility in distylous *Amsinckia douglasiana*. Evolution 42:248–253
Charlesworth B, Charlesworth D (1979) The maintenance and breakdown of heterostyly. Am Nat 114:499–513
Charlesworth D (1979) The evolution and breakdown of tristyly. Evolution 33:486–498
Charlesworth D (1989) Allocation to male and female function in hermaphrodites, in sexually polymorphic populations. J Theor Biol 139:327–342
Charlesworth D, Charlesworth B (1979) A model for the evolution of heterostyly. Am Nat 114:467–498
Charlesworth D, Charlesworth B (1981) Allocation of resources to male and female function in hermaphrodites. Biol J Linn Soc 15:57–74
Charlesworth D, Charlesworth B (1987) Inbreeding depression and its evolutionary consequences. Ann Rev Ecol Syst 18:237–268
Charnov EL (1982) The theory of sex allocation. Princeton Univ Press, Princeton, New Jersey
Coyne JA, Lande R (1985) The genetic basis of species differences in plants. Am Nat 126:141–145
Cronquist A, Holmgren AH, Holmgren NH, Reveal JL, Holmgren PK (1984) Intermountain flora, vol 4. N Y Bot Gard, New York
Crosby JL (1949) Selection of an unfavourable gene-complex. Evolution 3:212–230
Crowe LK (1964) The evolution of outbreeding in plants I. The angiosperms. Heredity 19:435–457
Curtis J, Curtis CF (1985) Homostyle primroses re-visited I. Variation in time and space. Heredity 54:227–234
Dahlgren RMT (1980) A revised system of classification of the angiosperms. Bot J Linn Soc 80:91–124
Darlington CD (1939) The evolution of genetic systems. Cambridge Univ Press, Cambridge
Darlington CD, Mather K (1949) The elements of genetics. Allen and Unwin, Lond
Darwin C (1865) On the sexual relations of the three forms of *Lythrum salicaria*. J Linn Soc Lond 8:169–196
Darwin C (1877) The different forms of flowers on plants of the same species. Murray, Lond
DeWinton D, Haldane JBS (1933) The genetics of *Primula sinensis* II segregation and interaction of factors in the diploid. J Genet 27:1–44

Donoghue MJ (1989) Phylogenies and the analysis of evolutionary sequences, with examples from seed plants. Evolution 43:1137–1156
Dulberger R (1964) Flower dimorphism and self-incompatibility in *Narcissus tazetta* L. Evolution 18:361–363
Dulberger R (1970) Floral dimorphism in *Anchusa hybrida* Ten. Isr J Bot 19:37–41
Dulberger R (1975a) *S*-gene action and the significance of characters in the heterostylous syndrome. Heredity 35:407–415
Dulberger R (1975b) Intermorph structural differences between stigmatic papillae and pollen grains in relation to incompatibility in Plumbaginaceae. Proc R Soc Lond Ser B 188:257–274
Eckenwalder JE, Barrett SCH (1986) Phylogenetic systematics of the Pontederiaceae. Syst Bot 11:373–391
Eckert CG, Barrett SCH (1992) Stochastic loss of style morphs from tristylous populations of *Lythrum salicaria* and *Decodon verticillatus* (Lythraceae). Evolution (in press)
Emerson S (1939) A preliminary survey of the *Oenothera organensis* population. Genetics 24:524–537
Feinsinger P, Busby WH (1987) Pollen carryover: experimental comparisons between morphs of *Palicourea lasiorrachis* (Rubiaceae), a distylous, bird-pollinated, tropical treelet. Oecologia 73:231–235
Felsenstein J (1988) Phylogenies from molecular sequences: inference and reliability. Annu Rev Genet 22:521–565
Fernandes A (1935) Remarque sur L'hétérostylie de *Narcissus triandrus* et de *N. reflexus*. Brot Bol Soc Broteriana Ser 2, 10:278–288
Fisher RA (1958) The genetical theory of natural selection, 2nd edn. Dover Publications, New York
Fisher RA, Martin VC (1948) Genetics of style length in *Oxalis*. Nature (Lond) 177:942–943
Ford EB (1964) Ecological genetics. Methuen, Lond
Ganders FR (1974) Disassortative pollination in the distylous plant *Jepsonia heterandra*. Can J Bot 52:2401–2406
Ganders FR (1975a) Mating patterns in self-compatible distylous populations of *Amsinckia* (Boraginaceae). Can J Bot 53:773–779
Ganders FR (1975b) Heterostyly, homostyly and fecundity in *Amsinckia spectabilis* (Boraginaceae). Madroño 23:56–62
Ganders FR (1979) The biology of heterostyly. N Z J Bot 17:607–635
Ganders FR, Denny SK, Tsai D (1985) Breeding system variation in *Amsinckia spectabilis* (Boraginaceae). Can J Bot 63:533–538
Gibbs PE (1986) Do homomorphic and heteromorphic self-incompatibility systems have the same sporophytic mechanism? Plant Syst Evol 154:285–323
Glover DE, Barrett SCH (1983) Trimorphic incompatibility in Mexican populations of *Pontederia sagittata* Presl. (Pontederiaceae). New Phytol 95:439–455
Glover DE, Barrett SCH (1986) Variation in the mating system of *Eichhornia paniculata* (Spreng.) Solms (Pontederiaceae). Evolution 40:1122–1131
Goldblatt P, Bernhardt P (1990) Pollination biology of *Nivenia* (Iridaceae) and the presence of heterostylous self-compatibility. Isr J Bot 39:93–111
Gottlieb LD (1984) Genetics and morphological evolution in plants. Am Nat 123:681–709
Graham SW, Barrett SCH (1990) Pollen precedence in *Eichhornia paniculata* (Pontederiaceae): a tristylous species. Am J Bot 77:54-55 (abstract)
Grant V (1975) Genetics of flowering plants. Columbia Univ Press, New York
Hamilton CW (1990) Variations on a distylous theme in Mesoamerican *Psychotria* subgenus *Psychotria* (Rubiaceae). Mem N Y Bot Gard 55:62–75
Haring V, Gray JE, McClure BA, Anderson MA, Clarke AE (1990) Self-incompatibility: a self-recognition system in plants. Science 250:937–941
Heuch I (1979) Equilibrium populations of heterostylous plants. Theor Popul Biol 15:43–57
Heuch I (1980) Loss of incompatibility types in finite populations of the heterostylous plant *Lythrum salicaria*. Hereditas 92:53–57
Hildebrand F (1866) Ueber den Trimorphismus in der Gattung *Oxalis*. Monatsber Königl Preuss Akad Wiss Berl 1866:352–374
Holsinger KE (1991) Mass action models of plant mating systems I. The evolutionary stability of mixed mating systems. Am Nat (in press)

Holsinger KE, Feldman MW, Christiansen FB (1984) The evolution of self-fertilization in plants: a population genetic model. Am Nat 124:446–453
Husband BC, Barrett SCH (1991a) Pollinator visitation patterns in populations of tristylous *Eichhornia paniculata* in northeastern Brazil. Oecologia (in press)
Husband BC, Barrett SCH (1991b) Colonization history and population genetic structure of *Eichhornia paniculata* in Jamaica. Heredity 66:287–296
Jernstedt JA (1982) Floral variation in *Chlorogalum angustifolium* (Liliaceae). Madroño 29:87–94
Karron JD, Marshall DL, Oliveras DM (1990) Numbers of sporophytic self-incompatibility alleles in populations of wild radish. Theor Appl Genet 79:457–460
Kohn JR, Barrett SCH (1992) Experimental studies on the functional significance of heterostyly. Evolution (in press)
Lack A, Kevan PG (1987) The reproductive biology of a distylous tree, *Sarcotheca celebica* (Oxalidaceae) in Sulawesi, Indonesia. Biol J Linn Soc 95:1–8
Lande R, Schemske DW (1985) The evolution of self-fertilization and inbreeding depression in plants I. Genetic models. Evolution 39:24–40
Leach CR (1983) Fluctuations in heteromorphic self-incompatibility systems. Theor Appl Genet 66:307–312
Lemmens RHMJ (1989) Heterostyly in Connaraceae. Acta Bot Neerl 38:224–225
Levin DA (1974) Spatial segregation of pins and thrums in populations of *Hedyotis nigricans*. Evolution 28:648–655
Lewis D (1982) Incompatibility, stamen movement, and pollen economy in a heterostylous tropical forest tree, *Cratoxylum formosum* (Guttiferae). Proc R Soc Lond Ser B 214:273–283
Lloyd DG (1979) Evolution towards dioecy in heterostylous populations. Plant Syst Evol 131:71–80
Lloyd DG (1980) Demographic factors and mating patterns in angiosperms. In: Solbrig OT (ed) Demography and evolution in plant populations. Blackwell, Oxford, pp 67–88
Lloyd DG, Webb CJ, Dulberger R (1990) Heterostyly in species of *Narcissus* (Amaryllidaceae), *Hugonia* (Linaceae) and other disputed cases. Plant Syst Evol 172:215–227
Lovett Doust J, Lovett Doust L, Cavers PB (1981) Fertility relationships in closely related taxa of *Oxalis*, section Corniculatae. Can J Bot 59:2603–2609
Martin FW (1967) Distyly, self-incompatibility, and evolution in *Melochia*. Evolution 21:493–499
Mather K (1950) The genetical architecture of heterostyly in *Primula sinensis*. Evolution 4:340–352
Mather K, DeWinton D (1941) Adaptation and counter-adaptation of the breeding system in *Primula*. Ann Bot II 5:297–311
Morgan MT, Barrett SCH (1988) Historical factors and anisoplethic population structure in tristylous *Pontederia cordata*: a re-assessment. Evolution 42:496–504
Morgan MT, Barrett SCH (1989) Reproductive correlates of mating system variation in *Eichhornia paniculata* (Spreng.) Solms (Pontederiaceae). J Evol Biol 2:183–203
Morgan MT, Barrett SCH (1990) Outcrossing rates and correlated mating within a population of *Eichhornia paniculata* (Pontederiaceae). Heredity 64:271–280
Muenchow GE (1982) A loss-of-alleles model for the evolution of distyly. Heredity 49:81–93
Muenchow GE, Grebus M (1989) The evolution of dioecy from distyly: reevaluation of the hypothesis of the loss of long-tongued pollinators. Am Nat 133:149–156
Mulcahy DL (1964) The reproductive biology of *Oxalis priceae*. Am J Bot 51:915–1044
Murray BG (1990) Heterostyly and pollen-tube interactions in *Luculia gratissima* (Rubiaceae). Ann Bot 65:691–698
Nicholls M (1987) Pollen flow, self-pollination and gender specialization: factors affecting seed set in the tristylous species *Lythrum salicaria* (Lythraceae). Plant Syst Evol 156:151–157
O'Brien SP, Calder DM (1989) The breeding biology of *Epacris impressa*. Is this species heterostylous? Austr J Bot 37:43–54
Olesen JM (1979) Floral morphology and pollen flow in the heterostylous species *Pulmonaria obscura* Dumort (Boraginaceae). New Phytol 82:757–767
Olmstead RG (1989) The origin and function of self-incompatibility in flowering plants. Sex Plant Reprod 2:127–136
Opler PA, Baker HG, Frankie GW (1975) Reproductive biology of some Costa Rican *Cordia* species (Boraginaceae). Biotropica 7:243–247

Ornduff R (1966) The origin of dioecism from heterostyly in *Nymphoides* (Menyanthaceae). Evolution 20:309–314
Ornduff R (1976) The reproductive system of *Amsinckia grandiflora*, a distylous species. Syst Bot 1:57–66
Ornduff R (1978) Features of pollen flow in dimorphic species of *Lythrum* section Euhyssopifolia. Am J Bot 65:1077–1083
Ornduff R (1980) Heterostyly, population composition, and pollen flow in *Hedyotis caerulea*. Am J Bot 67:95–103
Ornduff R (1986) Comparative fecundity and population composition of heterostylous and non-heterostylous species of *Villarsia* (Menyanthaceae) in western Australia. Am J Bot 73:282–286
Ornduff R (1988) Distyly and monomorphism in *Villarsia* (Menyanthaceae): some evolutionary considerations. Ann Mo Bot Gard 75:761–767
Palmer JD, Jansen RK, Michaels HJ, Chase MW, Manhart JR (1988) Chloroplast DNA variation and phylogeny. Ann Mo Bot Gard 75:1180–1206
Philipp M, Schou O (1981) An unusual heteromorphic incompatibility system: distyly, self-incompatibility, pollen load and fecundity in *Anchusa officinalis* (Boraginaceae). New Phytol 89:693–703
Piper JG, Charlesworth B, Charlesworth D (1984) A high rate of self-fertilization and increased seed fertility of homostyle primroses. Nature (Lond) 310:50–51
Piper JG, Charlesworth B, Charlesworth D (1986) Breeding system evolution in *Primula vulgaris* and the role of reproductive assurance. Heredity 56:207–217
Ray PM, Chisaki HF (1957) Studies of *Amsinckia* I. A synopsis of the genus, with a study of heterostyly in it. Am J Bot 44:529–536
Richards AJ (1986) Plant breeding systems. Allen and Unwin, Lond
Richards AJ, Ibrahim H (1982) The breeding system in *Primula veris* L. II. Pollen tube growth and seed set. New Phytol 90:305–314
Riveros M, Arroyo MTK, Humana AM (1987) An unusual kind of heterostyly in *Quinchamalium chilense* (Santalaceae) on Volcán Casablanca, southern Chile. Am J Bot 74:1831–1834
Sampson DR (1967) Frequency and distribution of self-incompatibility alleles in *Raphanus raphanistrum*. Genetics 56:241–251
Schaeppi H (1935) Kenntnis der Heterostylie von *Gregoria vitaliana* Derby. Ber Schweiz Bot Ges 44:109–132
Schou O, Philipp M (1984) An unusual heteromorphic incompatibility system III. On the genetic control of distyly and self-incompatibility in *Anchusa officinalis* L. (Boraginaceae). Theor Appl Genet 68:139–144
Schwarz-Sommer Z, Huijser P, Nacken W, Saedler H, Sommer H (1990) Genetic control of flower development by homeotic genes in *Antirrhinum majus*. Science 250:931–936
Scribailo RW, Barrett SCH (1991) Pollen-pistil interactions in tristylous *Pontederia sagittata* (Pontederiaceae) II. Patterns of pollen tube growth. Am J Bot (in press)
Scribailo RW, Barrett SCH (1992) Effects of compatible and incompatible pollen mixtures on seed set in tristylous *Pontederia sagittata* (Pontederiaceae). Unpubl. manuscript
Shore JS, Barrett SCH (1984) The effect of pollination intensity and incompatible pollen on seed set in *Turnera ulmifolia* (Turneraceae). Can J Bot 62:1298–1303
Shore JS, Barrett SCH (1985) The genetics of distyly and homostyly in *Turnera ulmifolia* L. (Turneraceae). Heredity 55:167–174
Shore JS, Barrett SCH (1986) Genetic modifications of dimorphic incompatibility in the *Turnera ulmifolia* L. complex (Turneraceae). Can J Genet Cytol 28:796–807
Shore JS, Barrett SCH (1990) Quantitative genetics of floral characters in homostylous *Turnera ulmifolia* var. *angustifolia* Willd. (Turneraceae). Heredity 64:105–112
Stirling J (1932) Studies of flowering in heterostyled and allied species Part 1. The Primulaceae. Publ Hartley Bot Lab 8:3–42
Stirling J (1933) Studies of flowering in heterostyled and allied species Part 2. The Lythraceae: *Lythrum salicaria*. Publ Hartley Bot Lab 10:3–24
Stirling J (1936) Studies of flowering in heterostyled and allied species Part 3. Gentianaceae, Lythraceae, Oxalidaceae. Publ Hartley Bot Lab 15:1–24
Tseng HP (1938) Pollen-tube competition in *Primula sinensis*. J Genet 35:289–300

Vasudevan Nair R (1975) Heterostyly and breeding mechanism of *Nymphoides cristatum* (Roxb.) O. Kuntze. J Bombay Nat Hist Soc 72:677–682
Webb CJ, Lloyd DG (1986) The avoidance of interference between the presentation of pollen and stigmas in angiosperms II Herkogamy. N Z J Bot 24:163–178
Wedderburn F, Richards AJ (1990) Variation in within-morph incompatibility sites in heteromorphic *Primula* L. New Phytol 116:149–162
Weller SG (1976a) Breeding system polymorphism in a heterostylous species. Evolution 30:442–454
Weller SG (1976b) The genetic control of tristyly in *Oxalis* section *Ionoxalis*. Heredity 37:387–393
Weller SG (1981) Pollination biology of heteromorphic populations of *Oxalis alpina* (Rose) Kunth (Oxalidaceae) in southeastern Arizona. Bot J Linn Soc 83:189–198
Weller SG, Ornduff RO (1977) Cryptic self-incompatibility in *Amsinckia grandiflora*. Evolution 31:47–51
Wells H (1979) Self-fertilization: advantageous or disadvantageous? Evolution 33:252–255
Wolfe LM, Barrett SCH (1989) Patterns of pollen removal and deposition in tristylous *Pontederia cordata* L. (Pontederiaceae). Biol J Linn Soc 36:317–329
Wyatt R (1983) Pollinator-plant interactions and the evolution of plant breeding systems. In: Real L (ed) Pollination biology. Academic Press, Orlando Florida
Yeo PF (1975) Some aspects of heterostyly. New Phytol 75:147–153

Chapter 2

Historical Perspectives on Heterostyly

R. ORNDUFF[1]

1 Early Observations of Heterostyly

Heterostyly was recognized as a morphological feature of certain groups of flowering plants as early as the 16th century, when it was reputedly noted in *Primula* by Clusius (van Dijk 1943). Few, if any, attempts were made to interpret the adaptive significance of this floral heteromorphism until Charles Darwin and Friedrich Hildebrand studied the phenomenon just after the middle of the 19th century. In 1862 Darwin published his first paper on dimorphic primulas and their "remarkable sexual relations", and in 1863 Hildebrand published his initial observations on *Primula sinensis*. In his 1862 paper, Darwin introduced, but then abandoned, the horticultural terms "pin-eyed" and "thrum-eyed" (erroneously as "thumb-eyed") to refer to the long- and short-styled floral morphs of Primulas. Many subsequent authors, for the sake of brevity, have adopted the terms pin and thrum, although these are applicable only to the flowers of dimorphic species and not of trimorphic ones. Thus, if both distylous and tristylous species are under discussion, referring to the morphs as long-, mid-, and short-styled or Longs, Mids, and Shorts is the most economical terminology.

In his 1862 paper Darwin noted that there appeared to be a general consensus among botanists that heterostyly[1] represented a case of "mere variability" in length and position of floral parts, which, he observed "is far from the truth", since he had "never the slightest doubt under which form [of the two] to class a plant." After discussing the floral morphology of several species of *Primula*, Darwin concluded "that the existence of two forms is very general" in the genus. "But what, it may be asked, is the meaning of these several differences?" He admits that his initial conclusion was that dimorphic primulas "were tending" toward dioecism, with Longs tending to be "more feminine" in nature and Shorts "more masculine". But, as he pointed out, his observations of seed production by the two morphs demonstrated that his initial surmise was "exactly the reverse of the truth." Subsequently, Darwin carried out a crossing program and measured seed-set following intramorph self- and cross-pollinations and intermorph pollinations in cultivated specimens of various species of *Primula*. He concluded that the floral dimorphism of *Primula* represents a physiological-morphological adaptation to promote outcrossing between the two

[1] Department of Integrative Biology, University of California, Berkeley, CA 94720, USA

[2] Darwin (1877) credited Hildebrand with coining the term heterostyly (as heterostylie) and adopted it in preference to the term heterogonous, also in use at the time, and preferred by the American botanist Asa Gray.

Monographs on Theoretical and Applied Genetics 15
Evolution and Function of Heterostyly (ed. by S.C.H. Barrett)

morphs. Seed-set data led to this conclusion, as did the reciprocal positions of anthers and stigmas in the two morphs. Darwin viewed the floral dimorphism of *Primula* as a "novel and curious aid for checking continued self-fertilization." He queried the general significance of sex, a subject "as yet hidden in darkness." Although Darwin was perhaps not the discoverer of incompatibility systems in plants, he was at least unaware of previous knowledge of this phenomenon, and noted that the existence of healthy plants which produce no seeds when self-pollinated "is one of the most surprising facts which I have ever observed."

Three contemporaries of Darwin were also important in assisting in the development of his own ideas on heterostyly by providing their own observations on the phenomenon. One of these was John Scott (1838–1880) who, while corresponding with Darwin, was head gardener at the Royal Botanic Garden, Edinburgh. Scott provided Darwin with his own observations on heterostylous (and other) plants in the Edinburgh collection; Darwin characterized Scott as someone with "remarkable talent, astonishing perseverance... and what I admire, determined difference from me on many points." A second individual of importance in Darwin's early work on heterostyly was the American botanist Asa Gray (1810–1888). Although not an experimentalist, Gray was a close observer of plants and provided Darwin with numerous new examples of heterostyly, particularly in North American taxa, and corresponded with him on numerous other topics as well. Indeed, as early as 1842 Gray noted distyly in the eastern American *Hedyotis purpurea* (Rubiaceae) and observed that both morphs of this species "appeared to be equally fertile." Darwin's debt to Gray is acknowledged by the dedication of *Forms of Flowers* (1877) to this American colleague and supporter.

Another contemporary of Darwin who played a key role in the early development of knowledge of the significance and functioning of heterostyly was the German Friedrich H. G. Hildebrand (1835–1915), Professor of Botany at Freiburg. Darwin and Hildebrand began corresponding as early as 1863, the year in which Hildebrand published a brief note on dimorphism in *Primula sinensis.* Hildebrand apparently initiated correspondence with Darwin, mentioning his own interest in the pollination of European orchids and work in progress on experimental studies of distyly in *Linum perenne.* Darwin encouraged Hildebrand in his investigations of heterostyly, in part because French botanists had viewed his own first publication on *Primula* to be "the work of imagination[3]." Darwin apparently believed he needed an advocate on the continent. Correspondence between the two was not voluminous, but was friendly, covered diverse topics, and continued for several years. Just as Gray was in some respects Darwin's American botanical bulldog, so was Hildebrand his German botanical bulldog, since Hildebrand published favorable reviews of Darwin's work in German journals. Hildebrand published the results of his own crossing programs and other related studies with *Primula sinensis* (1863, 1864) *Linum perenne* (1864), *Pulmonaria officinalis* (1865a), *Lythrum salicaria* (1865b), several species of *Oxalis*

[3] Coutance, in his preface to Heckel's translation into French of *Forms of Flowers* (1878), remarked that if Darwin were not known to be a "serious and conscientious" observer, one would be tempted to view this work as "a romance or a fantasy' in large part. Bonnier (1884), in his comments on the second edition of *Forms of Flowers*, doubted the existence of heterostyly, noting that in *Lythrum salicaria* there are not only three forms of flowers, but an "infinity" of them.

(1867, 1871, 1887, 1899), and *Forsythia* (1894). Although Hildebrand continued publishing botanical articles until he was almost 80, his last major published paper on heterostyly appeared in 1899 and was concerned with further studies of tristyly in species of *Oxalis*.

Darwin, Gray, and Hildebrand all worked with herbaceous species of north temperate latitudes. Although not an experimentalist, W. Burck was an important 19th century contributor to our knowledge of the distribution of heterostyly in tropical groups, particularly woody ones (Burck 1883, 1884). Working in Buitenzorg (now Bogor), he described distyly in various southeastern Asian genera of woody Rubiaceae, Connaraceae, and Oxalidaceae. Burck (1887) was perhaps the first to speculate on the evolutionary derivation of distyly from tristyly in the Oxalidaceae, a pattern that was later documented in this family as well as in the Lythraceae.

Another important correspondent of Darwin who resided on the fringe of the Brazilian tropics was the schoolmaster Fritz (J.F.T.) Müller, who reported on the occurrence there of heterostyly in the Pontederiaceae (1871). Fritz was the brother of Hermann Müller, whose *Die Befruchtung der Blumen durch Insekten und die gegenseitigen Anpassungen beider* is one of the early classics of pollination biology.

2 Darwin's Publications

After his first paper on heterostyly in 1862, Darwin published a second one in 1864, concerned with the nature of distyly and crossing relationships in *Linum* species. The same year, he also published an extensive paper on tristyly in *Lythrum salicaria* (Darwin 1865). In 1869, a paper *On the Character and Hybrid-Like Nature of the Offspring from the Illegitimate Unions of Dimorphic and Trimorphic Plants* was published. In 1877 he published a general review of the subject of heterostyly in the book *The Different Forms of Flowers on Plants of the Same Species*. The same year, an American edition was published by Appleton and a German translation was issued by Schweizerbart. In 1878, a French version was published by Reinwald but apparently converted only some of the French skeptics. Editions in still other languages followed.

In 1880, a second edition of *Forms of Flowers* was published, but the text was identical to the first edition except for correction of errors; Darwin added a few pages of prefatory material which updated the subject matter on the basis of additional published articles and correspondence he had received. In a reprint of the second edition published in 1884, Darwin's son Francis added a second preface, again bringing material up to date based on articles published by several workers.

Forms of Flowers contains eight chapters. Two deal with dimorphism in *Primula* and two other genera of Primulaceae, one with several miscellaneous distylous taxa, one with tristylous taxa, one with the "illegitimate" offspring of heterostyled plants ("illegitimate" was Darwin's term for pollinations between stigmas and anthers not in equivalent positions, and the progeny obtained from such pollinations), and a sixth chapter with "concluding" remarks on heterostyled plants. Chapter 7 is concerned with various combinations of unisexual flowers in a number of species and the final chapter deals with cleistogamy. Chapter 5 is perhaps anachronistic when read with a modern perspective; in it, Darwin discusses the intraspecific incompatibility characteristic of illegitimate pollinations between the two or three morphs of heterostylous plants with that commonly observed when attempts are made to make interspecific crosses of nonheterostylous plants. However, Darwin did discuss observations indicating the occurrence of pollen tube, or gametophytic, competition in styles of flowers that received mixed pollen loads. He also puzzled over the significance of the pollen-size heteromorphism that characterizes so many unrelated heterostylous species, a matter which has not yet been clearly resolved. It is ironic that although *Forms of Flowers* is one of Darwin's lesser known books, it summarizes work about which he wrote "I do not think anything in my scientific life has given me so much satisfaction as making out the meaning of the structure of these plants."

3 Opposition to Darwin's Views

While Darwin's emphasis in his studies of heterostylous and nonheterostylous plants was on floral mechanisms promoting xenogamy, his views met with considerable opposition. The views of French botanists have already been mentioned. In the United States, Thomas Meehan (1876) maintained that even flowers with conspicuous and complex perianths "are self-fertilizers," although he later changed his mind and agreed, at least, that "cross-fertilization will result in a plant better fitted for the struggle for life" than one that is self-fertilized (Meehan 1897). Even as late as 1899, the Englishman Edward Bell published a lengthy rebuttal to Darwin's primrose work, a tirade which was expanded and published (anonymously) in 1902. Despite Darwin's observations on *Primula* as well as those of other 19th century naturalists, Bell maintained that self-fertilization of heterostylous plants "is the natural fertilization." These strenuous arguments seem quaint in a modern context, but reveal views that were widely held less than a century ago.

4 Early Field Studies

Despite the fact that Darwin had relatively easy access to natural populations of heterostylous species of *Primula, Menyanthes trifoliata, Hottonia palustris,* and *Lythrum salicaria*, he chose to conduct his work and observations solely on cultivated specimens of the heterostylous species with which he worked.

Although the birth of population biology was many decades in the future, Darwin's work inspired others to conduct a number of interesting field studies of heterostylous plants. A particularly interesting and generally overlooked example of such a study was that published in 1884 by Christy. In this lengthy paper, *On the Species of the Genus Primula in Essex; with Observations on Their Variation and Distribution, and the Relative Number and Fertility in Nature of the Two Forms of Flower* in what even then was probably an obscure journal (Transactions of the Essex Field Club), Christy acknowledges that Darwin's *Forms of Flowers* "led me to pay special attention to several points in the Natural History... of the genus *Primula*."

Unlike Bell, Christy accepted the notion that the floral structure of distylous Primulas is a mechanism designed to promote intermorph pollinations. Christy confirmed Darwin's observations on the several differences between the two morphs of *Primula* species, adding one or two additional features he noted himself, and also confirmed the occasional occurrence of homostyled (and thus self-compatible) plants in some species. Beyond this, Christy dissected very young buds of two *Primula* species to "enquire how early in the development of the bud heterostylism exists", finding that in "very young" buds "the two forms were totally indistinguishable." Similar studies, using modern techniques, continue in this decade (see Chap. 4). Christy also investigated morph ratios of pins and thrums in natural populations of various British Primulas, taking issue with Darwin's observation that "the two forms [of *P. veris*] exist in a wild state in about equal numbers". Christy sampled 22 populations of the primrose (*P. vulgaris*), finding that in 14 of the samples Longs predominated (with a few homostyles present in three populations). Taking all samples, he concluded that Longs constitute 53% of the total sample. I applied the χ^2 statistic to his raw figures, however, and in only one population he examined do the figures deviate from 1:1 Long:Short in the direction of an excess of Long.

Similar figures are presented for *P. veris* and *P. elatior*; despite the possibility that Christy's interpretation of his data may not stand up under simple statistical scrutiny, he made a very interesting observation on *P. elatior*, namely, that under shaded conditions Shorts may prevail, but under open conditions, Longs prevailed [compare this with Levin's (1975) observations on differential distribution of pins and thrums in *Hedyotis nigricans*.] Christy also suggested that under open conditions, Longs produce more flowers per plant than Shorts. These observations should be explored today. Christy expressed surprise that, considering the compatibility relationships between the two morphs of the various *Primula* species, Longs and Shorts — according to his interpretation of his data — do not occur in equal numbers. Christy translated Darwin's measurement of pollen sizes of the two morphs of *P. vulgaris* and *P. veris* into pollen production figures, with Longs producing ca. 1.3 times more pollen per anther than Shorts (again, these sorts of calculations have been revived by those working with pollen flow in heterostylous plants in recent decades), and interpreted the significance of these differences in pollen production to the requirement of Short stigmas for more pollen than Long stigmas because of greater seed production of the former. Although other interpretations of these observations are possible, Christy's line of thinking strikes me as clever and having a remarkably modern ring.

After further discussion of differential seed production by the two morphs of *Primula*, Christy invoked fitness, suggesting that while Shorts produce more seeds than Longs, the "quality" of the seeds produced by the latter is superior. Modern plant demographers should find this surmise of interest. Although Christy's paper possesses quaint Victorianisms and some surmises that Mendelian genetics would later resolve, it represents the activities of an original intellect and an observant fieldworker. Christy's interest in *Primula* appeared to subside for nearly four decades, but in 1922 he published a paper on the pollination of British primulas, a subject about which there has been much controversy.

Darwin died in 1882, 5 years after the publication of *Forms of Flowers*. The latter portion of the 19th century saw a few dozen papers published on heterostyly by other workers in several countries. These included mostly reports of heterostyly in groups in which it was previously unrecorded, or field studies of insect visitors to the flowers of heterostylous flowers. Thus, the 19th century can be viewed largely as a time when the morphological nature of heterostyly was described, its functional significance suggested, its occurrence documented, and some field studies conducted. The numerous field observations of floral visitors did not explore the role of these visitors in effecting legitimate pollen transfer. Many workers, including Darwin, puzzled over the "transmission" of heterostylous traits, but no progress was made in understanding inheritance patterns until the advent of Mendelian genetics in the first year of the 20th century.

5 Early Genetic Work

Although the term "genetics" was not invented until later in the decade, the field was established in 1900 with the independent "rediscovery" of Mendel's work by Correns, von Tschermak, and de Vries. Interestingly, two of these workers also later investigated features of heterostylous plants (Correns 1921; von Tschermak 1935). Bateson, who coined the term genetics in 1906, collaborated in a joint investigation of the inheritance of distyly in *Primula sinensis* published the preceding year (Bateson and Gregory 1905). Although Mawe and Abercrombie (1778) specifically indicated that Short primulas were preferred by horticulturists for esthetic reasons, a little over a century later taste in flower patterns had evidently changed, since Bateson and Gregory state that "Fashion has decreed that *P. sinensis* shall be exhibited in the long-styled form alone." These workers begin their report by paying homage to their predecessors: "In view of the results obtained by Darwin, Hildebrand and others, it seems likely that the characters long-style and short-style... might have a Mendelian inheritance." Short *P. sinensis* was shown to be dominant, Long recessive (both genetic terms used by those workers). Since *P. sinensis* has a high degree of intramorph compatibility, Bateson and Gregory were able to obtain *SS* Shorts via illegitimate pollinations as well as the usual *Ss* Shorts; Longs are *ss*. The two workers suggested that this simple Mendelian inheritance also occurs in *P. vulgaris*, though this species was somewhat more difficult to deal with than *P. sinensis* because of its strong intramorph incompatibility. Bateson and Gregory also investigated the genetics of homostyly ("equal-styled plants") in *P. sinensis*; these are short-homostyles, with low anthers of the Long morph and small pollen and short styles of the Short morph combined in a single flower.

The genetics of tristyly was first worked out in *Oxalis valdiviana* and *Lythrum salicaria* by Darwin's granddaughter, Nora Barlow (Barlow 1913, 1923). In most tristylous species that have been investigated, Longs are *ssmm*, Mids are *ss M__*, and Shorts are *S__*. Other schemes are known (see Chaps. 1 and 5), but all involve the action of two loci with two alleles at each locus. The influence of polyploidy on inheritance of tristyly in *L. salicaria* was discussed by Fisher and Mather (1943) without altering Barlow's basic conclusions for this species. Important contributions to an understanding of the genetics of heterostyly include those of von Ubisch (1925), Bodmer (1927), Dowrick (1956), East (1927), Ernst (e.g., 1925, 1928, 1943), Gregory (1911), Laibach (1929), Mather (1950), and Mather and De Winton (1941). Indeed, the voluminous work of Ernst deserves to be more widely known, and it is good that Lewis and Jones refer to it in this Volume (see Chap. 5).

6 Conclusions

Probably the most significant advances in knowledge and understanding of heterostyly prior to 1950 were in documenting its distribution in flowering plants, in describing morphological and phenological differences between the morphs of heterostylous plants, in recording (although not studying behavior) visitors to flowers of such plants, in determining morph ratios of a few species in the field, and in working out the genetic bases of distyly and tristyly in a number of species. In recent years, important advances have been made in understanding the physiological and biochemical basis of heteromorphic incompatibility, in pollen flow patterns under natural conditions and other features of the population biology of heterostylous plants, and in the comparative organogenesis of heterostylous flowers. There also has been attention given to various theoretical aspects of the population biology and evolutionary modification of heterostylous species. Important general reviews that range far more widely than this one include those of Whitehouse (1959), Vuilleumier (1967), Ganders (1979), and Richards (1986), as well as the papers that follow in this Volume, all by workers who have made important contributions to our understanding of the nature and functioning of heterostyly and the population biology of heterostylous plants.

References

Barlow N (1913) Preliminary note on heterostylism in *Oxalis* and *Lythrum*. J Genet 3:53–65

Barlow N (1923) Inheritance of the three forms in trimorphic species. J Genet 13:133–146

Bateson W, Gregory RP (1905) On the inheritance of heterostylism in *Primula*. Proc R Soc Lond Ser B 76:581–586

Bell EA [A field naturalist] (1899) The primrose and Darwinism. Lond Q Rev 92:14–235

Bell EA [A field naturalist] (1902) The primrose and Darwinism. Grant Richards, Lond

Bodmer H (1927) Beiträge zum Heterostylie-Problem bei *Lythrum salicaria* L. Flora 122:306–341

Bonnier G (1884) Sur les différentes formes des fleurs de la même espèce. Bull Bot Soc Fr 31:240–244

Burck W (1883) Sur l'organisation florale chez quelques Rubiacées. Ann Jard Bot Buitenzorg 3:105–119

Burck W (1884) Sur l'organisation florale chez quelques Rubiacées. (suite) Ann Jard Bot Buitenzorg 4:12–87

Burck W (1887) Notes biologiques. I. Relation entre l'hétérostylie dimorphe et l'hétérostylie trimorphe. Ann Jard Bot Buitenzorg 6:251–254

Christy RM (1884) On the species of the genus *Primula* in Essex; with observations on their variation and distribution, and the relative number and fertility in nature of the two forms of flower. Trans Essex Field Club 3:148–211

Christy M [RM] (1922) The pollination of British primulas. J Linn Soc Bot 46:105–139

Correns C (1921) Zahlen- und Gewichtsverhältnisse bei einigen heterostylen Pflanzen. Biol Zentralbl 41:97–109

Coutance AGA (1878) Preface to Darwin. Des différentes formes des fleurs dans les plantes de la même espèce. Reinwald, Paris

Darwin C (1862) On the two forms, or dimorphic condition, in species of *Primula*, and on their remarkable sexual relations. Proc Linn Soc Bot 6:77–96

Darwin C (1864) On the existence of two forms, and of their reciprocal sexual relation, in several species of the genus *Linum*. J Linn Soc Bot 7:69–83

Darwin C (1865) On the sexual relations of the three forms of *Lythrum salicaria*. J Linn Soc Bot 8:169–196

Darwin C (1869) On the character and hybrid-like nature of the offspring from the illegitimate unions of dimorphic and trimorphic plants. J Linn Soc Bot 10:393–437

Darwin C (1877) The different forms of flowers on plants of the same species. Murray, Lond

Darwin C (1878) Des différentes formes des fleurs dans les plantes de la même espèce. Reinwald, Paris

Dowrick VPJ (1956) Heterostyly and homostyly in *Primula obconica*. Heredity 10:219–236

East EM (1927) The inheritance of heterostyly in *Lythrum salicaria*. Genetics 12:393–414

Ernst A (1925) Genetische Studien über Heterostylie bei *Primula*. Arch Julius Klaus-Stift 1:13–62

Ernst A (1928) Zur Genetik der Heterostylie. Z Indukt Abstammungs Vererbungsl (Suppl) I:635–665

Ernst A (1943) Kreuzungen zwischen dimorphen und monomorphen *Primula*-Arten und ihre Aufschlüsse zum Heterostylieproblem. Planta 33:615–636

Fisher RA, Mather K (1943) The inheritance of style length in *Lythrum salicaria*. Ann Eugenics 12:1–23

Ganders FR (1979) The biology of heterostyly. NZ J Bot 17:607–635

Gray A (1842) Notes of a botanical excursion to the mountains of North Carolina. Am J Sci Arts 42:1–49

Gregory RP (1911) Experiments with *Primula sinensis*. J Genet 1:73–132

Hildebrand F (1863) Dimorphismus von *Primula sinensis*. Verh Naturh Vereins Rheinl Westf Sitzungber 20:183–184

Hildebrand F (1864) Experimente über Dimorphismus von *Linum perenne* und *Primula sinensis*. Bot Zeit 22

Hildebrand F (1865a) Experimente zur Dichogame und zum Dimorphismus. 2. Dimorphismus von *Pulmonaria officinalis*. Bot Zeit 23:13–15

Hildebrand F (1865b) Über den Trimorphismus von *Lythrum salicaria*. Verh Naturh Vereins Rheinl Westf Sitzungber 4–6

Hildebrand F (1867) Über den Trimorphismus in der Gattung *Oxalis*. Monatsber Königl Preuss Akad Wiss Berl Juni:353–374

Hildebrand F (1871) Experimente und Beobachtungen an einigen trimorphen *Oxalis*-Arten. Bot Zeit 29:415–425, 431–440

Hildebrand F (1887) Experimente über die geschlechtlichen Oxalisarten. Bot Zeit 45:1–6, 16–23, 33–40

Hildebrand F (1894) Ueber die Heterostylie und Bastardierungen bei *Forsythia*. Bot Zeit 52:191–20

Hildebrand F (1899) Einige weitere Beobachtungen und Experimente an *Oxalis*-Arten. Bot Centralbl 79:1–10, 35–44

Hildebrand F (1902) Einige systematische und biologische Beobachtungen. Beitr Bot Centralbl 13:333–340

Laibach F (1929) Die Bedeutung der homostylen Formen für die Frage nach der Vererbung der Heterostylie. Ber Dtsch Bot Ges 47:584–596

Levin D (1975) Spatial segregation of pins and thrums of *Hedyotis nigricans*. Evolution 28:648–655

Mather K (1950) The genetical architecture of heterostyly in *Primula sinensis*. Evolution 4:340–352

Mather K, de Winton D (1941) Adaptation and counter-adaptation of the breeding system in *Primula*. Ann Bot NS 5:297–311

Mawe T, Abercrombie J (1778) The universal gardener and botanist; or, a general dictionary of gardening and botany. Robinson and Cadell, Lond

Meehan T (1876) Are insects any material aid to plants in fertilization? Proc Am Assoc Adv Sci 24:243–251
Meehan T (1897) Contributions to the life histories of plants, no. XII. Proc Acad Natrl Sci Philadelphia 49:169–203
Müller F (1871) Über den Trimorphismus der Pontederien. Jena Z Med Naturwiss 6:74–78
Richards AJ (1986) Plant breeding systems. Allen and Unwin, Lond
van Dijk W (1943) Le decóuverte de l'hétérostylie chez *Primula* par Ch. de l'Écluse et P. Reneaulme. Ned Kruidkd Arch 53:81–85
von Ubisch G (1925) Genetisch-physiologische Analyse der Heterostylie. Bibliogr Genet 2:284–342
von Tschermak E (1935) Über die Genik des Dimorphismus und das Vorkommen von Homostylie von *Primula*. Anz Akad Wiss Wien 27:165–166
Vuilleumier B (1967) The origin and evolutionary development of heterostyly in the angiosperms. Evolution 21:210–226
Whitehouse HLK (1959) Cross- and self-fertilization in plants. In: Bell PR (ed) Darwin's biological work. Cambridge Univ Press, Cambridge, pp 207–261

Chapter 3

Floral Polymorphisms and Their Functional Significance in the Heterostylous Syndrome

RIVKA DULBERGER[1]

1 Introduction

The emblem of the heterostylous syndrome is the positioning at equivalent levels of the stigmas of one floral morph and the anthers of one or two alternate morphs. Of less prominence in the syndrome are distinguishing characters of the pistils and stamens of the morphs, particularly of the stigmas and pollen grains. At the core of heterostyly there is typically a sporophytically controlled, diallelic incompatibility system which prevents or reduces self- and intra-morph fertilizations. Thus, in distyly the entire syndrome is controlled by a "supergene", i.e., a tightly linked group of genes, each with two alleles and apparently controlling three groups of characters; in tristyly the control is by genes at two loci, each with two alleles and epistatic interaction.

Darwin (1877) suggested that reciprocal placement of stigmas and anthers serves as a mechanical device for promoting insect-mediated legitimate pollinations, i.e., pollinations between stigmas of one floral morph and anthers at the equivalent level of another morph. The same device would then reduce pollen wastage deriving from inefficient illegitimate pollinations. According to Darwin's interpretation, the morphological features of heterostyly and sterility in self- and intramorph cross-pollinations are two distinct outbreeding mechanisms. This interpretation has received wide acceptance among students of heterostyly (Baker 1966; Vuilleumier 1967; Ganders 1979; D. Charlesworth and B. Charlesworth 1979; Lewis 1982; Barrett 1988a).

Mather and De Winton (1941) suggested that "the real significance of the morphological differences shown by the pistils and stamens lies in the physiological differences which follow from them" and also that "the two kinds of differences must be developmentally connected." These authors based their suggestion on the obligatory association between heterostyly and illegitimacy and also on the reverse argument that if the two mechanisms were mutually independent, they would have been encountered either each separately or in combination with a different breeding mechanism.

Subsequently, however, it has been shown that heterostyly may also be accompanied by multiallelic incompatibility, as appears to be the case in *Narcissus tazetta* (Dulberger 1964), *Anchusa hybrida* (Dulberger 1970a), and *A. officinalis* (Philipp and Schou 1981; Schou and Philipp 1984). Or, it may be accompanied by self-com-

[1] Department of Botany, G.S. Wise Faculty of Life Sciences, Tel Aviv University, Tel Aviv 69978, Israel

Monographs on Theoretical and Applied Genetics 15
Evolution and Function of Heterostyly (ed. by S.C.H. Barrett)

patibility, as in *Amsinckia* species (Ray and Chisaki 1957; Ganders 1975b), *Eichhornia crassipes* (Barrett 1977a, 1979), and *E. paniculata* (Barrett 1985a), or in *Cryptantha flava* (Casper 1985).

In the course of the last two decades there has been a resurgence of interest in heterostyly, and studies of this phenomenon have been extended to taxa in which heterostyly was previously little known or investigated. Most of the articles appearing in the 1970s are included in the review of Ganders (1979). As for the pertinent literature of the 1980s, this includes articles on members of the Boraginaceae (Philipp and Schou 1981; Schou and Philipp 1983, 1984; Casper 1983, 1985; Casper et al. 1988; Olesen 1979; Weller 1980; Weller and Ornduff 1989; the Menyanthaceae (Barrett 1980; Ornduff 1982, 1986, 1987a, 1988a,b; Olesen 1986), Oxalidaceae (Weller 1979, 1981a,b, 1986; Ornduff 1987b), the Pontederiaceae (Barrett 1985a, 1988b; Anderson and Barrett 1986), and Turneraceae (Shore and Barrett 1984, 1985a,b; Barrett and Shore 1987).

For some taxa new observations related to heterostyly have been reported (Lewis 1982; Riveros et al. 1987; Ornduff 1987a, 1988b). The efficiency of pollen transfer has been examined in natural populations of numerous species (review by Ganders 1979; Ornduff 1979a, 1980a,b; Schou 1983; Barrett and Glover 1985; Piper and Charlesworth 1986; Piper et al. 1986; Barrett 1990). The role of heterostyly in enhancing pollen carryover has been explored (Nicholls 1985a; Feinsinger and Busby 1987; Wolfe and Barrett 1989). Development of heterostyly was traced in *Eichhornia* and *Pontederia* species (Richards and Barrett 1984, 1987) and in *Linum and Fagopyrum* (Dulberger 1984 and unpubl.). Theoretical models have been proposed for the evolution and breakdown of distyly and tristyly (D. Charlesworth and B. Charlesworth 1979; B. Charlesworth and D. Charlesworth 1979; Charlesworth 1979). Ecological genetics and evolutionary modifications of the heteromorphic mating systems have been studied (Barrett 1985b; Shore and Barrett 1986). Two prominent new hypotheses have been proposed on the significance of heterostyly and its maintenance in plant populations. The first concerns sexual selection and optimal allocation of sexual resources (Willson 1979; Casper and Charnov 1982), while the second postulates that mutual interference between the presentation of pollen and stigmas can be obviated by reciprocal herkogamy (Lloyd and Yates 1982; Webb and Lloyd 1986).

The majority of the above studies have focused on the functioning of heterostyly sensu stricto, i.e., of reciprocal herkogamy (Webb and Lloyd 1986) as a floral mechanism for promoting pollinations between morphs and avoiding those within each morph. To a great extent this approach has continued to eclipse an elucidation of the possibility that the morphological syndrome functions to prevent inbreeding through incompatibility.

To date pollen and stigma polymorphisms have been the least investigated component of heterostyly, and their functional significance and evolution have remained obscure. Generally, these polymorphisms have been treated as scorable characters in the heterostyly syndrome (Vuilleumier 1967; Dulberger 1974; Ganders 1979; Charlesworth and Charlesworth 1979; Bir Bahadur et al. 1984a,b; Richards 1986; Barrett 1988a; Barrett and Richards 1989). Apart from the style and stamen lengths, the only polymorphism to which adaptive significance has been ascribed is the disparate pollen production by the morphs and this again, only in its relevance to the promotion of disassortative pollination (Ganders 1979; Piper and Charlesworth 1986).

The detection of structural and cytochemical dimorphisms of the stigmas in species of *Linum* and in various Plumbaginaceae, coupled with dissimilar behavior of the male gametophyte in the two types of incompatible pollinations, have led to the suggestion that dimorphisms of stigmas and of the pollen exine function in the physiology of the incompatibility mechanism (Dulberger 1974, 1975a,b, 1987a,b,c, 1991). This conclusion has been extended also to polymorphisms of pollen size and of style and stamen lengths, thus leading to an hypothesis on the developmental relationships between incompatibility and morphological and structural traits of heterostyly (Dulberger 1975b).

In parallel to the above-mentioned studies on heterostyly, much progress has also been made in the past two decades towards an understanding of pollen-pistil interactions, particularly in plants having homomorphic incompatibility systems. Pollen grains and tubes, the receptive surface, and the transmitting tract have been investigated with electron microscopy, cytochemical and immunological methods and, more recently, by approaches and techniques employed in molecular biology. These investigations have expanded our knowledge of homomorphic incompatibility systems considerably and are the beginnings of an insight into the molecular basis of self-incompatibility (Heslop-Harrison 1978, 1983; review by Knox 1984; Clarke et al. 1985, 1989; Nasrallah et al. 1985, 1988; Nasrallah and Nasrallah 1986; Gaude and Dumas 1987; Bernatzky et al. 1988; Cornish et al. 1987, 1988; McClure et al. 1989, 1990; Broothaerts and Vending 1990).

Pollen-pistil relationships have also been investigated in heterostylous plants, e.g., in species of *Linum* (Dulberger 1974, 1975b, 1981a,b, 1987a, 1989, 1990, 1991; Ghosh and Shivanna 1980a,b, 1983; Murray 1986), in members of the Plumbaginaceae (Dulberger 1975a, 1987b; Mattsson 1983), *Primula* species (Heslop-Harrison et al. 1981; Shivanna et al. 1981, 1983; Richards and Ibrahim 1982; Stevens and Murray 1982; Schou 1984; Richards 1986) and Rubiaceae (Bawa and Beach 1983), *Pontederia* species (Glover and Barrett 1983; Barrett and Anderson 1985; Scribailo and Barrett 1986, 1989; Scribailo 1989), and in *Anchusa officinalis* (Schou and Philipp 1983). These and additional investigations have yielded much information on stigmas and pollen grains and on the behavior of the male gametophyte, thereby considerably enhancing our understanding of the relationships between the male and female partners after compatible and incompatible pollinations. Most of the above studies are based on the premise that the morphological characters of heterostyly and the physiology of incompatibility are independent phenomena.

In this chapter the essential pistil and stamen polymorphisms encountered in heterostylous plants are surveyed. Morph-specific characters of styles, stamens, pollen, and stigmas are considered in relation to the behavior of the male gametophyte in the different incompatible pollinations. Some of the polymorphisms are also discussed with regard to disassortative pollen transfer. Available information is presented, which supports the view (Dulberger 1975b) that most polymorphic characters participate in the incompatibility mechanism. It is postulated that recognition of some of the polymorphisms as components of incompatibility mechanisms can lead to a better comprehension of the nature, origins, and evolution of heterostyly. It must, of course, be taken into account that polymorphic characters are also involved in other functions. An integrative concept of morphological, structural, developmental, and physiological facets of heterostyly is therefore proposed.

Although the present chapter is focused primarily on distylous plants, it is assumed that the governing principles and conclusions drawn concerning distyly are also essentially valid and applicable for tristyly.

2 Polymorphisms

2.1 Style and Stamen Lengths

Heterostylous taxa display an array of characters that enable discrimination between pistils and stamens of the morphs. The one and only heteromorphic trait which is common to almost all species with diallelic incompatibility is the presence of two or three modal classes of style and stamen lengths. In distyly there are long-styled (pin) and short-styled (thrum) flowers, and in tristyly- also mid-styled flowers. The exceptions are found among members of the Staticeae (Baker 1966), and these will be dealt with later.

The long-style to short-style length ratio varies, e.g., 4:1 in *Primula auricula*, 3:1 in *P. elatior*, 2:1 in *Limonium vulgare*, 1.8:1 in *Linum flavum* and 1.3:1 in *Acantholimon glumaceum*. In tristylous species, the Long: Mid: Short style-length ratios are 6:3:1 for *Lythrum salicaria*, 4:2:1 for *Oxalis valdiviensis* (Schill et al. 1985) and 12:6:1 for *Pontederia sagittata* (Scribailo 1989). The style length may vary to different degrees in each of the morphs. Considerable intramorph variation in the length of the style and stamens occurs particularly in large flowers, as in *Plumbago capensis* and *Narcissus tazetta*.

In the vast majority of heterostylous plants the stamens of the morphs differ in anther level and reciprocal herkogamy is obvious, although it often lacks precision. Exceptions are rare, e.g., *Linum grandiflorum* (Darwin 1877) and some populations of *L. suffruticosum* (Rogers 1979 but see also Nicholls 1985b). As for *L. grandiflorum*, our measurements revealed a slight difference in anther level between the two floral morphs (Table 1), which was also noted by Heitz (1980).

Tristyly is restricted to taxa with two distinct whorls of stamens in a flower. In some tristylous species of *Eichhornia* and *Pontederia*, the three stamens within each of the two levels in a flower originate from two whorls (Richards and Barrett 1987).

Some species which possess two whorls of stamens are distylous and not tristylous, e.g., species of *Jepsonia*, *Fagopyrum esculentum*, *Byrsocarpus coccineus*, and *Erythroxylon coca*.

A dissimilar shape of the style has been reported for *Linum suffruticosum* where the long style is ribbon-like (Rogers 1979; Nicholls 1985b), and for *Waltheria viscosissima* in which the long style is repeatedly curved and bears large verrucae and long hairs, whereas the short style is only slightly curved, has smaller verrucae and bears shorter hairs (Köhler 1973, 1976).

In some species style length polymorphism is associated with differences in style structure. Long and Mid styles of *Lythrum salicaria* have a compact transmitting tissue devoid of intercellular spaces, while in Short styles a canal leads from the stigma to the ovary (Schoch-Bodmer 1945). A similar situation has been found in *L. junceum* (Dulberger and Cohen, unpubl. data). A much smaller area of transmitting

Table 1. Dimensions of floral parts in the two floral morphs of *Linum grandiflorum*

	Pin		Thrum	
Character	N	Mean ± SD (range)	N	Mean ± SD (range)
Length of pistil (mm)	73 Flowers	11.86 ± 1.08 (10.0–14.8)	73 Flowers	8.11 ± 0.83 (7–11)
Length of stigma (mm)	10 Flowers	4.35 ± 0.48 (3.64–5.04)	10 Flowers	2.07 ± 0.24 (1.68–2.52)
Length of stigmatic papillae (µm)	60 Papillae	55.88 ± 12.48 (32–88)	60 Papillae	47.6 ± 10.16 (28–80)
Max. width of stigmatic papillae (µm)	60 Papillae	27 ± 4.40 (16–36)	60 Papillae	20.52 ± 2.6 (16–24)
Length of stamen (mm)	73 Flowers	12.74 ± 1.01 (10.0–15.0)	73 Flowers	14.27 ± 1.16 (11.5–18)
Length of stamen filament (mm)	73 Flowers	9.48 ± 0.75 (8–12)	73 Flowers	10.74 ± 0.9 (9–13.5)

All the dimensions were significantly different between the morphs at the 0.1% (0.001) significance level according to Student's *t* test.

tissue in pin than thrum styles has been detected in *Primula obconica* (Dowrick 1956) and *Linum pubescens* in which the total area of cross-sections is also considerably smaller in pin than thrum styles (Dulberger and Cohen, unpubl. data). In *Pontederia sagittata* short styles have a narrower canal than long and mid styles (Scribailo 1989). Structural differences between pin and thrum styles have also been observed in *Anchusa hybrida* and in an unidentified distylous *Lythrum* species (Dulberger and Cohen, unpubl. data).

Sporadically occurring polymorphisms in the color or pubescence of the style or stamen filaments, or in corolla size of the morphs are beyond the scope of the present review and are not dealt with here.

There is an additional category of species in which the flowers display style length dimorphism not accompanied by a reciprocal placement of stigmas and anthers in the morphs. These species also deviate from conventional heterostylous ones by their mating system, which is not based on diallelic incompatibility. Examples are: *Narcissus tazetta* (Dulberger 1964), *A. officinalis* (Philipp and Schou 1981) and *Quinchamalium chilense* (Riveros et al. 1987). Stylar dimorphisms in these and additional species are reviewed by Barrett and Richards (1989). Although they do not display the entire syndrome known as heterostyly, these species are included here for methodological reasons given in Section 5.1.

2.2 Pollen

2.2.1 Pollen Size

Differences in size constitute the most widespread polymorphism of pollen (Darwin 1877; Vuilleumier 1967; Dulberger 1974). Pollen size heteromorphism occurs in most distylous and tristylous species belonging to diverse genera and families. Generally, pollen grains from short-styled flowers are larger than those from long-styled flowers. In the examples of distylous plants listed in Table 2, the ratios of thrum to pin pollen size vary from 1.06 to 1.80. In *Lythrum californicum* (Ganders 1979) and *Hedyotis caerulea* (Lewis 1976), some populations show pronounced pollen dimorphism while others do not. Considerable variation in pollen size has been detected among populations of *Amsinckia vernicosa* (Ganders 1979). Pin pollen was larger than thrum pollen in one of two populations of *Fauria crista-galli* (Ganders 1979).

Among distylous species, absence of size polymorphism has been reported in a few species of *Cordia* and *Nesaea* (Darwin 1877), *Amsinckia spectabilis* var. *spectabilis* (Ray and Chisaki 1957), *Pauridiantha* and 29 species of *Linum* (Darwin 1877; East 1940; Verdcourt 1958; Hallé 1961, all quoted from Vuilleumier 1967), *Byrsocarpus coccineus* (Baker 1962), nine rubiaceous species (Darwin 1877; Burck 1887; Baker 1958; Hallé 1961 and Bremekamp 1963, all quoted from Bir Bahadur 1968), and *Linum pubescens* (Dulberger 1973).

Table 3 reveals that in many additional distylous species no significant size differences occur between pollen of the morphs, or the typical size relations are reversed, with pin grains larger than thrum grains.

In several tristylous species pollen from anthers of the three levels shows three modal classes of grain size. Generally, size trimorphism is positively correlated with

Table 2. Ratios of thrum to pin pollen size in some distylous species

Species	Thrum/pin ratio		Reference
	Polar axis	Equatorial axis	
Cratoxylum formosum	1.29		Lewis (1982)
Fagopyrum esculentum	1.68	1.33	Dulberger (unpubl. data)
Hedyotis caerulea	1.20		Ornduff (1977)
Jepsonia parryi	1.47		Ornduff (1970a)
Lithospermum caroliniense	1.41	2.00	Levin (1968)
Lythrum californicum	1.51		Ornduff (1978)
Nivenia binata	1.30		Mulcahy (1965)
Oldenlandia scopulorum	1.06	1.00	Bir Bahadur (1966)
O. umbellata	1.14	1.11	Bir Bahadur (1964)
Primula auricula	1.30	1.33	Schill et al. (1985)
P. elatior	1.63	1.57	Schou (1983)
P. farinosa	1.80	1.27	Schill et al. (1985)
P. malacoides	1.41		Pandey and Troughton (1974)
P. veris	1.62		Ornduff (1980b)
P. vulgaris	1.43		Ornduff (1979a)
Pulmonaria obscura	1.22		Olesen (1979)
Psychotria vogeliana	1.29		Baker (1958)
Rudgea jasminoides	1.43	1.43	Baker (1956)
Turnera hermannioides	1.14	1.20	Barrett and Shore (1985)
Uragoga nimbana	1.17		Baker (1958)
Villarsia congestiflora	1.23		Ornduff (1988a)
Waltheria viscosissima	1.23	1.33	Köhler (1973)

anther height. Measurements of pollen dimensions have been given for *Lythrum salicaria* (Darwin 1877; Schoch-Bodmer 1945; Dulberger 1967), *L. junceum* (Dulberger 1970b), *Biophytum sensitivum* (Mayura Devi 1964), *B. intermedium* (Bir Bahadur et al. 1984b), *Oxalis suksdorfii* (Ornduff 1964), *O. tuberosa* (Gibbs 1976), *O. alpina* (Weller 1976), *O. pes-caprae* (Ornduff 1987b), *Eichhornia crassipes* (Barrett 1977a), *E. paniculata* (Barrett 1985a), and six additional *Eichhornia* species (Barrett 1988b), *Pontederia rotundifolia* (Barrett 1977b), *P. cordata* (Price and Barrett 1982), and *P. sagittata* (Glover and Barrett 1983). In the tristylous species examined, a wide gamut has been found, ranging from accentuated pollen size trimorphism at one extreme, as in *P. cordata*, to considerable overlap in size of pollen from different anther levels in *E. crassipes* and *E. paniculata*.

2.2.2 Exine Sculpturing

Dissimilar pollen size in the morphs may be associated with slight differences in exine sculpture. Thus, a coarse reticulation of the exine in larger thrum grains as opposed to a more delicate one in pin grains has been reported for *Nivenia binata* (Mulcahy 1965, but see Goldblatt and Bernhardt 1990), *Oldenlandia umbellata, O. scopulorum, Pentas lanceolata, Hedyotis nigricans,* and *Schismatoclada psychotrioides* (Bir Bahadur 1968), *Jepsonia parryi* (Ornduff 1970a), *Forsythia intermedia, Primula malacoides* (Pandey and Troughton 1974), and *P. vulgaris* (Heslop-Harrison et al. 1981), and for *Hedyotis caerulea, H. procumbens, Morinda tomentosa, Neanotis montholoni, N. indica, Fagopyrum esculentum, Polygonum chinense,* and *Turnera subulata* (Bir Bahadur et al. 1984b). In some of these species dimorphic sculpturing of the sexine muri and/or lumina has been observed (Bir Bahadur et al. 1984b).

The afore-mentioned differences in exine sculpturing may be a developmental outcome of size dimorphism. However, in *Armeria maritima*, grains of extremely aberrant size display the exine pattern characteristic of the morph to which they belong (Philipp 1974).

In distylous and tristylous species of *Lythrum*, size differences between pollen from anthers of dissimilar levels are associated with differences in sculpturing of nonreticulate exine (R. Dulberger and S.C.H. Barrett, unpubl. data).

Size polymorphism of grains may also be accompanied by differing numbers or shape of apertures, with larger grains of high-level anthers having more apertures. Instances reported are: *Schismatoclada psychotrioides* (Lewis 1965, as cited by Bir Bahadur 1968), species of *Waltheria* (Köhler 1973, 1976), *L. junceum* (Zavala 1978), *Primula obconica*, and *Turnera subulata* (Bir Bahadur et al. 1984b). A similar difference has also been observed in pollen of some polyploid *Linum* species when compared to their diploid relatives (Ockendon 1968).

Remarkable, though less frequent, are dissimilarities in major characters of pollen in the morphs, such as pronounced dimorphism of exine ornamentation or of shape of grains. In some distylous species of *Linum* the exine of thrum grains bears blunt processes of more or less uniform size, each with a marginal ring of small verrucae. The exine of pin grains, on the other hand, bears two kinds of processes, namely a few large processes, provided with a central spinule as well as with a peripheral ring of spinules or buttresses, which are interspersed among numerous

Table 3. Ratios of thrum to pin or B type to A type pollen size in species of *Linum*, Plumbaginaceae, and Boraginaceae

Species	Ratios		Reference
	Thrum/pin or B/type/A type[a]		
	Polar axis	Equatorial axis	
Plumbaginaceae			
Acantholimon androsaceum	1.17	1.19	Weber (1981)
A. glumaceum	1.04	1.01	Schill et al. (1985)
Armeria alpina	0.98	1.08	Schill et al. (1985)
A. maritima	1.02	0.95	Weber (1981)
A. plantaginea	1.35	1.25	Weber (1981)
Dyerophytum africanum	0.86	1.03	Weber – El Ghobary (1986)
D. indicum	0.93	1.03	Weber (1981)
Goniolimon tataricum	1.01	1.01	Weber (1981)
	1.01	0.97	Schill et al. (1985)
Limoniastrum feei	0.97	0.96	Weber (1981)
L. guyonianum	0.87	0.90	Weber (1981)
L. meygandiorum	0.76	0.78	Weber (1981)
Limonium binervosum	0.83	0.86	Weber (1981)
L. meyeri	0.97	0.99	Dulberger (unpubl. data)
L. minutum	1.05	1.00	Weber (1981)
L. sinuatum	0.99	0.93	Dulberger (unpubl. data)
L. vulgare	1.01	1.01	Weber (1981)
	1.06	0.95	Schill et al. (1985)
Plumbago capensis	1.03	1.00	Dulberger (unpubl. data)
Linum			
Linum altaicum	1.06	1.03	Schill et al. (1985)
L. austriacum	0.95	0.98	Punt and den Breejen (1981)
	1.00	1.00	Heitz et al. (1971)
L. flavum	1.05	0.98	Schill et al. (1985)
L. flavum	0.90	0.98	Punt and den Breejen (1981)
L. grandiflorum	0.98	0.97	Dulberger (1981a)
L. perenne	1.00	1.12	Nicholls (1986)
L. pubescens	0.98	0.98	Dulberger (1981a)
L. suffruticosum	0.92	1.08	Punt and den Breejen (1981)
Boraginaceae			
Amsinckia douglasiana	1.27		Ray and Chisaki (1957)
	1.25		Ganders (1976)
A. grandiflora	1.28		Ornduff (1976)
A. spectabilis	1.00		Ray and Chisaki (1957)
A. vernicosa var. *furcata*	1.23		Ganders (1976)
Anchusa officinalis	1.05	0.97	Philipp and Schou (1981)
Cryptantha flava	No noticeable size dimorphism		Casper (1983)
C. flavoculata	No noticeable size dimorphism		Casper (1983)

[a]In *Limonium vulgare* and *L. meyeri* pins have pollen of A type and thrums of B type.

smaller processes, each terminating in a single long spinule (Saad 1961; Ockendon 1968, 1971; Heitz et al. 1971; Rogers 1979, 1980; Dulberger 1981a; Punt and den Breejen 1981). A similar exine dimorphism occurs in *Reinwardtia indica* (Bir Bahadur et al. 1984b).

In species of several genera in the tribe Staticeae (Plumbaginaceae), two kinds of pollen grains occur. In *Armeria maritima* pollen grains of type A have a coarsely reticulate sexine, with prominent buttressed muri surrounding deep polygonal areoles, whereas grains of type B have a considerably finer reticulation, with buttresses reduced to minute spines. Similar exine dimorphisms are known in other species of *Armeria, Limonium, Goniolimon, Acantholimon*, and *Limoniastrum*. In distylous species of *Limonium*, grains of type A are produced by long-styled flowers and those of type B are produced by short-styled flowers (Macleod 1887; Erdtman 1940; Iversen 1940; Baker 1953, 1966; Dulberger 1975a).

Some species of *Dyerophytum* and *Plumbago* of the tribe Plumbagineae display another kind of pollen dimorphism. Here the exine of thrum pollen has blunt processes with rounded tips, whereas pin pollen has processes that are pointed and terminate in spinules (Erdtman 1970; Weber 1981; Weber-El Ghobary 1986).

As can be seen in Table 3, the exine dimorphism in *Linum* and in the Plumbaginaceae frequently occurs in almost equal-sized grains.

Striking differences between the morphs in exine sculpturing, shape, or in overall size of the grains are not necessarily mutually exclusive. This inference can be made on the basis of the following examples. In *Rudgea jasminoides*, Rubiaceae, the pollen exine is spiniferous in thrum flowers but granulate in pin flowers, yet the thrum to pin pollen size ratio is about 1.5 (Baker 1956). A second example are *Litho- spermum canescens* and *L. caroliniense* of the Boraginaceae, which have thrum pollen grains that are ovoidal and only slightly constricted in the middle, and pin grains that are oblong and much constricted in the middle (Baker 1961; Weller 1980). This notwithstanding, the thrum to pin grain size ratio is 1.4 in *L. caroliniense* (Levin 1968).

The most extreme disparity between pollen types has been reported for *Waltheria viscosissima*, (Köhler 1973) and for 22 additional species of the same genus of the Sterculiaceae (Köhler 1976). Here, pollen grains of the morphs differ in shape, size (up to 32% difference, with an average difference of 18%), exine sculpturing, number and shape of apertures, length of the ectoapertures, and other traits as well. Thrum grains are oblate spheroidal, and their exine is of the spinulose type and has 5–7 very short, unclear colpi. The smaller pin grains are spheroidal, their exine is of the suprareticulate type and has 3–4 long colpi. Furthermore, the two morphs differ in the shape of the ora, in supratectal elements and the fine structure of the tectum.

The differences reviewed above between pollen characters of the floral morphs in *Rudgea, Lithospermum* and *Waltheria* are of a magnitude similar to that distinguishing pollens of taxa ranking from genus and above.

2.2.3 Pollen Colour

Rarely, pollen grains of the morphs differ in their colour. In *Linum grandiflorum*, for example, thrum pollen is dark blue whereas pin pollen is dark gray. Long-styled flowers of *L. tenuifolium* have brick-red anthers with pale brick-red pollen, whereas short-styled flowers have cream or yellow anthers and pollen (Rogers 1979; Nicholls 1985b).

Pollen from long stamens of *Lythrum salicaria* is olive green (due to flavones and anthocyanin) and anthers are dark purple; in contrast, pollen from mid- and short-level stamens is yellow and anthers are light green (Darwin 1877; Schoch-Bodmer 1938, 1945). In distylous *Lythrum californicum* and *L. alatum*, anthers and pollen grains of the two sets of stamens show similar differences in color.

Pollen grains from long stamens of *L. salicaria* (Tischler 1917) and *L. junceum* (Dulberger 1970b) characteristically stain for starch, while pollen from mid and short stamens does not show such a reaction. Presence of starch in high-level anthers is known also in *Hypericum aegypticum* (Ornduff 1979b) and *Jepsonia* (Ganders 1979).

2.2.4 *Anther Size*

Stamens may also differ in anther size. Anthers of long stamens are smaller than those of short stamens in *Linum flavens* and *Forsythia suspensa*, but larger in *Hottonia palustris, Limnanthemum indicum*, six rubiaceous genera, *Nymphoides indica* and *Pulmonaria angustifolia* (Darwin 1877), *Lithospermum* (Johnston 1952), *Amsinckia vernicosa* var. *furcata* and *A. spectabilis* var. *microcarpa* (Ganders 1979), and tristylous *Pontederia cordata* (Price and Barrett 1982), *P. sagittata* (Glover and Barrett 1983), *Lythrum salicaria*, and *L. junceum*. In *Amsinckia vernicosa* and the tristylous species of *Pontederia* and *Lythrum*, differences in anther size are associated with polymorphisms in pollen size and amounts of grains produced.

2.2.5 *Pollen Production*

Differences between pollen production of the morphs have been found in most species examined, with stamens of long-styled flowers generally producing more pollen than those of short-styled flowers (Ganders 1979). Table 4 shows pollen production ratios of pin to thrum flowers varying from 0.56 to 3.18. It is interesting that in many species, thrums produce more pollen than pins, while in other species only small differences in the number of pollen grains were detected. In five of these species the grains of long and short-styled flowers are of almost equal size (Table 3).

In one of four populations of *Hedyotis caerulea*, pin flowers were found to produce considerably more pollen than thrum flowers, whereas in three populations pollen production was similar (Ornduff 1980a).

2.3 Stigmas

2.3.1 *Size, Shape, and Color of Stigmas*

Size polymorphism of stigmas is widespread among heterostylous species, with the receptive surface of the long-styled morph typically larger than that of the short-styled one. In *Jepsonia parryi* the area of pin stigmas is 50% greater than that of

Table 4. Pollen production per flower in some heterostylous plants

Species	Pin	Thrum	P/T ratio	Reference
Amsinckia douglasiana	17 490	15 380	1.13	Ganders (1976)
A. grandiflora	32 953	28 650	1.15	Ornduff (1976)
A. spectabilis var. *microcarpa*	7 700	13 600	0.56	Ganders (1975a)
A. vernicosa var.	18 200	24 000	0.76	Ganders (1976)
furcata	16 507	23 947	0.69	
Anchusa officinalis	2 790	2 540	1.09	Philipp and Schou (1981)
Cryptantha flava	109 938	102 581	1.07	Casper (1983)
C. flavoculata	219 731	230 313	0.95	Caspter (1983)
Fagopyrum esculentum	1 830	1 080	1.69	Ganders (1979)
	1 354 ± 164	1 088 ± 339	1.24	Dulberger (unpubl. data)
Gelsemium sempervirens	75 511	63 966	1.18	Ornduff (1970b)
Hedyotis caerulea	13 814	13 486	1.02	Ornduff (1980a)
Jepsonia parryi	206 000	78 000	2.64	Ornduff (1970a)
Limonium meyeri	1 566 ± 360 (type A)	1 430 ± 386 (type B)	1.09	Dulberger (unpubl. data)
Linum grandiflorum	21 131 ± 423	23 257 ± 4 104	0.91	Dulberger
	19 609 ± 3 862	23 697 ± 3 975	0.82	(unpubl. data)
L. perenne	2 140	2 640	0.81	Nicholls (1986)
L. pubescens	11 495 ± 2 546	12 041 ± 1 836	0.95	Dulberger (unpubl. data)
Lithospermum caroliniense	160 800	64 360	2.50	Weller (1980)
Lythrum californicum	30 552	10 988	2.78	Ornduff (1978)
L. curtisii	14 421	7 099	2.03	Ornduff (1978)
Plumbago capensis	1 760 ± 130	1 636 ± 187	1.07	Dulberger (unpubl. data)
Primula elatior	138 710	86 645	1.60	Schou (1983)
P. veris	211 000	87 000	2.42	Ornduff (1980b)
P. vulgaris	283 000	89 000	3.18	Ornduff (1979a)
Pulmonaria obscura	51 460	33 405	1.54	Olesen (1979)
Turnera hermannioides	15 200	11 800	1.29	Barrett and Shore (1985)

thrum stigmas (Ornduff 1970a). The length ratio of pin to thrum stigmas is 1.55 in *Plumbago capensis* and about 2.0 in *Linum grandiflorum*, *L. pubescens*, and *L. mucronatum* (Table 1 and unpubl. data). In contrast, thrum stigmas are larger than pin stigmas in *Gregoria vitaliana* (Schaeppi 1935), *Amsinckia grandiflora* (Ornduff 1976), *Primula malacoides* (Pandey and Troughton 1974), *Hedyotis caerulea* (Ornduff 1980a), *Palicourea lasiorrachis* (Feinsinger and Busby 1987), *Gelsemium sempervirens*, and *Neanotis montholoni* (Bir Bahadur et al. 1984b).

Pronounced differences between the morphs in the shape and size of stigmas have also been described. In *Rudgea jasminoides* thrum stigmas are long, narrow, and considerably curled, whereas pin stigmas are short and broad (Baker 1956). Pin stigmas of *Waltheria viscosissima* are densely branched with many delicate lobes;

thrum stigmas are shorter and less branched, but the branches proper are longer and sturdier (Köhler 1973). In *Villarsia capitata* pin stigmas are bilobate, and thrum stigmas are irregular, and lacerated crateriform (Ornduff 1982). Slight differences in stigma shape are also reported in *Anchusa hybrida* (Dulberger 1970a), *A. officinalis* (Philipp and Schou 1981), *Primula vulgaris* (Heslop-Harrison et al. 1981), *P. elatior* (Schou 1983), *P. obeonica*, and *Neanotis montholoni* (Bir Bahadur et al 1984b).

Heteromorphism of stigma color occurs in *Linum grandiflorum*, where the short thrum stigmas are dark red with more crowded papillae compared to the light red to pink pin stigmas. Similarly, in *L. pubescens* thrum stigmas are yellowish, whereas pin stigmas are white, probably as a result of cytochemical differences (Dulberger 1974, 1987a). Differences between the morphs in density of papillae may stem from differential cell growth of underlying stigmatic tissue.

2.3.2 Size and Shape of Papillae

Perhaps the most frequently reported polymorphism of stigmas is that of papillar size, with papillae in long-styled flowers larger than in short-styled flowers (Vuilleumier 1967; Dulberger 1974). Measurements of papilla size are documented in relatively few plants: *Lythrum salicaria* (Dulberger 1967), *L. junceum* (Dulberger 1970b), *L. curtisii* (Ornduff 1978), *Pulmonaria obscura* (Olesen 1979), *Primula vulgaris* (Heslop-Harrison et al. 1981) and *Linum pubescens* (Dulberger 1987a), and *L. grandiflorum* (Table 1). Thrum papillae are reported to be larger than pin papillae in *Anchusa officinalis* (Schou and Philipp 1984) *Reinwardtia indica* (Bir Bahadur et al. 1984b).

Shape dimorphism of unicellular papillae is known in *Armeria, Limonium* (Macleod 1887; Erdtman 1940; Iversen 1940; Baker 1948, 1953, 1966), *Acantholimon* and *Goniolimon* (Bokhari 1972; Schill et al. 1985). In flowers having type A pollen grains the stigmas are cob-like, having a bullate surface, while in those having type B pollen grains stigmas are distinctly papillate.

Dimorphisms in the shape of multicellular clusters of papillae, size and number of clusters on a stigma lobe, and papilla size and number of papillae in a cluster, occur in the two morphs in *Plumbago, Ceratostigma* (Dahlgren 1918, 1923; Dulberger 1975a, 1987b), and *Dyerophytum* (Dahlgren 1970). The clusters are larger and less numerous in long-styled flowers than in short-styled flowers.

2.3.3 Dimorphic Structure of the Papilla Wall

In the Plumbaginaceae, morphological dimorphisms of stigmas are combined with structural and cytochemical differences between stigmatic papillae of the morphs. Thus, in *Limonium meyeri, L. sinuatum* and *Armeria maritima*, the cuticle of each papilla is uniformly thin in the cob-like stigmas, while in the papillate type it is considerably thickened and thimble-shaped at the papillar tip (Dulberger 1975a).

Pin papillae of *Ceratostigma willmottianum* have a subcuticular space filled with pectins, which is absent in thrum papillae (Dulberger 1975a). Thrum papillae of *Plumbago capensis* produce a secretion; their surface readily stains with Coomassie blue, especially at the clefts between papillae, where osmiophilic plugs are visible using TEM. The larger pin papillae have a thicker cuticle, particularly at the clefts; they are devoid of osmiophilic secretion and of protein plugs and, unlike thrum papillae, their cuticle is impermeable to neutral red solution. Only a thin epicuticular film stains with Coomassie blue (Dulberger 1987b and unpubl. data).

The Plumbaginaceae are the only taxon investigated in which the morphs differ in shape and structure of the stigmatic papillae, as well as in exine sculpturing.

In *Linum grandiflorum, L. pubescens,* and *L. mucronatum* the cuticle of thrum papillae is torn at anthesis and secretory material staining intensely for proteins and lipids is released within a pectinaceous matrix. The cuticle of the larger pin papillae is continuous and overlaid by only an extremely thin film of proteins. The cuticle of the upper third of the papilla is thicker than in thrum papillae and is impermeable to water. On the basis of their structure and staining responses, pin stigmas belong to the dry type while thrum stigmas resemble the wet type. Electrophoretic protein profiles of leachates of the two stigma types showed both qualitative and quantitative differences (Dulberger 1974, 1987a; Ghosh and Shivanna 1980a,b).

The converse situation exists in *Primula obconica.* Here smaller thrum papillae show certain signs of secretory activity but remain "dry" at maturity with only a superficial pellicle, as typical of the dry type of stigmas. In pin papillae, on the other hand, the cuticle becomes disrupted at maturity, and an exudate containing lipids, carbohydrates, and protein is released onto the papilla surface (Schou 1984).

It is reasonable to assume that stigma polymorphisms, particularly those in the size of the receptive area and of individual papillae, are much more widespread than hitherto reported. In most studies of heterostylous species for which sizes of pollen grains in the morphs are given, stigma characters are not described. Differences between characters of the stigma of the floral morphs are more difficult to assess than characters of pollen grains. To begin with, it is often difficult to measure the receptive area. Secondly, unlike differences in size of individual pollen grains of the morphs, those in size of the receptive cells are not conspicuous and, consequently, are easily overlooked, particularly in nonpapillate stigmas. Thirdly, in papillate stigmas, the papillae usually display considerable variation in size and are crowded with their bases concealed, all of which render measurements difficult (albeit in some *Linum* species a procedure for spreading the papillae for measurements has been employed — see Dulberger 1987a). Lastly, unlike pollen characters, attributes of the stigma cannot be determined accurately in herbarium material.

The main physiological and histological types of the receptive surface recognized by Heslop-Harrison and Shivanna (1977) are represented in heterostylous plants. The prevailing type is that of the dry unicellular papillate stigma. Nonpapillate stigmas of the wet type with copious secretion on the receptive surface are not encountered in heterostylous species.

The polymorphic nature of pollen grains and stigmas is usually characteristic at least at the level of the genus, although fine details of such pollen and stigma polymorphisms may vary among species of a genus. For example, in distylous species of *Linum* there is interspecific variation in fine structure of the papilla wall of each morph (Dulberger 1974, 1987a) and in details of the exine sculpturing of the morphs (Ockendon 1968; Rogers 1979; Dulberger 1981a).

3 Functional Significance of the Polymorphisms

In order to assess the functional significance of the polymorphisms displayed by heterostylous plants, it is necessary to consider lines of evidence pertaining to two distinct functions of heterostyly, namely, the promotion of legitimate (intermorph) pollinations while obviating illegitimate (self- and intramorph) pollinations, and the prevention of self- or intramorph fertilization through incompatibility.

3.1 Pollen Flow and Pollen Grain Size and Production

An inverse correlation between polymorphism in pollen grain size or anther size and pollen production within the same species was first inferred on the basis of developmental considerations (Dulberger 1975b). Measurements and counts of pollen grains in the morphs have substantiated this conclusion for many heterostylous plants. By plotting thrum to pin pollen size ratios against pin to thrum number of grains per flower in 15 distylous species, Ganders (1979) found an inverse correlation between pollen size and pollen production of the morphs and suggested that the polymorphism in pollen production has adaptive value in regulating disassortative pollen flow.

Studies of pollen loads on stigmas of many distylous species have revealed that the deposition of pollen is asymmetric and that stigmas of long-styled flowers capture larger numbers of pollen grains than short-styled flowers. However, short-styled flowers experience greater levels of legitimate pollination. This asymmetry appears to be a general phenomenon (Ganders 1979), and has been confirmed in recent reports for *Primula veris* (Ornduff 1980b), *Hedyotis caerulea* (Ornduff 1980a), *Lithospermum caroliniense* (Weller 1980), *Cratoxylum formosum* (Lewis 1982), *Primula elatior* (Schou 1983), *Cryptantha flava* (Casper 1983), *Linum tenuifolium and L. perenne* (Nicholls 1985a,b, 1986), *Pontederia cordata* (Price and Barrett 1984; Barrett and Glover 1985; Glover and Barrett 1986), and *Primula vulgaris* (Piper and Charlesworth 1986). The dissimilar capture of pollen by flowers of the two morphs is attributed to the greater accessibility to contact by insects of pin than thrum stigmas (Ganders 1974, 1975a,b, 1976) and of thrum than pin anthers (Piper and Charlesworth 1986).

According to Ganders (1979), disparities in pollen production of the morphs serve to compensate for the asymmetrical pollen deposition. He suggested that greater pollen production by pins counteracts the dissimilar pollen flow and different pollen sizes in the morphs may be a developmental means to increase pollen production in pins. By this view, the adaptive significance of pollen-size polymorphism ultimately entails enhancement of disassortative (i.e., compatible) pollen flow, and/or dissimilar requisites for storage product, as imposed by differences in style length.

Most of the afore-mentioned investigations of pollen transfer have tested the Darwinian hypothesis on the function of heterostyly in promoting legitimate pollinations. Pollen loads on open-pollinated stigmas have been analyzed in comparison with pollen loads expected from random pollination. The results revealed various degrees of efficiency with respect to intermorph pollen transfer. Large amounts of illegitimate pollen were found in the pollen loads, especially on stigmas of the long-styled morph, but there was no deficiency of pin pollen on thrum stigmas.

To obviate masking of the intermorph promotional effect by intrafloral or geitonogamous pollen deposition on stigmas, studies of pollen flow have also been conducted with emasculated flowers of distylous *Jepsonia heterandra* (Ganders 1974) and tristylous *Pontederia cordata* (Barrett and Glover 1985). The results obtained were found to support the Darwinian hypothesis. This and additional aspects of the function of heterostyly in promoting intermorph pollinations and reducing intramorph pollinations have been discussed by Ganders (1974, 1975a,b, 1976, 1979), Schou (1983), Nicholls (1985a,b), Barrett and Glover (1985), Glover and Barrett (1986), Piper and Charlesworth (1986), Piper et al. (1986), Barrett (1988a, 1990).

It is not yet clear to what extent the compensation by excess pin pollen for lower pollen capture by thrums is indispensable for disassortative pollination and maximum seed set of thrums. In most cases it is difficult to assess the role of excess pin pollen in ensuring maximum potential seed set in the short-styled morph. Nevertheless, in 13 out of 14 species examined (Lewis 1982), compatible pollen grains on thrum stigmas exceeded the number of potential seeds. The only exception was *Cratoxylum formosum*: in this species the short-styled morph was at risk of receiving fewer compatible pollen grains than the number of potential seeds (Lewis 1982).

As noted earlier, the available data regarding pollen dimensions and amounts of pollen in flowers of the morphs in heterostylous species indeed show that, in most instances, pollen from high-level anthers of short-styled flowers is larger than from low-level anthers of long-styled flowers. Generally, the latter are also found to produce more pollen grains. There are, however, exceptions that contravene the view that pollen size polymorphism is primarily a device for achieving excess pin pollen and thereby more efficient pollen transfer. As can be seen in Table 3, such exceptions occur mainly among three groups of plants.

In some *Linum* species pin and thrum grains are equal-sized or thrum grains are smaller (Table 3) yet, in *Linum grandiflorum*, *L. pubescens*, and *L. perenne* (Table 4), the two morphs differ in number of grains produced, probably as a result of unequal anther size. Intriguingly, however, in these three species thrum rather than pin flowers produce more pollen.

The Plumbaginaceae are another group in which the relationships between pollen size and production frequently deviate from the norm. In some members of the Staticeae pollen and stigma dimorphisms occur in the absence of heterostyly. Among these species, pollen size in the morphs varies widely: from B type pollen larger than A in *Armeria plantaginea*, through near equal pollen size in the majority of species, to A type larger than B in *Limonium binervosum*. These instances suggest that, in the absence of reciprocal herkogamy, pollen flow may be symmetrical. This, in turn, implies that polymorphisms in pollen size and production are unlikely to be devices to counteract asymmetrical pollen flow in these species. However, *Acantholimon glumaceum, Goniolimon tataricum* (Bokhari 1972; Schill et al. 1985), *Limonium meyeri*, and *Plumbago capensis* are distylous, yet pollen of the two morphs is almost equal-sized (Table 3) and pollen counts in the latter two species show similar numbers in short- and long-styled flowers (Table 4).

The third exceptional group of distylous plants wherein atypical ratios of pollen size and/or amounts occur between the pin and thrum flowers includes some boraginaceous species and certain populations of *Hedyotis caerulea. Amsinckia grandiflora* and *A. douglasiana* have the typical excess of pin grains with larger thrum grains. In *A. spectabilis* pollen grains of the two morphs are equal-sized, but in *A. vernicosa*, thrum grains are larger than pin grains. Yet in *Amsinckia spectabilis* and *A. vernicosa* thrum anthers are larger and produce considerably more pollen than pin anthers. Other species of Boraginaceae in which there are no appreciable differences between the morphs in pollen size or production include *Anchusa officinalis* and *Cryptantha flava.*

A question that naturally arises is how asymmetrical pollen flow is compensated for in heterostylous species in which anthers of the morphs produce either almost equal amounts of pollen or an excess of thrum pollen. We have no answer to this question. Similarly intriguing are instances in which pin stigmas capture more total pollen than thrum stigmas, even though thrum flowers produce more pollen than pin flowers or the pollen production is similar in the two morphs. Examples are *Linum perenne* (Nicholls 1986), *Amsinckia spectabilis* var. *microcarpa, A. vernicosa* var. *furcata* (Ganders 1975a,b, 1976), and two *Cryptantha* species (Casper 1983).

So far as differential pollen deposition is concerned, a decisive factor not taken into account in most distylous plants is the disparity in size of the receptive area in the morphs. Analyses of pollen loads have mostly been based on acetolyzed total bulk samples of pollinated stigmas rather than on assessment of pollen distribution on individual pin and thrum stigmas. It appears that pin stigmas receive more total pollen than thrum stigmas not only because they are more accessible to insect visitors but, presumably to a great extent, also because their receptive area is larger than that of thrum flowers. Exceptions are *Amsinckia grandiflora* and *Hedyotis caerulea*, in which pin stigmas are smaller, but capture more pollen than thrum stigmas (Ornduff 1976, 1980a).

Although thrum stigmas and pin anthers are less accessible to visiting pollinators than the exposed pin stigmas and thrum anthers, deposition of pollen on thrum stigmas is more precise, apparently owing to positioning of the insect during visitation and localization of pin pollen on a smaller part of its body (Ganders 1974, 1975a,b, 1976). Disassortative pollination of thrums has been found to be more efficient than that of pins (Ganders 1979; Lewis 1982; Barrett and Glover 1985). The actual pollen loads found on thrum stigmas may in some cases reflect greater pollen production by pins.

It therefore appears that disparities between the morphs in pollen loads on stigmas are the result of differences in stigma size, accessibility of stigmas and anthers to insects, and in the amounts of pollen produced. Hence, the widespread phenomenon of pollen size polymorphism cannot invariably be interpreted solely as a means for achieving efficient disassortative pollen transfer. In fact, the selective advantage of increased pollen production by pin anthers for counteracting the asymmetrical pollen flow is debatable in the cases in which thrums produce as much or more pollen than pins.

Why then, is pollen heteromorphism, with more and smaller pollen in long-styled flowers, the prevailing condition in heterostylous plants? The following points in particular merit consideration. Why is greater pollen production in long-styled flowers generally attained by reduced grain size and not by increased size of anthers? Polymorphism in anther size has been reported in few species, in contrast to the widespread pollen size polymorphism reported in several dozens of species. How is one to explain the significance of "aberrant" ratios of pollen production and/or pollen size in the morphs? Why, in cases of greater pollen production by thrums, does one not observe thrum grains which are significantly smaller than pin grains? As argued below, these questions can be answered by assuming that polymorphisms in pollen size and production are involved in other functions of heterostyly and not only in the pollination process.

3.2 Heteromorphic Incompatibility

The incompatibility response is grounded on interaction between the male and female partners. Yet, to better assess possible participation in the incompatibility mechanism of each of the various polymorphisms, this chapter treats pollen, stamens, styles, and stigmas separately.

3.2.1 Pollen Size

Morph-specific pollen size seems to be involved in regulating pollen-pistil interactions in incompatible as well as compatible pollinations.

Darwin (1877) suggested that size differences between pollen grains of the morphs are related to requirements for larger energy reserves in pollen from high-level anthers. In legitimate pollinations, the latter type of grains produce longer tubes to traverse longer styles than grains from low-level anthers, which need only to traverse short styles. Darwin, however, failed to find any marked correlation between inequality of the grains and that of the styles. He therefore ascribed this lack of correlation to physiological differences and so also the absence of size polymorphism he encountered among certain heterostylous species: in some of these species the tubes purportedly developed mainly by energy contained within the grains, while in others by energy from the pistil.

In a study of 23 distylous species, Ganders (1979) also failed to find any correlation between the ratios of thrum to pin pollen volumes and those of pin to thrum style length. Both studies mentioned related the volume of pollen grains to the length of tubes they must produce to grow down the styles in *compatible* pollinations.

There are several neglected early studies which provide valuable information on the relations between pollen size and style length polymorphisms in compatible and incompatible pollinations. From these and from comparative data on pollen-tube growth in *Fagopyrum esculentum* (Schoch-Bodmer 1930, 1934; Dulberger, unpubl. data), *Primula obconica* and *P. sinensis* (Lewis 1942) the following common patterns emerge.

In legitimate pollinations, the tubes of thrum pollen grow down the pin style nearly twice as fast as tubes of the smaller pin pollen down the thrum style. In contrast, in illegitimate pollinations, larger thrum pollen is inhibited earlier and more strongly than pin pollen. In thrum × thrum pollinations, the tubes grow at a rate and attain a length which is only half or much less that in pin × pin pollinations. In pin × pin pollinations of *F. esculentum* and *P. sinensis*, the pollen tubes grow longer than in thrum × pin pollinations, thus contrasting with thrum × thrum pollinations in which tubes were shorter than in pin × thrum pollinations.

From his study of the two *Primula* species, Lewis (1942) concluded that larger thrum grains are adapted to grow faster in compatible longer styles but are inhibited earlier in the shorter own-morph style. Since the thrum style is shorter, there is a selective advantage to earlier inhibition of thrum than pin pollen. Under equal inhibition or equal growth rates of the tubes, there would be penetration into the ovary and fertilization.

Observations on the rate of pollen-tube growth in tristylous *Lythrum salicaria* (Schoch-Bodmer 1942, 1945; Esser 1953) and *L. junceum* (Dulberger 1970b) reveal features that are basically similar to those described in *Fagopyrum* and *Primula*. In legitimate pollinations, tubes of the large grains from long stamens grow two to three times faster in compatible long styles than tubes of smaller grains from mid- or short-level stamens in their respective compatible Mid or Short styles. Also, in illegitimate pollinations, large grains are generally inhibited at shorter distances in incompatible Short styles than small grains in incompatible Long styles.

The extent to which these patterns are shared by other species cannot be evaluated at present. The rate of pollen tube growth in vivo has been monitored in only a few additional heterostylous plants. Yet, gynoecial sites at which the male gametophyte is inhibited in incompatible pollinations have been determined in many heterostylous species belonging to different genera and families. Table 5 summarizes information of which only fragments have thus far been considered sporadically in the literature on heteromorphic incompatibility.

As can be seen in Table 5, the development of the male gametophyte may be arrested at numerous stages and sites. The table also implies that the site of inhibition may be a familial characteristic, as in the case of the Plumbaginaceae. More frequently, however, the patterns vary within the same family at all levels, from intergeneric to intramorph variation.

Hence, the pattern of the inhibitory response as shared by different taxa displaying the same kind of polymorphism merits special attention, because it helps to clarify the role played by the polymorphic character.

Contrary to accepted views, the site of inhibition in the majority of species examined, in at least one of the morphs, is not the stigmatic surface but rather the stigmatic zone sensu Knox (1984) or the transmitting tract of the style or ovary.

Table 5 shows that in 9 out of 11 species belonging to six genera of the Rubiaceae, pollen tubes are inhibited within the style in intramorph pin × pin pollinations, but the inhibition is stigmatic in intramorph thrum × thrum pollinations, as was pointed out by Bawa and Beach (1983). Presumably, these species possess dimorphic pollen size, similarly to most distylous Rubiaceae; at least they are not included by Bir Bahadur (1968) among rubiaceous species with no pollen size dimorphism.

Within the genus *Primula* there is also considerable intramorph variation in site, stage, and rigor of the incompatibility response, in addition to intersectional and interspecific variation (Richards 1986). Some of this variation is shown in Table 5. Richards (1986) distinguished four types of incompatibility responses among *Primula* species, pertaining to the behavior of the male gametophyte in incompatible pollinations. In three of the four types the larger thrum pollen is inhibited at an earlier

Table 5. Sites of inhibition of the male gametophyte in intramorph pollinations of some heterostylous plants

Species	Site of inhibition		Reference
	Pin × pin pollinations	Thrum × thrum pollinations	
Plumbaginaceae			
Acantholimon androsaceum	Stigma surface	Stigma surface	Baker (1966)
Armeria maritima	Stigma surface	Stigma surface	Baker (1966)
Ceratostigma willmottianum	Stigma surface	Stigma surface	Dulberger (1975a)
Goniolimon tataricum	Stigma surface	Stigma surface	Baker (1966)
Limonium meyeri	Stigma surface	Stigma surface	Dulberger (1975a)
L. sinuatum	Stigma surface	Stigma surface	Dulberger (1975a)
L. vulgare	Stigma surface	Stigma surface	Baker (1966)
Plumbago europaea	Stigma surface	Stigma surface	Dulberger (1975a)
P. capensis	Stigma surface	Stigma surface	Dulberger (1975a)
Rubiaceae			
Cephaelis elata	Base of the style; stigma	Stigma	Bawa and Beach (1983)
Coussarea sp.	Base of the style	Stigma	Bawa and Beach (1983)
Coussarea sp.	Half the length of style	Stigma	
Faramea suerrensis	Stigma	Stigma	Bawa and Beach (1983)
Faramea sp	Within stigma	Within stigma	Bawa and Beach (1983)
Psychotria acuminata	Base of the style	Base of the style	Bawa and Beach (1983)
P. chiapensis	Half the length of the style or stigma	Stigma	Bawa and Beach (1983)
P. officinalis	Half the length of the style	Stigma	Bawa and Beach (1983)
P. suerrensis	Junction of stigma and style	Stigma	Bawa and Beach (1983)
Rudgea cornifolia	Base of the style	Stigma	Bawa and Beach (1983)
Palicourea lasiorrachis	Half to 3/4 of the style	Stigma surface	Feinsinger and Busby (1987)
Primula			
Primula elatior	Stigma surface	Stigma surface	Schou (1984)
P. obeoconica	Stigma surface, within the stigma or style	Stigma surface	Lewis (1942); Dowrick (1956); Stevens and Murray (1982); Schou (1984); Richards (1986);
P. veris	Style	Stigma surface	Richards and Ibrahim (1982); Richards (1986);

Table 5. *(Continued)*

Species	Site of inhibition		Reference
	Pin × pin pollinations	Thrum × thrum pollinations	
P. vulgaris	Stigma surface, within the stigma head, or in the style	Stigma surface, within the stigma head, or in the style	Shivanna et al. (1981)
Fagopyrum esculentum	Base or the style	Within stigma	Schoch-Bodmer (1934); Dulberger (unpubl. data)
Linum			
Linum grandiflorum	Stigma surface	Within stigma	Darwin (1877); Lewis (1942)
L. pubescens	Stigma surface	Within stigma	Dulberger (1973)
L. mucronatum	Stigma surface	Within stigma	Dulberger (1973)
L. austriacum	Top of the style	Top of the style	Darwin (1877); Baker (1975)
L. perenne	Top of the style	Top of the style	Darwin (1877); Baker (1975)
L. maritimum	Base of the stigma	Base of the stigma	Dulberger (1987a)
Turnera ulmifolia	Junction of stigma and style	Junction of stigma and style	Martin (1965)
T. ulmifolia	Half the length of the style	Stigma-style interface	Shore and Barrett (1984)
Melochia tomentosa	Within stigma or style	Within stigma or style	Martin (1967)
Oxalis alpina	Upper part of the style	Upper part of the style	Weller (pers. commun.)
Boraginaceae			
Amsinckia douglasiana		Cryptic self-incompatibility	Casper et al. (1988)
A. grandiflora		Cryptic self-incompatibility	Weller and Ornduff (1977, 1989)
	Mid-stylar region	Upper part of the style	
A. spectabilis	Self-compatible		Ray and Chisaki (1957)
A. vernicosa var. *furcata*	Self-compatible		Ganders (1975b)
Anchusa officinalis	Ovule	Ovule	Schou and Philipp (1983)
Cryptantha flava	Self-compatible		Casper (1985)
Narcissus tazetta	Ovule	Ovule	Dulberger (1964)

stage and more strongly in thrum × thrum pollinations than the smaller pin pollen in pin × pin pollinations.

An overall relationship between pollen size and tube length has been noted in tristylous *Pontederia cordata* (Anderson and Barrett 1986) and *P. sagittata* (Glover and Barrett 1983; Scribailo and Barrett 1986, 1989; Scribailo 1989; Barrett 1990). In these species, after pollinations of Long or Mid styles with pollen of short stamens and of Long styles with pollen of mid stamens (L × s, M × s and L × m), pollen tubes grew to the same length as in legitimate pollinations with the same pollen type. This finding suggested that the inability of pollen of small size to produce tubes capable of traversing longer styles may be due to inadequate storage reserves in small grains.

On the other hand, in *Oxalis alpina, Melochia tomentosa,* some populations of *Turnera ulmifolia*, three *Linum*, and two *Faramea* species, intermorph differences in pollen size are associated with inhibition at similar sites in the pistils of the two morphs (Table 5). Arrest of pollen tube growth in the stigma head or upper style has also been reported for *Cordia curassavica, Linum narbonense, Oxalis valdiviensis*, and *Epacris impressa* (Gibbs 1986), but it is not clear whether this disparity in sites of inhibition has been observed between the morphs or within each floral morph.

Despite the intriguing exceptions of *Melochia, Oxalis,* and *Faramea* species, it is clear that in many other taxa, such as *Primula*, Rubiaceae, *Fagopyrum, Lythrum*, and *Pontederia*, morph-specific pollen size generally correlates with dissimilar behavior of the male gametophyte and differing sites of inhibition. The physiological basis, if any, for such correlation has yet to be elucidated.

The large thrum and smaller pin pollen grains of *Primula vulgaris* have been found to differ in their ability to rehydrate and germinate in vitro (Shivanna et al. 1983). But a similar difference has been inferred from the behavior in vitro of equal-sized grains of *Linum grandiflorum* (Lewis 1943), *Limonium meyeri*, and *Plumbago capensis* (see Sect. 3.2.5).

As mentioned, differences in storage reserves related to pollen size have been invoked repeatedly to explain the different lengths of pollen tubes in styles of the morphs. However, in *Lythrum junceum* the present author found that the polymorphism of pollen in starch content does not appear to be involved in incompatibility. In legitimate pollinations of Longs, with starch-containing pollen grains from long stamens of Mids or Shorts, pollen tubes reacted positively to IKI test as far down as their style base. On the other hand, in illegitimate pollinations of Mids or Shorts with pollen from long stamens, tubes were inhibited within the stigma or at the upper part of the style, even though they were still densely filled with starch (Dulberger 1967 and unpubl. data).

Thus, polymorphism in starch content of pollen does not appear to be operative in incompatibility. It is possible that in *Lythrum junceum* and *L. salicaria* the presence of starch in pollen of high-level anthers, and in the latter species also the differing coloration of anthers and of their pollen grains, serve as reward and attractant for pollinators rather than being involved in the process of differential tube growth.

A different growth rate of pollen tubes in the two morphs is not necessarily dependent on pollen size dimorphism. In *Linum grandiflorum* and *L. pubescens*, for instance, pin and thrum grains are of equal size, and yet in pin × thrum pollinations the tubes grow twice as fast as in thrum × pin pollinations (Lewis 1942; Dulberger 1973 and unpubl. data). Even in species displaying pollen size polymorphism, the dissimilar behavior of the male gametophyte in incompatible pollinations is not dependent on size differences between grains of the morphs as such. Thus, in *Lythrum salicaria* and *Lythrum junceum*, tubes of the large pollen grains from high level anthers of Mids or Shorts are unable to traverse styles of Mids or Shorts, even though they do traverse styles of Longs. Secondly, in many distylous species there is overlap between the dimensions of pin and thrum pollen grains and, in some tristylous species, also between the pollen from the two anther levels of the same morph. Yet in these cases, the incompatibility response is characteristic for pollen of the corresponding morph or anther level and is not dependent on magnitude of the size disparity or similarity.

Barrett (1978) and Richards and Barrett (1984) noted a relevant contrasting condition among tristylous members of the Pontederiaceae. Self-incompatible *Eichhornia azurea, Pontederia rotundifolia, P. cordata,* and *P. sagittata* display strongly developed pollen size trimorphism. In contrast, *Eichhornia crassipes* and *E. paniculata* are highly self-compatible with only weakly developed size trimorphism of pollen grains from different anther levels. Similarly, *Quinchamalium* (Riveros et al. 1987) and *Nivenia* species (Goldblatt and Bernhardt 1990) have weak pollen-size dimorphism and are also self-compatible.

At this point it should be stressed that the incompatibility reaction and morph-specific pollen size have not been found separated genetically (Lewis 1949; Ernst 1936a,b, 1955; Charlesworth 1979). Breakdown of heteromorphic self-incompatibility is associated with a loss in pollen size heteromorphism (Ornduff 1972; Barrett 1979, 1988b).

If the correlation between heteromorphic pollen dimensions and the differing behavior of pollen has a functional basis, then the factor responsible for pollen behavior is clearly neither the storage product nor the grain size proper, although it must be strongly linked with morph-specific size or dry weight. What we are looking for could be a protein or perhaps a growth factor deriving from an early premeiotic stage of the microsporangium, during differential growth of the sporogenous cells or of the pollen mother cells.

If morph-specific pollen size is involved in incompatibility, how is one to account for equal-sized pollen in the morphs or reversed thrum/pin size ratios of species listed in Table 3? To answer this question a comparison could be instructive. The incidence of taxa with equal or similar-sized pollen in both morphs should be compared with that of taxa in which the reaction of the male gametophyte deviates from the usual pattern of intrastigmatic or stylar inhibition in one of the morphs.

The Plumbaginaceae with equal-sized pollen are the only group in which the inhibition response occurs on the surface of the stigma in both morphs (Iversen 1940; Baker 1966; Dulberger 1975a).

In several *Linum* species possessing similar-sized pollen, rejection of incompatible pollen also occurs on the stigma surface prior to germination, at the early stage of adherence, albeit in only one of the morphs.

In the distylous Boraginaceae in which pollen of the two morphs does not differ appreciably in size, a gamut of deviations both from the usual sites of inhibition and from typical diallelic incompatibility occurs. In *Anchusa hybrida* and *A. officinalis* incompatibility is apparently under multiallelic control and thus genetically independent of floral dimorphism (Dulberger 1970a; Philipp and Schou 1981; Schou and

Philipp 1983, 1984). Moreover, in *A. officinalis*, self-pollen tubes are arrested in the ovules, similarly to *Narcissus tazetta*. In distylous *Cryptantha flava* and *C. flavoculata*, absence of differences in pollen size and pollen production are associated with self-compatibility (Casper 1983, 1985). *Amsinckia grandiflora* and *A. douglasiana* display typical dimorphisms of pollen size and production and in both species cryptic self-incompatibility has been reported (Weller and Ornduff 1977, 1989; Casper et al. 1988). Conflicting results obtained with *A. grandiflora* (Weller and Ornduff 1977, 1989; Ganders 1979) may be due to presence of pollen size differences in some but not other populations or to differences in the reactivity of styles. *Amsinckia spectabilis* var. *microcarpa* has equal-sized pollen grains and is self-compatible (Ray and Chisaki 1957). In *A. vernicosa* var. *furcata* thrum grains are larger than pin grains and plants are self-compatible (Ganders 1976).

From the foregoing it is evident that in distylous species of Plumbaginaceae, *Linum*, and Boraginaceae, male gametophytes react in a variety of ways, ranging from rejection at the stigma surface in one or both morphs, through cryptic incompatibility or inhibition within the ovary, to acceptance in self-pollinations. In some of these species incompatibility is under multiallelic control, or the plants are self-compatible, but in no case is the rejection response expressed as an arrest of pollen tubes at two morph-specific sites as is typical of most distylous plants with pollen size polymorphism. It seems therefore, that in taxa in which morph-specific pollen size does not function in diallelic incompatibility, typical pollen size polymorphism is absent. Consequently, pollen grains in these species may be equal-sized, or may display size differences which are insignificant, inconsistent, and perhaps too small to be functional, or else the typical size relations are reversed. It is noteworthy that most of the species that lack pollen-size heteromorphism have an excess of thrum rather than pin pollen grains.

A recognition of the role of pollen-size polymorphisms in interactions of the male gametophyte with the pistil in incompatible pollinations provides answers to some of the questions raised earlier. Polymorphism in anther size is considerably less widespread than polymorphism in pollen size, because in heterostyly morph-specific anther size as such has no selective value. In instances where it occurs, namely where grains in the morphs are equal-sized and thrum anthers produce more pollen than pin anthers, incompatibility is not affected by pollen size. An example are *Linum* species in which differences in pollen production are perhaps important per se, but the inhibition occurs on the stigma, without differential pollen tube growth operating in incompatible pollinations. Another example is *Amsinckia spectabilis* with higher number of thrum grains, similar pollen size and self-compatibility.

Thrum pollen grains that are smaller than pin grains occur only rarely (Table 3), and apparently only in plants in which the stigma is the site of inhibition. Examples are *Limonium binervosum*, *Dyerophytum africanum*, *Linum austriacum* and *L. suffruticosum*, and *Fauria crista-galli*. In the first two inhibition is probably on the stigma, as in other members of the Plumbaginaceae; in *Linum* species, inhibition is within the stigma in one or both morphs. In none of the instances in which thrum grains are smaller than pin grains is incompatibility manifested by differential pollen tube growth in the morphs.

The assumption that pollen size heteromorphism functions in incompatibility accounts for the normal presence of larger grains in long than in short stamens and is not disproved by exceptions listed in Table 3, where morphs were found to produce equally sized pollen or pin grains larger than thrum grains.

3.2.2 *Stamen Length*

There is no evidence to indicate a direct involvement of stamen length in incompatibility. There can be no doubt, however, that the function of stamen length polymorphism is to create a reciprocity between the levels of anthers and stigmas in the morphs. Stamen-length polymorphism is directly correlated with pollen-size polymorphism and inversely correlated with polymorphism of pollen production. The first two attributes have been found to be genetically independent (Ernst 1936a,b, 1955; Dowrick 1956).

Absence of a marked stamen length dimorphism is associated with an absence of pollen-size dimorphism, as is true for *Linum grandiflorum*, members of the Staticeae and plants with atypical heterostyly, with no diallelic incompatibility (Sect. 5.).

3.2.3 *Style Length*

There are grounds for believing that in most heterostylous species, style length contributes to the physiological control of incompatibility, in addition to the polymorphisms of pollen and stigmas described above.

Baker (1964) noted that both efficient mechanical promotion of pollination between the morphs and reduction of pollen wastage in intramorph pollinations are dependent on an adequate dimensional relationship between a flower and its pollinator. He also pointed out that in flowers of the open type, lacking a tube, as in *Fagopyrum esculentum*, there is little advantage to the complementary placement of stigmas and anthers. Neither is it likely that heterostyly functions to increase disassortative pollination in flowers such as those in distylous *Linum grandiflorum*, where there is no reciprocity in positioning of stamens and stigmas of the morphs. Yet, style length polymorphisms do occur in variously shaped flowers, and not only in flowers of an optimal shape for promotion of intermorph pollen transfer.

The ubiquity of style-length polymorphism in plants with diallelic incompatibility strongly suggests that morph-specific style length is a sine qua non for the functioning of heteromorphic incompatibility systems.

It is customary to compare the length attained by pollen tubes in incompatible pollinations with that reached by tubes of the same pollen type in the compatible gynoecia. Upon such comparisons it is found that in incompatible pollinations pollen tubes usually cease growth before attaining the length reached by tubes of the same pollen type in the compatible styles to which they are adapted.

As noted earlier (Sect. 3.2.1), in the majority of heterostylous plants pollen tubes of the short-styled morph are inhibited on or just underneath the stigma surface, within the stigmatic tissue, at the transition between stigma and style, or in the upper part of the style.

On the other hand, pollen tubes of long-styled flowers may in some instances be inhibited in styles of their own morph after having traversed a distance similar or exceeding that traversed in compatible thrum styles. In *Fagopyrum esculentum*, thrum pollen tubes reach a mere 200 µm in thrum styles, vs. 1850 µm in pin styles. In contrast, pin pollen tubes reach 1440 µm in pin styles vs. 650 µm in thrum styles (Schoch-Bodmer 1930, 1934). In *Primula sinensis*, pin tubes reach a greater length in their own styles than in thrum styles, and Lewis (1942), who reported this finding, suggested that this in itself could suffice to arrest the growth of pin grains in long styles, obviating the need for a physiological inhibition mechanism in this morph.

In incompatible pollinations of short-styled flowers the inhibitory effect of the style is fairly obvious. In long-styled flowers, on the other hand, the inefficiency of intramorph pollinations can be ascribed to the very difference in style length, and also to the inability of tubes formed by smaller grains of this morph to descend to the base of the long style. Thus it is often reasoned that the style length itself becomes the female factor responsible for cessation of pollen tube growth.

In recent years a similar interpretation was proposed for *Pontederia cordata* (Glover and Barrett 1983) and *P. sagittata* (Anderson and Barrett 1986; Scribailo 1989). These authors emphasized that in certain incompatible pollinations there is an apparent correlation between pollen-size heteromorphism and style length, perhaps without the need for an inhibitory effect of the transmitting tract in long-styled and mid-styled flowers. Accordingly, in these morphs the incompatibility mechanism might be largely "passive" and not involve rejection specificities.

An alternative interpretation, however, could be that long styles do exert an inhibitory effect, but in some instances this effect is delayed till tubes attain a length which is greater than that attained in the compatible styles to which they are adapted.

That this is the correct interpretation for *Fagopyrum esculentum* is evident from a convincing though overlooked experiment of Schoch-Bodmer (1934), in which she pollinated pin and thrum styles of *F. esculentum* with pollen of a non-heterostylous species. The results revealed that pollen-tube growth was dissimilar in styles of the two morphs. Thus, pollen tubes of *Polygonum aviculare* attained 1100 µm in pin styles and only 200 µm in thrum styles (of *Fagopyrum*), while those of *Plantago lanceolata* reached 1100 µm and 360 µm in pin and thrum styles, respectively. Hence, conditions for tube growth are clearly different in long and short styles. It should also be mentioned that pollen of the two morphs germinated similarly on stigmas of *Veronica*.

That this difference could be due to a stimulatory action of the style in compatible pollinations is ruled out by results of another experiment by Schoch-Bodmer (1934), in which the two pollen types of *Fagopyrum* were found to form in styles of *Veronica chamaedrys* or *V. tournefortii* tubes of similar length, namely, pin tubes of 630 µm and thrum tubes of 850 µm. Thus, for growth in a short style of *Fagopyrum*, both types of pollen found the necessary conditions in each of the two foreign styles.

The elegant observations of Schoch-Bodmer demonstrate irrefutably that in *F. esculentum* not only the short style but also the long style exerts an inhibitory effect. The cessation of pollen tube growth in long styles is not a direct outcome of the greater style length proper, nor is it attributable exclusively to a smaller volume of the grain or to a difference in storage reserves.

Within the style, pollen tubes grow in the intercellular spaces of the transmitting tissue. Histochemical tests have shown that these spaces contain substances which are secreted by the surrounding cells and stain for proteins and polysaccharides. This has been observed in distylous *Primula vulgaris* (Heslop-Harrison et al. 1981) and *Linum grandiflorum* and *L. pubescens* (Dulberger, unpubl. data). In the style, the rejection response probably occurs between the tip of the pollen tube and substances present in the interstitial matrix of the transmitting tract (Knox 1984).

Attempts to detect and determine recognition substances in styles of the morphs have thus far been limited to *Primula obconica* (Golynskaya et al. 1976) and *P. vulgaris* (Shivanna et al. 1981). They have shown that extracts obtained from thrum and pin stigmas and styles inhibit the growth of self-pollen tubes in vitro differentially. At certain concentrations, thrum extracts proved more inhibitory to thrum pollen tube growth than to the growth of pin pollen tubes, and vice versa.

The disparity in sites of inhibition could also be an effect of morphological and structural dissimilarities, without incompatibility determinants being the product of style length polymorphism. For a better assessment of the role of style length polymorphism in the incompatibility response, it is therefore crucial to elucidate whether morph-specific stylar recognition factors inhibitory to the male gametophyte are generated by the different style lengths of the morphs.

It is noteworthy in this connection that, although style-length polymorphism is the most characteristic feature of heterostyly, basic aspects such as the comparative anatomy, development and physiology of the style in the floral morphs have thus far received relatively little attention. The cause for this neglect is probably the prevailing opinion that di- or trimorphism in style length are relevant only in the context of pollination.

The possible involvement of morph-specific style-length differences in generating an incompatibility specificity receives greater plausibility when one collates five apparently unrelated lines of evidence. The different sources of evidence pertain mainly to the mode of growth and histology of the style in the morphs.

First, there are comparative observations which have evinced disparities in structural details of the transmitting tract of the respective morphs in all species examined (Sect. 2.1.1).

Second, it has been shown that the style develops mostly by secondary elongation via intercalary growth (Takhtajan 1948; Eames 1961). Furthermore, Linskens (1964, 1974) established that in the very young style of *Petunia*, the transmitting tissue grows mainly by cell divisions localized in the basal meristematic zone. Subsequently, during the second phase, style growth entails elongation of cells without change in their number, and occurs with a strong dependence on the presence of the androecium.

Third, there are early works on developmental stages of heterostyly in various species. In young buds of *Primula* species and *Gregoria vitaliana*, (Rubiaceae), the length of styles and stamens has been shown to be initially equal in the floral morphs, but subsequently, differences in length begin to appear as a result of differential cell elongation in the style and stamens of the morphs (Ernst 1932; Schaeppi 1935; Bräm 1943). Accordingly, in the long style and stamens there is induction of growth, while in the short style and stamens growth is repressed. Again, Ernst (1932) described and depicted elongated cells in pin styles and short cells in thrum styles of *Psychotria malayana*. Observations by Schoch-Bodmer (1930) in *Fagopyrum esculentum* indicate a 3:1 ratio between the pin to thrum style length as well between the lengths of epidermal cells of the styles in the two morphs. A similar condition has been reported for *Primula vulgaris* (Heslop-Harrison et al. 1981). In *Primula sinensis* (Stirling 1932) and *Eichhornia paniculata* both cell divisions and cell expansion contribute to differences in style length (Richards and Barrett 1984).

Fourth, there is an edifying study by Russell (1986) of pollen tube growth in relation to the transmitting tract of monomorphic *Plumbago zeylanica*. Russell found that pollen tubes display a strongly biphasic pattern of growth, related to changes in the organization of the stylar transmitting tissue, rather than to specific changes in pollen physiology. Thus, in the upper part of the style cells are axially aligned, long and narrow, having a length to width ratio of 45:1 and walls rich in pectins. In the lower part of the style cells are shorter and broader, have a length to width ratio varying from 11.4 to 1.5, and have thinner walls. The orientation of the latter cells becomes skewed and may provide greater resistance to the passage of pollen tubes. They clearly do not elongate similarly to the upper cells during the second phase detected by Linskens (1964, 1974) in *Petunia*.

The fifth line of evidence entails studies on wall extension (Lamport 1970; Sadava and Chrispeels 1973; Sadava et al. 1973), which have shown that differences in the rate of cell extension and cessation of elongation growth are associated with differences in wall proteins.

The above evidence strongly suggests that the morph-specific mode of growth and length of the style, the organization of the transmitting tract, the substances present in the interstices, and the differential sites of inhibition are all interrelated in one way or another.

3.2.4 Size of Stigmatic Papillae

Lewis (1949) conjectured that dimorphism of papillar size and style length may be "two ways of measuring the same effects." He reached this conclusion in the light of Ernst's (1932) finding that the style and stamen lengths in *Psychotria malayana* are proportional to the lengths of their cells. Lewis also noted that no separate genetic control of papilla size and style length was known. In *Primula vulgaris*, however, in which the pin/thrum ratios of style length and epidermal cell lengths are similar, the corresponding ratios for papilla length are significantly different (Heslop Harrison et al. 1981).

Larger stigmatic papillae in long- than short-styled flowers may be a consequence of differential style elongation in the morphs; style and stigma extension may have a common physiological basis. Whether this condition is an outcome of developmental constraint, or a disassociation of style and stigmatic extension is possible, is not known. If elongation of style cells and of stigmatic papillae are physiologically separable, then the reverse condition of thrum papillae larger than pin papillae should be encountered.

An inverse correlation between stigma size and style length of the morphs has been reported in few species (Sect. 2.3.1). In *Amsinckia grandiflora* and *Anchusa officinalis*, thrum papillae appear to be larger than pin papillae. In *Primula malacoides*, however, in which thrum stigmas are larger than pin stigmas, pin papillae are longer (Pandey and Troughton 1974), thus indicating that stigma size and papilla size can be independent characters.

In *Linum pubescens* and *L. grandiflorum*, the highest rates of style elongation occur on the final 2 days of preanthesis, after papillae have attained most of their mature size (Dulberger 1984 and unpubl. data).

In floral buds of nonheterostylous *Lupinus pilosus*, a marked extension of the brush hairs was observed prior to the final extension of the papillae; in some Compositae, extension growth of the stylar branches, brush hairs, and adjacent receptive cells of the stigma occur at distinct stages of ontogeny (Dulberger, unpubl. data). Thus, extension growth of the style and stigma cells is not necessarily concomitant; separate timings for cell extension may occur in adjacent tissues that perform different functions in the stigma area. It is not known whether the same or different hormonal controls are involved in extension growth of these tissues.

How can one account for the prevailing condition in most heterostylous plants of larger papillae in long- than short-styled morphs? Is it due to paucity of hormonal control enabling the reverse condition or to the fact that control ensuring the prevailing condition has been selected for its functional advantage? The fact that papilla-size polymorphism may be the consequence of differential style growth does not rule out functionality of morph-specific papilla size.

Morphological types of stigmas are determined by factors of various nature such as the mode of carpel closure and fusion, placentation, and the pollination mechanism. Cytological features of the receptive surface, however, are only rarely molded by any mechanism of pollen transfer; interrelations with pollen grains appear to be more relevant for the characters of papillae. Striking examples of this can be found in stigma papillae of the Staticeae (Dulberger 1975a) and the Boraginaceae (Heslop-Harrison 1981).

Early opinion holds that the larger papillae of pin styles accomodate the larger grains of thrum flowers. Shivanna et al. (1981), who studied the physiology of rehydration of pin and thrum pollen on the two types of stigmas of *Primula vulgaris*, suggested that the slight structural differences between the papillae of the two morphs may be involved in control of pollen hydration, chemotropic guidance of emerging tubes, and tube penetration. Observations on pollen-stigma interactions in distylous *Linum* species support this suggestion.

3.2.5 Pollen-Stigma Interactions in Linum Species and Plumbaginaceae

In *Linum grandiflorum, L. pubescens*, and *L. mucronatum* the usual papilla-size dimorphism is associated with dimorphic wall structures, (Sect. 2.3.3) almost equal-sized pollen grains (Table 3), and dimorphic exine sculpturing(Sect. 2.1.2.2).

After thrum × thrum pollinations, stigmas become covered by hundreds of pollen grains but pollen tubes are arrested within the stigma. In pin × pin pollinations most grains fail to adhere to the stigma. In pollinations of thrum stigmas the secretory product released onto the disrupted cuticle of the papillae provides an adhesive for both types of pollen grains (Dulberger 1974, 1975b, 1987a; Ghosh and Shivanna 1980a,b, 1983; Murray 1986).

On the pollen side, dimorphic exine sculpturing provides a differential contact area between the proteinaceous-lipoidal coating of the grains and the surface of stigma papillae. The uniformly long, blunt processes of the exine in thrum grains offer a large area for contact between the exine coating and substantial secretory materials on the surface of thrum papillae, as well as with the thin proteinaceous film overlying pin papillae.

In contrast, the contact area of pin pollen grains is considerably smaller, being restricted to tips of the longest, pointed processes. The area here is sufficient for establishing intimate contact with thrum papillae, but not for adherence of pin pollen to the extremely thin film overlying the smooth surface of pin papillae. A brief contact of pin pollen grains and pin papillae elicits a repulsion response, presumably of an electrostatic nature. Although the physicochemical basis of this rejection is not known, there can be little doubt that the structural dimorphism of papillae and exine play an important part in the recognition process (Dulberger 1991).

In the Plumbaginaceae, the site of inhibition in both incompatible pollinations is the surface of the stigma. In *Armeria maritima* the lengths of styles and stamens are similar in the two morphs. In pollen loads of cob stigmas both pollen types occur, whereas in pollen loads of papillate stigmas almost exclusively A grains were present (Iversen 1940). Contrary to Mattsson's (1983) assertion, Iversen attributed the absence of B pollen on papillate stigmas to failure of B type grains to germinate, rather than to failure to adhere to these stigmas.

Dulberger (1975a) observed that in *Limonium sinuatum* and distylous *Limonium meyeri*, in the combination of papillate stigmas with B type pollen, the grains are rejected prior to germination, namely at the stage of adherence to the receptive surface. In pollinations of cob stigmas with pollen of the A type, the grains adhere but do not germinate. She suggested that a morphological complementarity between the stigma papillae and pollen exine in the papillate × A combination is absent in the papillate × B combination and that this lock and key mechanism reinforces chemical recognition. Mattsson's (1983) observations on structural, physical, and chemical factors responsible for adherence vs. nonadherence of pollen in the two incompatible combinations of *Armeria maritima* lend strength to this suggestion.

Recently, stigma dimorphism has also been encountered in *Goniolimon tataricum* and *Acantholimon glumaceum* (Schill et al. 1985). In *Goniolimon tataricum, G. callicomum,* and *Acantholimon androsaceum* the two types of plants, with A or B pollen, respectively, have been found to be self-incompatible and cross-compatible (Baker 1966). It is therefore reasonable to assume that in *Goniolimon* and *Acantholimon* too, incompatibility operates on the basis of both pollen and stigma dimorphisms.

Following intermorph pollinations of *Plumbago capensis*, pollen becomes hydrated and swollen within minutes of deposition. In pin × pin pollinations considerably fewer pollen grains adhere to the stigma than in thrum × thrum pollinations. In both types of incompatible pollinations, the grains fail to take up water and maintain the shape characteristic of their desiccated state.

The failure of pin pollen to swell on pin stigmas can be explained by the absence of secretory products on their surface, as well as by the morph-specific structure of the papilla wall and cuticle, and their impermeability. Nevertheless, thrum pollen does become hydrated on pin stigmas, despite impermeability. On the other hand, thrum pollen does not hydrate on its own-morph stigma, despite presence on the latter of a secretory product on which pin grains do germinate. It must be concluded that pin and thrum pollen grains differ in their ability to extract water from the stigma.

Pollen grains of both morphs are coated with lipids and proteins and there is an epi-exinous veil-like layer which masks the exine sculpturing. This layer constitutes the contact area of the pollen grain with the stigma (Dulberger 1987b). It is not clear to what extent differences in exine sculpturing of pin and thrum pollen are relevant for incompatibility in *Plumbago capensis*. Presumably, both the patterning of the pollen exine and its coating are functionally significant in the acceptance and rejection responses.

3.2.6 Patterns of Heteromorphic Incompatibility

From the foregoing, two major categories are discernible in the relationships between polymorphisms and the physiological mechanisms of diallelic incompatibility.

1. A category in which the majority of taxa display di- or trimorphism in pollen size, and the morph-specific behavior of the male gametophyte seems to depend on a factor strongly linked with pollen size.

Three subcategories are clearly differentiated here:

a) Inhibition of pollen tubes in the different incompatible pollinations takes place at two or three dissimilar sites in the style and/or stigma. Examples are *Primula sinensis, Fagopyrum esculentum, Lythrum salicaria* and *L. junceum*.
b) The rejection response occurs at the same site in both morphs, usually at the transition between stigma and style or at the upper part of the style, but may also take place on/or within the stigma in both morphs. Examples are some *Linum* and *Oxalis* species and *Primula elatior*.
c) The incompatibility response occurs in the ovary at least in one of the two or three morphs. The style is of the hollow type. This pattern occurs in tristylous *Pontederia* and atypically distylous *Narcissus tazetta* but presumably also in *Anchusa officinalis*; in *Anchusa hybrida* at least, there is a canal in the lower half of the gynobasic style (Dulberger and Cohen unpubl. data).

2. A second major category is represented in taxa where the pollen grains display dimorphic exine sculpturing combined with dimorphic shape and/or structure of the stigmatic papillae. Examples are *Limonium meyeri* and *Plumbago capensis*. Here the male gametophyte is rejected on the stigma in both or one of the morphs.

Regrettably, no information is available on stigma structure and on the behavior of the male gametophyte in instances where pollen of the morphs differs in size, shape, and exine sculpturing.

4 Constraints Limiting the Distribution of Heterostyly

The distribution of heterostyly may become more understandable when factors limiting its appearance are considered. Such factors may be morphological, developmental, or physiological in nature but may also relate to the pollination system, life form, pre-existing breeding system or to combinations of these.

Heterostyly occurs almost exclusively in hermaphrodite, entomophilous flowers, with rare exceptions such as the ornithophilous *Palicourea lasiorrachis* (Feinsinger and Busby 1987). The phenomenon is absent in wind-pollinated and unisexual flowers. The importance of tubular flowers for intermorph pollen delivery and deposition has been noted by Baker (1964) and Ganders (1979). Heterostyly is rare in nontubular flowers, or in very small flowers which are usually bowl-shaped or flat. A notable exception is *Fagopyrum esculentum*. Heterostyly is rare in large flowers. To the examples given by Ganders (1979) one can add *Plumbago capensis* with a corolla tube attaining 35 mm and a style up to 38 mm in length and various *Narcissus* species.

Only a few heterostylous species have zygomorphic flowers, e.g., distylous species of *Gelsemium* and tristylous species of *Lythrum, Eichhornia,* and *Pontederia.* This paucity may partly be explained by the requirement for achieving reciprocal herkogamy despite the frequent positioning of anthers at different levels on two opposing sides of the same flower. The degree of constraint imposed by zygomorphy is illustrated by tristylous *Pontederia* and *Eichhornia* species. In these flowers there are two whorls of three stamens each, but because of zygomorphy the anthers at each level originate from both whorls and not from one (Richards and Barrett 1987; Scribailo 1989).

In most taxa in which heterostyly occurs, the stamen filaments are separate from their bases up to much of their distal part. In taxa where the filaments are adnate to the corolla along most of their length, stamen elongation in the course of bud development also entails elongation of the corolla tube. This constraint explains why thrum flowers are sometimes larger than pin flowers: the greater extension of their stamens involves greater extension of the corolla tube. It is perhaps significant that most instances in which a difference in size of the flowers of the morphs has been reported (Ganders 1979) belong to the Boraginaceae.

Thus, differential elongation of stamens in the floral morphs may be limited by their adnation to the corolla and may not be possible without changes in the dimensional relationships between the tube and limb of the corolla. This could influence the search image of the morphs. Significantly perhaps, in *Quinchamalium chilense*, where anthers are adnate to the perianth tube, to the extent of being almost sessile (Riveros et al. 1987), there is no reciprocal herkogamy and anthers are situated at the same level in the two morphs. Nevertheless, reciprocal herkogamy is widely represented in the Rubiaceae, where various degrees of adnation of stamens to the corolla occur.

In distylous flowers with numerous stamens, the bases of stamen filaments are either free or connate forming a tube, as in *Melochia, Hypericum, Waltheria,* and *Cratoxylum* species. Hence, differential elongation of stamens does not entail differential elongation of the corolla.

Apocarpous gynoecia are extremely rare in heterostylous plants, an example being *Byrsocarpus coccineus*. The only genus of the Saxifragaceae in which heterostyly is represented is *Jepsonia* in which the carpels are fused at the ovary level. Flowers with coenocarpous or monomerous gynoecia but separate styles are likewise rare among heterostylous plants. Examples include *Linum grandiflorum*, *L. suffruticosum*, and distylous species of *Jepsonia*, *Limonium*, and *Fagopyrum*.

Most families with heterostylous members have hypogynous flowers. Among the dicotyledons, however, the family with the most numerous heterostylous genera and species are the Rubiaceae, which have epigynous flowers. It seems that in the Rubiaceae there is a background of morphological, developmental, and physiological traits particularly favorable for the evolution of heterostyly.

As mentioned earlier, the Boraginaceae include species in which distyly is combined with an array of mating systems. In some species the levels of anthers do not differ appreciably between the morphs. It is not known whether possession of a gynobasic style constitutes a factor limiting the appearance of diallelic incompatibility in some genera. It is worth noting that incompatibility has not been found in the related Lamiaceae, another family with a gynobasic style.

Among families lacking heterostyly are the Apiaceae. They have evolved an array of sexual systems and dichogamy, perhaps because of their very short styles, inferior ovary, stylopodium and small, flat and crowded dialypetalous flowers. Absence of heterostyly in other families of dicotyledons with an inferior ovary may be due to an anatomical or developmental constraint. The Lythraceae have a hypanthium. Heterostylous monocotyledons have epigynous flowers.

Heterostylous flowers are short-lived, anthesis usually lasting 1 or 2 days. Usually they are homogamous, with both male and female functions operating from the beginning of anthesis. The appearance of heterostyly is also determined by the pre-existing mating system which is, presumably, self-compatibility.

Most heterostylous plants are perennials, exceptions being the annual species *Fagopyrum esculentum*, *Cryptantha flava*, *C. flavoculata*, *Linum pubescens*, and *L. grandiflorum*. The presence of distyly, even though atypical, in the root parasite *Quinchamalium chilense* is unique.

The absence of heterostyly in primitive Magnoliidae may illustrate to an extreme a combination between various constraints. Heterostyly is absent in flowers having elongated receptacles, apocarpous gynoecia and spirally arranged, numerous and indefinite numbers of carpels and stamens. With such a background, reciprocal herkogamy becomes inconceivable. Furthermore, many species of primitive Magnoliidae are cantharophilous. Indiscriminate pollen delivery by the body of beetles precludes any efficiency in the operation of heterostyly as a mechanical device for promoting intermorph pollinations.

5 Atypical Heterostyly

About a dozen species display style-length dimorphism without most of the other characteristics of the heterostylous syndrome. The variation in length of the style and/or stamens is less marked than in distylous species and even where length is bimodal, as for example in *Quinchamalium chilense*, there is no reciprocal herkogamy. In reviews of these polymorphisms, Barrett (1988a) and Barrett and Richards (1989) designated them "anomalous" heteromorphisms. Here belong species of *Cordia, Suteria,* and *Lipostoma* (Darwin 1877; Opler et al. 1975), *Narcissus tazetta* (Dulberger 1964), *Anchusa hybrida* (Dulberger 1970a), *Anchusa officinalis* (Philipp and Schou 1981; Schou and Philipp 1983, 1984), *Guettarda scabra* (Richards and Koptur 1987), *Quinchamalium chilense* (Riveros et al. 1987), and apparently also several *Nivenia* species (Goldblatt and Bernhardt 1990).

Remarkably, in most of these species anthers of short-styled flowers are situated at a level more or less equivalent to stigmas in long-styled flowers. However, anthers of long-styled flowers are situated much higher than stigmas of short-styled flowers. In this connection, the report of dimorphism in several *Nivenia* species is inconclusive, for the style lengths are not indicated.

Some of these species also share other characteristics with distylous plants. In *Nivenia* species a statistically measurable pollen-size dimorphism has been detected; in *Quinchamalium chilense* dimorphism both in grain size and pollen production is reported.

The feature of greatest relevance for the divergence of these plants from conventional distyly is the absence of diallelic incompatibility. Instead, the plants are either self-compatible, as in *Q. chilense, Guettarda scabra,* and *Nivenia* species, or have apparently multiallelic incompatibility, as in *N. tazetta* and the two *Anchusa* species.

The occurrence of stylar dimorphism in the absence of diallelic incompatibility leads to the conclusion that style length polymorphism is irrelevant for this type of incompatibility. However, possibly pollen specificity and not style specificity is missing here. The possibility that in these plants the control of style elongation differs from that in typical heterostylous plants is also not ruled out.

It is difficult to categorize these species. They belong to taxonomically unrelated groups and are heterogenous in the expression of heteromorphic traits and in their breeding systems. But for that matter, neither do conventional heterostylous plants represent a monolithic conglomerate.

Barrett and Richards (1989) distinguished between species with stylar dimorphisms and species like *Linum grandiflorum*, which show deviations from typical heterostyly but are related to species displaying the entire syndrome. This distinction is, however, not invariably applicable. Thus, *Quinchamalium chilense* is to date the only species of the Santalaceae with reported stylar dimorphism, so that its possible affinities to fully distylous species in this genus are so far uncertain. *Narcissus tazetta* has only stylar dimorphism, but in other species of the genus reciprocal herkogamy has been reported (see Lloyd et al. 1990). Provided there is reciprocal herkogamy in flowers of *Nivenia* species, they may be classified as heterostylous, like species of *Cryptantha* or *Eichhornia*, despite their self-compatibility. If, however, herkogamy proves to be absent or incomplete, these species would then belong to the anomalous group.

Generally, these polymorphisms are so far poorly understood. Nothing is known about the pollination process or about the rate of pollen tube growth in the gynoecia of the two phenotypes, so that we are left guessing whether the dimorphisms of style and stamens here are adaptive, or are manifestations of developmental noise or of a "loose" genetical switch.

Whatever our approach to classifying these polymorphisms, they provide instructive analogs to those occurring in typical heterostyly. Total exclusion of these species from the category of heterostylous plants (e.g., Ganders 1979) seems fruitless. Including them as atypical cases within a broader array of heterostyly seems to be more reasonable. In this context the following forecast is worth considering:

"Perhaps when we know more of the developmental basis of floral morphology, of the hormonal relations which control style and stamen elongation, and the nature of the incompatibility reactions which take place in the carpel, we shall be able to explain the persistence of style-length variability in *Mirabilis* and *Narcissus*. No simple explanation on a basis of mechanical promotion of cross-pollination or avoidance of self-pollination appears adequate, however. When we do have this answer, we may be in a better position to understand how it evolved and is controlled in those heterostylous species where it does have direct value in promoting outbreeding" (Baker 1964).

6 Reciprocal Herkogamy, Heteromorphic Incompatibility, and Evolution of Heterostyly

6.1 Hypothesis and Evidence

The viewpoint presented here does not contravene the widely accepted Darwinian one regarding the adaptive significance of the reciprocal arrangement of anthers and stigmas in the floral morphs. The virtual confinement of heterostyly to insect-pollinated flowers renders especially prominent its function as a mechanical device for promoting pollen transfer, as demonstrated in a few species. In certain cases also morph-specific pollen production may function in the regulation of pollen flow, as suggested by Ganders. The array of morphological polymorphisms can be synergistically involved in the incompatibility mechanism and in other functions.

The present hypothesis takes into account interactions between the male gametophyte and the pistil. In this it differs fundamentally from the Darwinian one, for it offers a more complete interpretation of the significance of most floral polymorphisms and of heterostyly in general.

It is theoretically possible that, as in homomorphic incompatibility systems, the male and female recognition specificities are not outcome of heteromorphisms or their products. The morphological and biochemical components of the syndrome may be developmentally and/or functionally independent of each other. If this is true, then the plethora of morphological, cytological, ultrastructural, and biochemical polymorphisms are rendered functionally meaningless, or else become useless relics of the past, devoid of selective value. The interpretation of the present author attributes to various polymorphisms not only past but also current functionality in incompatibility.

Admittedly, much of the evidence amassed in support of this hypothesis is circumstantial, as is true in numerous other instances where the value of morphological or structural adaptations is assessed. The virtual omnipresence of style length di- or trimorphism in plants with diallelic incompatibility strongly supports the view of a participation of this polymorphism in incompatibility. This conclusion has logical validity and can also be applied to other di- or trimorphic characters of apparently no value in the regulation of pollen transfer.

It has yet to be demonstrated experimentally that dimorphism of exine patterning and characteristics of the stigmatic papillae contribute to incompatibility in *Linum* or members of the Plumbaginaceae. Yet Mattsson's (1983) findings do not rule out participation of the morphological components; they actually strongly support it. As noted by Mattsson, in *Armeria maritima* it is difficult to distinguish between the structural and physiological features involved in the response.

Differing sites of inhibition in the morphs of a few heterostylous plants vs. one site in homomorphic incompatibility have been noted by various authors (Glover and Barrett 1983; Anderson and Barrett 1986; Gibbs 1986). Gibbs discussed the dissimilar incompatibility reactions in pin × pin vs. thrum × thrum pollinations of a few distylous species. He failed, however, to attribute or relate the physiological dissimilarities to polymorphisms of pollen and pistils.

The presence in most species of two or three sites of inhibition points to the functioning in incompatibility of pollen and pistil di- or trimorphisms. But perhaps the most convincing evidence for the participation in incompatibility of pollen and stigma dimorphisms is found in the Plumbaginaceae and some *Linum* and *Primula* species, rather than in taxa in which there are two or three sites of inhibition.

For instance, in *Limonium meyeri* and *Linum grandiflorum*, the nature of the rejection response differs in the two intermorph pollinations, even though the inhibition occurs on the stigma surface or underneath the surface in both floral types. In pollinations of papillate stigmas with B type pollen, or in pin × pin pollinations of *Linum*, the pollen does not adhere to the stigma, or is unable to extract water from the papillae, owing to the impermeable thick cuticle (Dulberger 1975a, 1987a). In contrast, in pollinations of cob stigmas with A type pollen, the grains adhere and swell, but fail to germinate, although the cuticle is thin. In thrum × thrum pollinations of *Linum*, the pollen grains germinate, but the tubes are arrested just underneath the stigma surface. In these species, stigmatic inhibition and the dimorphic properties of the receptive surface and pollen grains, all clearly point to these dimorphisms as being functional in incompatibility. Inhibition at the same site is obviously caused by different factors which are manifest also in pollen and stigma dimorphism. Whatever the biochemical or biophysical nature of the specificities, it is doubtful whether they can disprove the adaptiveness of the morphological or structural components.

An example of rejection at the same site but at dissimilar stages of development in taxa in which pollen is dimorphic in size is provided by *P. elatior*. In this species the incompatibility reaction takes place on the stigma surface in both morphs: pin pollen remains unhydrated on pin stigmas and the pollen tubes of germinated thrum pollen are incapable of penetrating into the thrum stigma (Schou 1984). Hence, both blockage at different sites and blockage at the same site but at different stages of pollen development point to functioning of polymorphic characters in incompatibility.

The hypothesis put forward here provides answers, necessarily incomplete, to some questions it poses, but also raises new questions. For instance: if morph-specific pollen size is significant in incompatibility, its mode of action is an enigma. It is not known whether the inhibitory effect of the long style is the result of a chemical specificity in all heterostylous species, or whether cessation of pollen-tube growth in styles of this morph is due, in few species, to the length of the style itself. It is not clear why in some *Linum* species the pollen tubes are arrested at the top of the style in both morphs and not on or within the stigma, as in *Linum grandiflorum*. It is not understood why in *Oxalis* species showing pollen-size polymorphism, the pollen tubes are inhibited in the upper style in both intramorph pollinations. The genetic and physiological control of the marked intramorph variation in site and stage of inhibition in several *Primula* species is obscure. These "exceptions" do not, however, invalidate the significance of the polymorphisms in incompatibility.

6.2 Evolution of Heterostyly

It has been suggested that stamen and pistil heteromorphisms are superimposed on pre-existing diallelic incompatibility. Historically, reciprocal stamen-style length differences are the latest acquisition in the syndrome (Baker 1966; Ganders 1979). Based on a similar assumption and on population genetics, Charlesworth D. and Charlesworth B. (1979) provided the first theoretical model of the evolution of heterostyly. These authors also proposed stages in the evolution of self-incompatibility. The physiology of incompatibility, however, was not considered in their models: floral polymorphism and incompatibility were regarded as independent phenomena. It would be interesting to analyze theoretical models of evolution based on the premise that polymorphic properties of the pistil and pollen grains participate in the physiology of incompatibility.

Theoretical models proposed by Lloyd and Webb (review by Barrett 1990) suggest that evolution of reciprocal herkogamy preceded self-incompatibility. The latter thus evolved by gradual adjustment of pollen-tube growth to the different stylar environments of the morphs. According to Scribailo (1989) and Barrett (1990), results obtained in tristylous *Pontederia* species are consistent with Lloyd and Webb's hypothesis.

There is general consensus that heterostyly has polyphyletic origins (Ganders 1979). This opinion is based mainly on the taxonomic distribution of reciprocal herkogamy. The patterns of heteromorphic incompatibility distinguished in Section 3.2.6 reflect a diversity of physiological mechanisms. It is reasonable to assume that mechanisms based on polymorphism in pollen shape or exine sculpturing have evolved independently of those based on pollen-size differences. Furthermore, in the Staticeae and *Linum*, the dimorphic characters of the pollen exine and stigmas and the behavior of pollen grains suggest independent evolution of the stigmatic rejection. Instances suggesting polyphyletic origins of incompatibility can also be found among taxa characterized by pollen-size polymorphism. Therefore, a single model does not suffice to explain the diverse modes by which the heterostylous syndrome originated.

Among taxa, differing facets, constraints, and selective forces may have assumed importance or accent in the evolvement of the heterostylous syndrome. Nevertheless, it seems that the same set of fundamental laws, presumably of developmental nature, must prevail in the repeated evolution of distylous and tristylous incompatibility.

Acknowledgement. I thank S. C. H. Barrett for advice.

References

Anderson JM, Barrett SCH (1986) Pollen tube growth in tristylous *Pontederia cordata* (Pontederiaceae). Can J Bot 64:2602–2607
Baker HG (1948) Dimorphism and monomorphism in the Plumbaginaceae. I. A survey of the family. Ann Bot N S 12:207–219
Baker HG (1953) Dimorphism and monomorphism in the Plumbaginaceae. II. Pollen and stigmata in the genus *Limonium*. Ann Bot 17:433–445
Baker HG (1956) Pollen dimorphism in the Rubiaceae. Evolution 10:23–31
Baker HG (1958) Studies in the reproductive biology of the West African Rubiaceae. J West Afr Sci Assoc 4:9–24
Baker HG (1961) Heterostyly and homostyly in *Lithospermum canescens* (Boraginaceae). Rhodora 63:229–235
Baker HG (1962) Heterostyly in the Connaraceae with special reference to *Byrsocarpus coccineus*. Bot Gaz 123:206–211
Baker HG (1964) Variation in style length in relation to outbreeding in *Mirabilis* (Nyctaginaceae). Evolution 18:507–512
Baker HG (1966) The evolution, functioning and breakdown of heteromorphic incompatibility systems. I. The Plumbaginaceae. Evolution 20:349–368
Baker HG (1975) Sporophyte-gametophyte interactions in *Linum* and other genera with heteromorphic self-incompatibility. In: Mulcahy DL (ed) Gamete competition in plants and animals. North Holland, Amsterdam, IX:288
Barrett SCH (1977a) Tristyly in *Eichhornia crassipes* (Mart.) (water hyacinth). Biotropica 9:230–238
Barrett SCH (1977b) The breeding system of *Pontederia rotundifolia* L., a tristylous species. New Phytol 78:209–220
Barrett SCH (1978) The floral biology of *Eichhornia azurea* (Swartz) Kunth (Pontederiaceae). Aquat Bot 5:217–228
Barrett SCH (1979) The evolutionary breakdown of tristyly in *Eichhornia crassipes* (Mart.) Solms (water hyacinth). Evolution 33:499–510
Barrett SCH (1980) Dimorphic incompatibility and gender in *Nymphoides indica* (Menyanthaceae). Can J Bot 58:1938–1942
Barrett SCH (1985a) Floral trimorphism and monomorphism in continental and island populations of *Eichhornia paniculata* (Spreng.) Solms. (Pontederiaceae). Biol J Linn Soc 25:41–60
Barrett SCH (1985b) Ecological genetics of breakdown in tristyly. In: Haeck I, Woldendorp JW (eds) Structure and functioning of plant populations, vol 2. North Holland, Amsterdam, pp 267–275
Barrett SCH (1988a) The evolution, maintenance and loss of self-incompatibility systems. In: Lovett Doust J, Lovett Doust L (eds) Plant reproductive ecology. Patterns and strategies. Oxford Univ Press, New York Oxford, 344 pp
Barrett SCH (1988b) Evolution of breeding systems in *Eichhornia* (Pontederiaceae): a review. Ann MO Bot Gard 75:741–760
Barrett SCH (1990) The evolution and adaptive significance of heterostyly. Trends Ecol Evol. 5:144–148
Barrett SCH, Anderson JM (1985) Variation in expression of trimorphic incompatibility in *Pontederia cordata* L. (Pontederiaceae). Theor Appl Genet 70:355–362
Barrett SCH, Glover DE (1985) On the Darwinian hypothesis of the adaptive significance of tristyly. Evolution 39:766–744
Barrett SCH, Richards JH (1989) Heterostyly in tropical plants. Mem N Y Bot Gard 55:35–61
Barrett SCH, Shore JS (1985) Dimorphic incompatibility in *Turnera hermannioides* Camb. (Turneraceae). Ann MO Bot Gard 72:259–263

Barrett SCH, Shore JS (1987) Variation and evolution of breeding systems in the *Turnera ulmifolia* L. complex (Turneraceae). Evolution 41:340–354

Bawa KS, Beach JH (1983) Self-incompatibility systems in the Rubiaceae of a tropical lowland wet forest. Am J Bot 70:1281–1288

Bernatzky R, Anderson MA, Clarke AE (1988) Molecular genetics of self-incompatibility in flowering plants. Dev Genet 9:1–12

Bir Bahadur (1964) Pollen dimorphism in heterostyled *Oldenlandia umbellata* L. Rhodora 66:56–60

Bir Bahadur (1966) Heterostyly in *Oldenlandia scopulorum*. Bull J Genet 59:267–272

Bir Bahadur (1968) Heterostyly in Rubiaceae: a review. J Osmania Univ Sci Golden Jubilee Vol:207–238

Bir Bahadur, Laxmi SB, Rama Swamy N (1984a) Pollen morphology and heterostyly – a historical review. Adv Pollen Spore Res 12:45-78

Bir Bahadur, Laxmi SB, Rama Swamy N (1984b) Pollen morphology and heterostyly. A systematic and critical account. Adv Pollen Spore Res 12:79-126

Bokhari MH (1972) A brief review of stigma and pollen types in *Acantholimon* and *Limonium*. Notes R Bot Gard Edinb 32:79–84

Bräm M (1943) Untersuchungen zur Phänanalyse und Entwicklungsgeschichte der Blüten von *Primula pulverulenta, P. cockburniana* und ihrer F1-Bastarde. Arch Julius Klaus-Stift Vererbungsforsch Sozialanthropol Rassenhyg 18:235–259

Bremekamp CEB (1963) On pollen dimorphism in heterostylous Psychotrieae, especially in the genus *Mapouria* Aubl. Grana Palynol 4:53–63

Broothaerts WJ, Vending JC (1990) Self-incompatibility proteins from styles of *Petunia hybrida* have ribonuclease activity. Physiol Plant 79:A43

Burck W (1887) Notes biologiques. I. Relation entre hétérostylie dimorphe et hétérostylie trimorphe. Ann Jard Bot Buitenzorg 6:251–254

Casper BB (1983) The efficiency of pollen transfer and rates of embryo initiation in *Cryptantha* (Boraginaceae). Oecologia 59:262–268

Casper BB (1985) Self-compatibility in distylous *Cryptantha flava* (Boraginaceae). New Phytol 99:149–154

Casper BB, Charnov El (1982) Sex allocation in heterostylous plants. J Theor Biol 96:143–149

Casper BB, Sayigh LS, Lee SS (1988) Demonstration of cryptic incompatibility in distylous *Amsinckia douglasiana*. Evolution 42:248–253

Charlesworth B, Charlesworth D (1979) The maintenance and breakdown of distyly. Am Nat 114:499–513

Charlesworth D (1979) The evolution and breakdown of tristyly. Evolution 33:489–498

Charlesworth D, Charlesworth B (1979) A model for the evolution of distyly. Am Nat 114:467–498

Clarke AE, Anderson MA, Bacic T, Harris PJ, Mau S-L (1985) Molecular basis of cell recognition during fertilization in higher plants. J Cell Sci (Suppl) 2:261–285

Clarke AE, Anderson MA, Atkinson A, Bacic A, Ebert PR, Jahnen W, Lush WM, Mau S-L, Woodward JR (1989) Recent developments in the molecular genetics and biology of self-incompatibility. Plant Mol Biol 13:267–271

Cornish EC, Pettitt JM, Bonig I, Clarke AE (1987) Developmentally controlled expression of a gene associated with self-incompatibility in *Nicotiana alata*. Nature (Lond) 326:99–102

Cornish EC, Anderson MA, Clarke AE (1988) Molecular aspects of fertilization in flowering plants. Annu Rev Cell Biol 4:209–228

Dahlgren KVO (1918) Heterostylie innerhalb der Gattung *Plumbago*. Svensk Bot Tidskr 12:362–372

Dahlgren KVO (1923) *Ceratostigma*, eine heterostyle Gattung. Ber Dtsch Bot Ges 41:35–38

Dahlgren KVO (1970) Heterostylie bei *Dyerophytum indicum* (Gibs ex White) O.K. (Plumbaginaceae). Svensk Bot Tidskr 64:179–183

Darwin C (1877) The different forms of flowers on plants of the same species. Murray, Lond

Dowrick VPJ (1956) Heterostyly and homostyly in *Primula obconica*. Heredity 10:219–236

Dulberger R (1964) Flower dimorphism and self-incompatibility in *Narcissus tazetta* L. Evolution 18:361–363

Dulberger R (1967) Pollination systems in plants of Israel: Heterostyly. PhD Thesis Hebrew Univ, Jerusalem

Dulberger R (1970a) Floral dimorphism in *Anchusa hybrida* Ten. Isr J Bot 19:37–41
Dulberger R (1970b) Tristyly in *Lythrum junceum*. New Phytol 69:751–759
Dulberger R (1973) Distyly in *Linum pubescens* and *Linum mucronatum*. Bot J Linn Soc 66:117–126
Dulberger R (1974) Structural dimorphism of stigmatic papillae in distylous *Linum* species. Am J Bot 61:238–243
Dulberger R (1975a) Intermorph structural differences between stigmatic papillae and pollen grains in relation to incompatibility in Plumbaginaceae. Proc R Soc Lond Ser B 188:257–274
Dulberger R (1975b) *S*-gene action and the significance of characters in the heterostylous syndrome. Heredity 35:407–415
Dulberger R (1981a) Dimorphic exine sculpturing in three distylous species of *Linum* (Linaceae). Plant Syst Evol 139:113–119
Dulberger R (1981b) Pollen-pistil interactions in *Linum grandiflorum*. XIII Int Bot Congr Sydney Australia (Abstr):60
Dulberger R (1984) Timing of stigma specificity relative to style elongation and incompatibility in buds of distylous *Linum grandiflorum*. Am J Bot 71:25
Dulberger R (1987a) Fine structure and cytochemistry of the stigma surface and incompatibility in some distylous *Linum* species. Ann Bot 59:203–217
Dulberger R (1987b) Incompatibility in *Plumbago capensis*: Fine structure and cytochemistry of the reproductive surface and pollen wall. XIV Int Bot Congr Berlin (Abstr):18
Dulberger R (1987c) The association of physiological incompatibility with heteromorphic stigma characters in distylous taxa. Isr J Bot 36:199–213
Dulberger R (1989) The apertural wall in pollen of *Linum grandiflorum*. Ann Bot 63:421–431
Dulberger (1990) Release of proteins from the pollen wall of *Linum grandiflorum*. Sex Plant Reprod 3:18–22
Dulberger R (1991) Exine dimorphism and incompatibility in *Linum grandiflorum*. Isr J Bot (in press)
Eames AJ (1961) Morphology of the angiosperms. McGraw-Hill, New York
East EM (1940) The distribution of self-sterility in flowering plants. Proc Am Philos Soc 82:449–518
Erdtman G (1940) Flower dimorphism in *Statice Armeria* L. Svensk Bot Tidskr 34:377–380
Erdtman G (1970) Über Pollendimorphie in *Plumbaginaceae* (unter besonderer Berücksichtigung von *Dyerophytum indicum*). Svensk Bot Tidskr 64:184–188
Ernst A (1932) Zur Kenntnis der Heterostylie bei tropischen Rubiaceae. Arch Julius Klaus-Stift Vererbungsforsch Sozialanthropol Rassenhyg 7:241–280
Ernst A (1936a) Heterostylie-Forschung. Versuche zur genetischen Analyse eines Organisations- und Anpassung-Merkmales. Z Abst Vererb 71:156–230
Ernst A (1936b) Weitere Untersuchungen zur Phänanalyse, zum Fertilitätsproblem und zur Genetik heterostyler Primeln. II. *Primula hortensis* Wettstein. Arch Julius Klaus-Stift Vererbungsforsch 11:1–280
Ernst A (1955) Self-fertility in monomorphic primulas. Genetica 27:391–448
Esser K (1953) Genomverdopplung und Pollenschlauchwaschstum bei Heterostylen. Z Indukt Abstammungs Vererbungsl 85:28–50
Feinsinger P, Busby WH (1987) Pollen carryover: experimental comparisons between morphs of *Palicourea lasiorrachis* (Rubiaceae), a distylous bird-pollinated tropical treelet. Oecologia 73:231–235
Ganders FR (1974) Disassortative pollination in the distylous plant *Jepsonia heterandra*. Can J Bot 52:2401–2406
Ganders FR (1975a) Heterostyly, homostyly, and fecundity in *Amsinckia spectabilis* (Boraginaceae). Madroño 23:56–62
Ganders FR (1975b) Mating patterns in self-compatible distylous populations of *Amsinckia* (Boraginaceae). Can J Bot 53:773–779
Ganders FR (1976) Pollen flow in distylous populations of *Amsinckia* (Boraginaceae). Can J Bot 54:2530–2535
Ganders FR (1979) The biology of heterostyly. NZ J Bot 17:607–635
Gaude T, Dumas C (1987) Molecular and cellular events of self-incompatibility. In: Giles KL, Parkash J (eds) Int Rev Cytol 107:333–366
Ghosh S, Shivanna KR (1980a) Pollen-pistil interaction in *Linum grandiflorum*. Planta 149:257–261

Ghosh S, Shivanna KR (1980b) Pollen-pistil interactions in *Linum grandiflorum*: stigma-surface proteins and stigma receptivity. Ind Natl Sci Acad 46:177–183

Ghosh S, Shivanna KR (1983) Studies on pollen-pistil interactions in *Linum grandiflorum*. Phytomorphology 32:385–395

Gibbs PE (1976) Studies on the breeding system of *Oxalis tuberosa* Mol. Flora 165:129–138

Gibbs PE (1986) Do homomorphic and heteromorphic self-incompatibility systems have the same sporophytic mechanism? Plant Syst Evol 154:285–323

Glover DE, Barrett SCH (1983) Trimorphic incompatibility in Mexican populations of *Pontederia sagittata* Presl. (Pontederiaceae). New Phytol 95:439–455

Glover DE, Barrett SCH (1986) Stigmatic pollen loads of *Pontederia cordata* from the southern U.S. Am J Bot 73:1607–1612

Goldblatt P, Bernhardt P (1990) Pollination biology of *Nivenia* (Iridaceae) and the presence of heterostylous self-compatibility. Isr J Bot 39:93–111

Golynskaya EL, Bashrikova NV, Tomchuk NN (1976) Phytohemagglutinins of the pistil of *Primula* as possible proteins of generative incompatibility. Sov Plant Physiol 23:169–176

Hallé F (1961) Contribution a l'étude biologique et taxonomique des Mussaendeae (Rubiaceae) d'Afrique tropicale. Adansonia1:266–298

Heitz B (1980) La pollinisation des Lins hétérostyles du groupe *Linum perenne* L. (Linaceae). CR Acad Sci Paris 290:811–814

Heitz MB, Jean R, Prensier G (1971) Observations de la surface du stigmate et des grains de pollen de *Linum austriacum* L. hétérostyle. C R Hebd Seances Acad Sci Ser D Sci Nat 273:2493–2495

Heslop-Harrison J (1978) Genetics and physiology of angiosperm incompatibility systems. Proc R Soc Lond Ser B 202:73–92

Heslop-Harrison J (1983) Self-incompatibility: phenomenology and physiology. Proc R Soc Lond Ser B 218:371–395

Heslop-Harrison Y (1981) Stigma characteristics and angiosperm taxonomy. Nord J Bot 1:401–420

Heslop-Harrison Y, Shivanna KR (1977) The receptive surface of the angiosperm stigma. Ann Bot 41:1233–1258

Heslop-Harrison Y, Heslop-Harrison J, Shivanna KR (1981) Heterostyly in *Primula*. 1. Fine-structural and cytochemical features of the stigma and style in *Primula vulgaris* Huds. Protoplasma 107:171–187

Iversen J (1940) Blütenbiologische Studien. I. Dimorphie und Monomorphie bei *Armeria*. K Dan Vidensk Selsk Biol Skr 15:1–39

Johnston IM (1950) Studies in the Boraginaceae, XIX. B. *Cordia* section *Gerascanthus* in Mexico and Central America. J Arnold Arbor 31:179–187

Johnston IM (1952) Studies in the Boraginaceae, XXIII. A survey of the genus *Lithospermum*. J Arnold Arbor 33:299–363

Knox BR (1984) Pollen-pistil interactions. In: Linskens HF, Heslop-Harrison J (eds) Encyclopedia of plant physiology, new ser, vol 17. Springer, Berlin Heidelberg New York, pp 508–608

Köhler E (1973) Über einen bemerkenswerten Pollendimorphismus in der Gattung *Waltheria* L. Grana 13:57–64

Köhler E (1976) Pollen dimorphism and heterostyly in the genus *Waltheria* L. (Sterculiaceae). In: Ferguson JK, Muller J (eds) The evolutionary significance of the exine. Linn Soc Symp Ser 1:147–162, Academic Press, Lond New York

Lamport DTA (1970) Cell wall metabolism. Annu Rev Plant Physiol 21:235–270

Levin DA (1968) The breeding system of *Lithospermum caroliniense*: adaptation and counteradaptation. Am Nat 102:427–441

Lewis D (1942) The physiology of incompatibility in plants. I. The effect of temperature. Proc R Soc Lond Ser B 131:13–26

Lewis D (1943) The physiology of incompatibility in plants. II. *Linum grandiflorum*. Ann Bot 7:115–117

Lewis D (1949) Incompatibility in flowering plants. Biol Rev 24:472–496

Lewis D (1982) Incompatibility, stamen movement and pollen economy in a heterostyled tropical forest tree, *Cratoxylum formosum* (Guttiferae). Proc R Soc Lond Ser B 214:273–283

Lewis WH (1965) Cytopalynological study of African Hedyotideae (Rubiaceae). Ann MO Bot Gard 52:182–211

Lewis WH (1976) Pollen size of *Hedyotis caerulea* (Rubiaceae) in relation to chromosome number and heterostyly. Rhodora 78:60–64

Linskens HF (1964) The influence of castration on pollen tube growth after self-pollination. In: Linskens HF (ed) Pollen physiology and fertilization. North-Holland Publ Co, Amsterdam, pp 230–236

Linskens HF (1974) Some observations on the growth of the style. Incomp Newsl 4:4–15

Lloyd DG, Yates JMA (1982) Intrasexual selection and the segregation of pollen and stigmas in hermaphrodite plants, exemplified by *Wahlenbergia albomarginata* (Campanulaceae). Evolution 36:903–913

Lloyd DG, Webb CJ, Dulberger R (1990) Heterostyly in species of *Narcissus* (Amaryllidaceae) and *Hugonia* (Linaceae) and other disputed cases. Plant Syst Evol 172:215–227

Macleod J (1887) Untersuchungen über die Befruchtung der Blumen. Bot Zentrabl 19:150–154

Martin FW (1965) Distyly and incompatibility in *Turnera ulmifolia*. Bull Torrey Bot Club 92:185–192

Martin FW (1967) Distyly, self-incompatibility, and evolution in *Melochia*. Evolution 21:493–499

Mather K, De Winton D (1941) Adaptation and counter-adaptation of the breeding system in *Primula*. Ann Bot 5:297–311

Mattsson O (1983) The significance of exine oils in the initial interaction between pollen and stigma in *Armeria maritima*. In: Mulcahy DL, Ottaviano E (eds) Pollen: biology and implications for plant breeding. Elsevier, New York Amsterdam, pp 257–264

Mayura Devi P (1964) Heterostyly in *Biophytum sensitivum* DC. J Genet 59:41–48

Mayura Devi P, Hashim M (1966) Homostyly in heterostyled *Biophytum sensitivum* DC. J Genet 59:245–249

McClure BA, Haring V, Ebert PR, Anderson MA, Simpson RJ, Sakiyama AE, Clarke AE (1989) Style self-incompatibility gene products of *Nicotiana alata* are ribonucleases. Nature (Lond) 342:955–957

McClure BA, Gray J, Haring V, Anderson MA, Mau S-L, Clarke AE (1990) Self-incompatibility in flowering plants. Physiol Plant 79:A1

Mulcahy DL (1965) Heterostyly within *Nivenia* (Iridaceae). Brittonia 17:349–351

Murray BG (1986) Floral biology and self-incompatibility in *Linum*. Bot Gaz 147:327–333

Nasrallah JB, Kao T-H, Goldberg ML, Nasrallah ME (1985) A cDNA clone encoding an S-locus-specific glycoprotein from *Brassica oleracea*. Nature (Lond) 318:263–267

Nasrallah JB, Yu S-M, Nasrallah ME (1988) Self-incompatibility genes of *Brassica oleracea*: expression, isolation, and structure. Proc Natl Acad Sci USA 85:5551–5555

Nasrallah ME, Nasrallah JB (1986) Molecular biology of self-incompatibility in plants. Trends Genet 2:239–244

Nettancourt D de (1977) Incompatibility in angiosperms. Monographs on theoretical and applied genetics, vol 3. Springer, Berlin Heidelberg New York

Nicholls MS (1985a) Pollen flow, population composition, and the adaptive significance of distyly in *Linum tenuifolium* L. (Linaceae). Biol J Linn Soc 25:235–242

Nicholls MS (1985b) The evolutionary breakdown of distyly in *Linum tenuifolium* (Linaceae). Plant Syst Evol 150:291–301

Nicholls MS (1986) Population composition, gender specialization, and the adaptive significance of distyly in *Linum perenne* (Linaceae). New Phytol 102:209–217

Ockendon DG (1968) Biosystematic studies in the *Linum perenne* group. New Phytol 67:787–813

Ockendon DG (1971) Cytology and pollen morphology of natural and artificial tetraploids in the *Linum perenne* group. New Phytol 70:599–605

Olesen JM (1979) Floral morphology and pollen flow in the heterostylous species *Pulmonaria obscura* Dumort. (Boraginaceae). New Phytol 82:757–767

Olesen JM (1986) Heterostyly, homostyly, and long-distance dispersal of *Menyanthes trifoliata* in Greenland. Can J Bot 65:1509–1513

Opler PA, Baker HG, Frankie GW (1975) Reproductive biology of some Costa Rican *Cordia* species (Boraginaceae). Biotropica 7:234–247

Ornduff R (1964) The breeding system of *Oxalis suksdorfii*. Am J Bot 51:307–314

Ornduff R (1970a) Incompatibility and the pollen economy of *Jepsonia parryi*. Am J Bot 57:1036–1041

Ornduff R (1970b) The systematics and breeding system of *Gelsemium* (Loganiaceae). J Arnold Arbor 51:1–17
Ornduff R (1972) The breakdown of trimorphic incompatibility in *Oxalis* section *Corniculatae*. Evolution 26:52–65
Ornduff R (1976) The reproductive system of *Amsinckia grandiflora*, a distylous species. Syst Bot 1:57–66
Ornduff R (1977) An unusual homostyle in *Hedyotis caerulea* (Rubiaceae). Plant Syst Evol 127:293–297
Ornduff R (1978) Features of pollen flow in dimorphic species of *Lythrum* section Euhyssopifolia. Am J Bot 65:1077–1083
Ornduff R (1979a) Pollen flow in *Primula vulgaris*. Bot J Linn Soc 78:1–10
Ornduff R (1979b) The genetics of heterostyly in *Hypericum aegypticum*. Heredity 42:271–272
Ornduff R (1980a) Heterostyly, population composition, and pollen flow in *Hedyotis caerulea*. Am J Bot 67:95–103
Ornduff R (1980b) Pollen flow in *Primula veris* (Primulaceae). Plant Syst Evol 13:89–93
Ornduff R (1982) Heterostyly and incompatibility in *Villarsia capitata* (Menyanthaceae). Taxon 31:495–497
Ornduff R (1986) Comparative fecundity and population composition of heterostylous and non-heterostylous species of *Villarsia* (Menyanthaceae) in Western Australia. Am J Bot 73:282–286
Ornduff R (1987a) The breakdown of heterostyly in *Villarsia* (Menyanthaceae): a unique scenario. Am J Bot 74:595–784
Ornduff R (1987b) Reproductive systems and chromosome races of *Oxalis pes caprae* L. and their bearing on the genesis of a noxious weed. Ann MO Bot Gard 74:79–84
Ornduff R (1988a) Distyly and incompatibility in *Villarsia congestiflora* (Menyanthaceae), with comparative remarks on *V. capitata*. Plant Syst Evol 159:81–83
Ornduff R (1988b) Distyly and monomorphism in *Villarsia* (Menyanthaceae): some evolutionary considerations. Ann MO Bot Gard 75:761–767
Pandey KK (1979) Overcoming incompatibility and promoting genetic recombination in flowering plants. NZ J Bot 17:645–663
Pandey KK, Troughton JH (1974) Scanning electron microscopic observations of pollen grains and stigma in the self-incompatible heteromorphic species *Primula malacoides* Franch. and *Forsythia intermedia* Zab., and genetics of sporopollenin deposition. Euphytica 23:337–344
Philipp M (1974) Morphological and genetical studies in the *Armeria maritima* aggregate. Bot Tidskr 69:40–51
Philipp M, Schou O (1981) An unusual heteromorphic incompatibility system: distyly, self-incompatibility, pollen load and fecundity in *Anchusa officinalis* (Boraginaceae). New Phytol 89:693–703
Piper JG, Charlesworth B (1986) The evolution of distyly in *Primula vulgaris*. Biol J Linn Soc 29:123–137
Piper JG, Charlesworth B, Charlesworth D (1986) Breeding system evolution in *Primula vulgaris* and the role of reproductive assurance. Heredity 56:207–217
Price SD, Barrett SCH (1982) Tristyly in *Pontederia cordata* (Pontederiaceae). Can J Bot 60:897–905
Price SD, Barrett SCH (1984) The function and adaptive significance of tristyly in *Pontederia cordata* L. (Pontederiaceae). Biol J Linn Soc 21:315–329
Punt W, den Breejen P (1981) Linaceae. Rev Palaeobot Palynol 33:75–115
Ray PM, Chisaki HF (1957) Studies on *Amsinckia* I. A synopsis of the genus, with a study of heterostyly in it. Am J Bot 44:529–536
Richards AJ (1986) Plant breeding systems. Allen and Unwin, Lond
Richards JH, Barrett SCH (1984) The developmental basis of tristyly in *Eichhornia paniculata* (Pontederiaceae). Am J Bot 71:1347–1363
Richards JH, Barrett SCH (1987) Development of tristyly in *Pontederia cordata* (Pontederiaceae). I. Mature floral structure and patterns of relative growth of reproductive organs. Am J Bot 74:1831–1841
Richards AJ, Ibrahim HBT (1982) The breeding system in *Primula veris* L. II. Pollen tube growth and seed-set. New Phytol 90:305–314
Richards JH, Koptur S (1987) Definition and development of unusual heterostyly in *Guettarda scabra* (L.) Vent. (Rubiaceae). Am J Bot 74:624. Abst #8

Riveros M, Arroyo MTK, Humana AM (1987) An unusual kind of distyly in *Quinchamalium chilense* (Santalaceae) on Volcan Casablanca, Southern Chile. Am J Bot 74:313–320
Rogers CM (1979) Distyly and pollen dimorphism in *Linum suffruticosum* (Linaceae). Plant Syst Evol 131:127–132
Rogers CM (1980) Pollen dimorphism in distylous species of *Linum* sect. *Linastrum* (Linaceae). Grana 19:19–20
Russell SD (1986) Biphasic pollen tube growth in *Plumbago zeylanica*. In: Mulcahy DL, Mulcahy GB, Ottaviano E (eds) Biotechnology and ecology of pollen. Springer, Berlin Heidelberg New York
Saad SI (1961) Pollen morphology and sporoderm stratification in *Linum*. Grana Palynol 3:109–129
Sadava D, Chrispeels MJ (1973) Hydroxyproline-rich cell wall protein (extensin): role in the cessation of elongation in excised pea epicotyls. Dev Biol 30:49–55
Sadava D, Walker F, Maarten J, Chrispeels MJ (1973) Hydroxyproline-rich cell wall protein (extensin): biosynthesis and accumulation in pea epicotyls. Dev Biol 30:42–48
Schaeppi H (1935) Kenntnis der Heterostylie von *Gregoria vitaliana* Derby. Ber Schweiz Bot Ges 44:109–132
Schill R, Baumm A, Wolter M (1985) Vergleichende Mikromorphologie der Narbenoberflächen bei den Angiospermen; Zusammenhänge mit Pollenoberflächen bei heterostylen Sippen. Plant Syst Evol 148:185–214
Schoch-Bodmer H (1930) Zur Heterostylie von *Fagopyrum esculentum*. Untersuchungen über das Pollenschlauchwachstum und über die Saugkräfte der Griffel und Pollenkörner. Ber Schweiz Bot Ges 39:4–15
Schoch-Bodmer H (1934) Zum Heterostylieproblem: Griffelbeschaffenheit und Pollenschlauchwachstum bei *Fagopyrum esculentum*. Planta 22:149–152
Schoch-Bodmer H (1938) Farbumschlag des Pollenanthocyans durch saure Narbensekrete bei *Lythrum salicaria*. Verh Schweiz Naturforsch Ges p 179
Schoch-Bodmer H (1942) Pollenbeschaffenheit und Fertilität bei *Lythrum salicaria* L. Ber Schweiz Bot Ges 52:317–352
Schoch-Bodmer H (1945) Zur Frage der "Hemmungsstoffe" bei Heterostylen. Arch Julius Klaus-Stift Vererbungsforsch Sozialanthropol Rassenhyg 20:403–416
Schou O (1983) The distyly in *Primula elatior* (L.) Hill (Primulaceae), with a study of flowering phenology and pollen flow. Bot J Linn Soc 86:261–274
Schou O (1984) The dry and wet stigmas of *Primula obconica*: ultrastructural and cytochemical dimorphism. Protoplasma 121:99–113
Schou O, Philipp M (1983) An unusual heteromorphic incompatibility system. II. Pollen tube growth and seed sets following compatible and incompatible crossing within *Anchusa officinalis* L. (Boraginaceae). In: Mulcahy DL, Ottaviano E (eds) Pollen: biology and implications for plant breeding. Elsevier, New York Amsterdam, pp 219–227
Schou O, Philipp M (1984) An unusual heteromorphic incompatibility system. 3. On the genetic control of distyly and self-incompatibility in *Anchusa officinalis* L. (Boraginaceae). Theor Appl Genet 68:139–144
Scribailo RW (1989) Structural studies of trimorphic incompatibility in *Pontederia sagittata* Presl. (Pontederiaceae). PhD Thesis, Univ Toronto, Toronto, Canada
Scribailo RW, Barrett SCH (1986) Sites of pollen tube inhibition in tristylous *Pontederia* L. (Pontederiaceae). Am J Bot 73:643
Scribailo RW, Barrett SCH (1989) Pollen-pistil interactions in tristylous *Pontederia sagittata* (Pontederiaceae). Am J Bot (Suppl) 76, 6 (Abstr):57
Shivanna KS, Heslop-Harrison J, Heslop-Harrison Y (1981) Heterostyly in *Primula*. 2. Sites of pollen inhibition, and effects of pistil constituents on compatible and incompatible pollen tube growth. Protoplasma 107:319–337
Shivanna KS, Heslop-Harrison J, Heslop-Harrison Y (1983) Heterostyly in *Primula*. 3. Pollen water economy: a factor in the intramorph-incompatibility response. Protoplasma 117:175–184
Shore JS, Barrett SCH (1984) The effect of pollination intensity and incompatible pollen on seed set in *Turnera ulmifolia* (Turneraceae). Can J Bot 62:1298–1302
Shore JS, Barrett SCH (1985a) Morphological differentiation and crossability among populations of *Turnera ulmifolia* L. complex (Turneraceae). Syst Bot 10:308–321

Shore JS, Barrett SCH (1985b) The genetics of distyly and homostyly in *Turnera ulmifolia* L. (Turneraceae). Heredity 55:167–174

Shore JS, Barrett SCH (1986) Genetic modifications of dimorphic incompatibility in the *Turnera ulmifolia* L. complex (Turneraceae). Can J Genet Cytol 28:796–807

Stevens VAM, Murray BG (1982) Studies on heteromorphic self-incompatibility systems. Physiological aspects of the incompatibility system of *Primula obconica*. Theor Appl Genet 61:245–256

Stirling J (1932) Studies of flowering in heterostyled and allied species. Part I. The Primulaceae. Publ Appl Genet 61:245–256

Takhtajan A (1948) Morphological evolution of the angiosperms. Moscow (in russian)

Tischler G (1917) Pollenbiologische Studien. Z Bot 9:417–488

Verdcourt B (1958) Remarks on the classification of the Rubiaceae. Bull Jard Bot Etat Brux 28:209–290

Vuilleumier BS (1967) The origin and evolutionary development of heterostyly in the angiosperms. Evolution 21:210–226

Webb CJ, Lloyd DG (1986) The avoidance of interference between the presentation of pollen and stigmas in angiosperms II. Herkogamy. NZ J Bot 24:163–178

Weber MO (1981) Pollen diversity and identification in some Plumbaginaceae. Mus Natl Hist Nat 23:321–348

Weber-El Ghobary MO (1986) Dimorphic exine sculpturing in two distylous species of *Dyerophytum* (Plumbaginaceae). Plant Syst Evol 152:267–276

Weller SG (1976) The genetic control of tristyly in *Oxalis* section Ionoxalis. Heredity 37:387–393

Weller SG (1979) Variation in heterostylous reproductive systems among populations of *Oxalis alpina* in southeastern Arizona. Syst Bot 4:57–71

Weller SG (1980) Pollen flow and fecundity in populations of *Lithospermum caroliniense*. Am J Bot 67:1334–1341

Weller SG (1981a) Fecundity in populations of *Oxalis alpina* in southeastern Arizona. Evolution 35:197–200

Weller SG (1981b) Pollination biology of heteromorphic populations of *Oxalis alpina* (Rose) Kunth (Oxalidaceae) in south-eastern Arizona. Bot J Linn Soc 83:189–198

Weller SG (1986) Factors influencing frequency of the mid-styled morph in tristylous populations of *Oxalis alpina*. Evolution 40:279–289

Weller SG, Ornduff R (1977) Cryptic self-incompatibility in *Amsinckia grandiflora*. Evolution 31:47–51

Weller SG, Ornduff R (1989) Incompatibility in *Amsinckia grandiflora* (Boraginaceae): distribution of callose plugs and pollen tubes following inter- and intramorph crosses. Am J Bot 76:277–282

Willson MF (1979) Sexual selection in plants. Am Nat 113:777–790

Wolfe LM, Barrett SCH (1989) Patterns of pollen removal and deposition in tristylous *Pontederia cordata* L. (Pontederiaceae). Biol J Linn Soc 36:317–329

Zavala ME(1978) Threeness and fourness in pollen of *Lythrum junceum*. Bot Soc Am Misc Ser Publ 156:163

Chapter 4

The Development of Heterostyly

J.H. RICHARDS[1] and S.C.H. BARRETT[2]

1 Introduction

Heterostyly is a genetic polymorphism in which the two (distyly) or three (tristyly) mating types in a population differ in floral morphology. The principal feature that distinguishes the floral morphs is that they differ in stigma and anther heights. The sex organs are reciprocally positioned with anthers in flowers of one morph at the same level as stigmas in flowers of the other morph(s). This structural difference is usually accompanied by a physiological self- and intramorph incompatibility that limits mating to crosses between organs at the same level. In this chapter we describe the diversity of organization in heterostylous flowers, consider the developmental bases for this diversity, and examine the genetic and environmental components of phenotypic variation in heterostylous breeding systems. We review the developmental implications of models for the genetic control and evolution of heterostyly and conclude with recommendations for future research.

2 Structure of Mature Heterostylous Flowers

2.1 Constraints on Morphology of Heterostylous Flowers

The evolution of heterostyly appears to occur within some general constraints on floral morphology (Ganders 1979a; Chaps. 3, 5 and 6). Heterostylous flowers are generally moderate sized and have a floral tube, a limited number of stamens, and a syncarpous ovary with few carpels. The reasons for these limitations are unclear but may depend on both developmental and functional constraints.

The length of heterostylous flowers is usually between 5 and 30 mm. In their study of breeding systems in *Cordia* species (Boraginaceae), Opler et al. (1975) found that dioecy is present in the smallest flowers and heterostyly in medium to large flowers. Similarly, the smallest-flowered species of *Melochia* (Sterculiaceae) in the Caribbean lack distyly or derived homostyly, conditions found in three larger-flowered species in the region (Martin 1966).

[1]Department of Biological Sciences, Florida International University, Miami, FL 33199, USA, and Fairchild Tropical Garden, 11935 Old Cutler Road, Miami, FL 33156, USA
[2]Department of Botany, University of Toronto, Toronto, Ontario, Canada M5S 3B2

Monographs on Theoretical and Applied Genetics 15
Evolution and Function of Heterostyly (ed. by S.C.H. Barrett)

Possible reasons for these size limits include both restrictions on pollination efficiency and developmental constraints imposed by the need for reciprocity in organ position. For example, the lower end of the size range may be determined by how precisely stigma-anther separation can be controlled developmentally, as well as whether pollen from small flowers can be effectively separated by morph on the bodies of pollinators. Small flowers usually have short stamens and styles. If distances of a few mm are all that separate stigmas and anthers within heterostylous flowers, selection must severely limit variation in sex organ level both within and between flowers of the morphs in order to preserve stigma-anther separation. The inherent variation in developmental processes may preclude maintenance of such small differences and thus set a lower limit on the size of heterostylous flowers.

The absence of heterostyly in large-flowered species may also have both ecological and developmental explanations. The feeding behavior of long-tongued moths, bats, and birds, which are common pollinators of large flowers, may make these pollinators less effective at mediating disassortative pollination in comparison with smaller pollinators, such as bees and butterflies that visit moderate-sized flowers. Some distylous plants, however, are hummingbird-pollinated, e.g., *Cephaelis* (Rubiaceae) (Bawa and Beach 1983) and *Palicourea* (Rubiaceae) (Feinsinger and Busby 1987). Upper limits on floral size could also result if variability in stigma-anther height exceeds the constraints imposed by the need for reciprocity among morphs in organ position. Long styles and stamens may show more absolute variability in stigma and anther levels because minor variations in early development are magnified during growth. The organs of large flowers may therefore be less efficient at promoting disassortative pollination.

In addition to size constraints, heterostylous species show limitations on floral architecture. Heterostylous flowers usually have few stamens. Distylous flowers commonly produce 2–12 (Ganders 1979a), while tristylous flowers have from 6 to 12. An increased number of stamens prolongs the time between the beginning and end of stamen initiation and increases the complexity of positional effects in development (cf. Hufford 1988). Both of these factors could increase the amount of developmental variability. The variability caused by timing and position of stamen origin probably accounts for variation in stamen length within flowers of distylous species that have two whorls or series of stamens, such as *Byrsocarpus coccineus* (Connaraceae) (Baker 1962) and *Erythroxylum coca* (Erythroxylaceae) (Ganders 1979b). Differences in length between stamen series of these species caused early researchers to speculate that these taxa are tristylous (Baker 1962; Ganders 1979b).

Because tristylous species have two stamen levels in each flower, Yeo (1975) proposed that two series of stamens are a necessary precondition for the evolution of tristyly. He assumed that one stamen series developed into one anther level, while the other produced the second level. Although all tristylous species have two series of stamens, in the Pontederiaceae each stamen level has members from both the first and second series (Richards and Barrett 1984, 1987). The organization of heterostyly in this family therefore negates the reason for assuming that two stamen series are a precondition for the evolution of tristyly, although a within-flower stamen dimorphism probably is a prerequisite (see Sect. 6).

Hypericum aegypticum and *Cratoxylum formosum* in the Guttiferae are unusual among heterostylous species because they have numerous stamens arranged in bundles (Ornduff 1975; Lewis 1982). In both species stamen length differs within a flower, and some overlap between stamen levels is found in *C. formosum*. Distyly is achieved in the latter species by an ephemeral bending of the filaments at anthesis in the long-styled morph (Lewis 1982). This bending places the stamens at the level of the short style. The occurrence of heterostyly in species of *Cratoxylum* and *Hypericum* indicates that a limited number of stamens is not a precondition for the evolution of heterostyly. The unusual stamen behavior in *Cratoxylum*, however, emphasizes that a large number of stamens presents different developmental problems in evolution of the syndrome.

The gynoecium of heterostylous species is usually syncarpous and has from one to five carpels [Chap. 6, this Vol.; heterostylous Sterculiaceae have five (*Melochia*) or one (*Waltheria*) carpel(s)]. Styles are free or united and the ovary can be superior or inferior. The floral structure of apocarpous species (many carpels spread over a broad or elongated receptacle, short styles, many stamens, and open flowers) may be inherently unsuitable for evolution of the heterostyly syndrome. Problems in regulating development of separate style lengths would be similar to those of regulating numerous stamens. In species with carpels on an elongated receptacle, the range of stigma heights could compromise legitimate pollen transfer between the floral morphs and lead to considerable pollen wastage.

2.2 Structural Basis for Style and Stamen Heteromorphisms

The characteristic that defines heterostyly is the reciprocal positioning of anthers and stigmas in flowers of different morphs. Stigma height depends on the size of the ovary and length and orientation of the style. Style length is the major source of differences in stigma height among heterostylous morphs, since ovary lengths are usually similar among morphs (Stirling 1932, 1936; Richards and Barrett 1984, 1987). Cases where this is not so have been associated with gender specialization, as in subdioecious *Cordia inermis* (Opler et al. 1975).

Bending of the style can make a significant contribution to final stigma position, especially in the short-styled morph. Darwin (1888) described bending in *Linum grandiflorum* (Linaceae), noting that short styles diverge much more than long styles and that bending allows the short styles to pass between the stamen filaments. A similar condition is found in *Oxalis* (Oxalidaceae) (Darwin 1888), *Piriqueta* (Turneraceae) (JH Richards pers. observ.), *Turnera* (Turneraceae) (Barrett and Shore 1987), and *Melochia* (Martin 1966). The contribution of stylar bending to stigma position has not been quantified nor has variation in this character among morphs been studied in detail. In *Oxalis* stylar bending in the short morph occurs prior to anthesis (JH Richards pers. observ.).

Anther height depends on filament length and orientation. Since many heterostylous plants are sympetalous (Ganders 1979a), both the position of filament insertion on the floral tube and the floral tube length can contribute to anther height. Filament length establishes anther height differences in some species, such as *Jepsonia* (Saxifragaceae) (Ornduff 1971), *Erythroxylum* (Ganders 1979b) and *Oxalis* (Ornduff 1964, 1972; Weller 1976a; and see Sect. 3.1), whereas insertion position determines anther height in other species, e.g., *Primula* spp. (Primulaceae) (Darwin

1888; A.J. Richards 1986) and *Hedyotis caerulea* (Rubiaceae) (Ornduff 1980). In many heterostylous species both filament length and position of insertion vary among morphs (e.g., *Pontederia* and *Eichhornia* (Pontederiaceae), Richards and Barrett 1984, 1987; species of *Psychotria, Cephaelis*, and *Faramea* (Rubiaceae), JH Richards unpubl. data).

In the unusual case of *Cratoxylum formosum*, anther height is not determined solely by filament length or insertion (Lewis 1982). Stamen bending positions the "short" anthers at a height comparable to the short stigma. Unlike stylar bending, this reorientation is reversible – the stamens erect themselves after about 6 h – and the developmental timing and, presumably, mechanism is different.

In studies of heterostylous flowers, stigma and anther height are usually measured from the base of the flower or some similar reference point. The biologically important measurements, however, are stigma-anther separation. Although separation can be calculated from stigma-anther height, this calculated value and the actual separation are not necessarily equal, especially if ephemeral changes in position, bending, or lateral displacement of sex organs are significant. To overcome this problem, careful observations of flowers in situ are needed in addition to the traditional measurements.

2.3 Ancillary Characters of Heterostylous Flowers

Heteromorphism of other reproductive characters is often associated with stigma-anther positions in heterostylous species (Darwin 1888; Ganders 1979a). Differences in anther length, pollen size and number, pollen shape and exine sculpturing, pollen color, stigma size and shape, stigma papillae size and shape, and stigma color have been reported (see Chap. 3). The pattern of heteromorphism is often similar in different heterostylous taxa. Long-level styles, for example, frequently have larger stigmas and longer stigmatic papillae than short-level styles, while long-level stamens usually produce larger but fewer pollen grains than short-level stamens (Dulberger 1975a; Ganders 1979a). Recurrent associations among characters in heterostylous species may reflect common developmental bases for the correlated characters. Larger stigmas with longer papillae, for example, can be a direct result of increased cell length in long styles.

Pollen-size dimorphism is so frequent in distylous taxa that Darwin (1888) considered it a basic feature of heterostyly. Number of pollen grains per anther also often varies among anther levels. Ganders (1979a) showed that the ratio of thrum:pin pollen volume is correlated with the ratio of pin:thrum pollen grain number. This correlation suggests that there is a developmental relationship between pollen size and pollen grain number that results in either many small grains or fewer large grains. Differences between morphs in pollen grain size could arise any time from sporogenous cell origin until the grains mature and could have a number of causes, such as size differences when sporogenous cells differentiate or variations in how long or rapidly the cells grow. The kinetics of pollen growth have not been studied in a heterostylous species, but premeiotic size heteromorphisms have been reported for sporogenous cells of *Primula* (Dulberger 1975a) and *Eichhornia paniculata* (Richards and Barrett 1984).

Differences among anther levels in pollen number probably originate during microsporogenesis. Although post-meiotic pollen abortion could also produce pollen number differences, a substantial number of cell deaths would be required to obtain the observed differences in pollen number [e.g., app. 4000 vs. 21 000 pollen grains in long-level vs. short-level anthers of *Pontederia cordata* (Price and Barrett 1982)], and evidence of such abortion has never been reported.

The developmental basis for pollen number differences may vary among species. Pollen production (P) is usually reported per anther level and can be approximated by the equation

$$P = (ALS)(2^n)(4),$$

where A is the number of anthers per stamen level, L is the number of locules per anther, S is the initial number of sporogenous cells per locule, and n is the number of sporogenous cell mitotic divisions prior to meiosis, which are multiplied by 4 to account for products of meiosis. Anther number and locule number per anther have not been observed to vary between morphs in heterostylous species, so the variables of interest are S and n. Ratios of short-level to long-level pollen number in most distylous species range from 1.13 to 3.12 (Ganders 1979a). Small differences in S could account for the lower end of this range and such small differences will be difficult to detect in developmental studies. The twofold or greater differences in sporogenous cell number required to achieve the upper end of this range, however, should be obvious.

Alternatively, differences between anther levels in n, caused by one or two additional divisions in some or all of the sporogenous cells, could account for the observed range of pollen number ratios. Morph-specific differences in the cell cycle or in the duration of microsporogenesis would produce different numbers of microspore divisions. The two anther levels within flowers of tristylous Pontederiaceae enter meiosis at different times, and differences in pollen mother cell numbers, as well as size, are present prior to meiosis (Richards and Barrett 1984; JH Richards unpubl. data), which indicates that the number of divisions in the sporogenous tissue contributes to pollen number heteromorphisms in the Pontederiaceae. Developmental studies in other heterostylous taxa are needed to define the processes responsible for pollen number differences and to understand the relation of pollen number to pollen size heteromorphisms.

While certain patterns of ancillary character heteromorphism predominate among heterostylous taxa, none of the correlations between primary and ancillary characters is invariant. Pollen and stigma heteromorphisms are lacking in some heterostylous genera (Vuilleumier 1967; Dulberger 1974). In addition, cases are known where the predominant pattern of heteromorphism is reversed. For example, the stigmas of short styles can be larger than those of long styles (e.g., *Palicourea lasiorrachis*, Feinsinger and Busby 1987), and long-level anthers can have more pollen than short-level anthers, as in *Amsinckia spectabilis* var. *microcarpa* and *A. vernicosa* var. *furcata* (Boraginaceae) (Ganders 1975, 1976). In the two *Amsinckia* species pollen grains from the long-level anthers are larger than grains from short-level anthers, so the increase in pollen number has occurred without change in the usual pollen size relations. The pattern of pollen size dimorphism is variable in *Fauria* (= *Nephrophyllidium*) *crista-galli* (Menyanthaceae), with long-level pollen larger than short-level pollen in one population but smaller in another (Ganders 1979a).

The diversity of expression in ancillary characters may reflect differences among species in development of the heteromorphisms. A species that produces long styles by increased cell division rather than increased cell elongation, for example, may not have a large stigma with long papillae. More detailed analysis of the structural and developmental basis of heterostyly are needed before the relation of these ancillary characters to the primary stamen-style heteromorphism can be understood.

3 Floral Development in Heterostylous Species

The structural diversity of stamen-style heteromorphisms and taxonomic diversity of heterostylous species suggests that different developmental mechanisms are involved in the evolution of the polymorphism in unrelated angiosperm families. Unfortunately, there are few detailed structural analyses or developmental studies with which to test this prediction. The most complete body of work is on tristylous species, and even among the small group of plants that share this polymorphism, several patterns of development are evident. Below we review what is known about the development of tristylous and distylous flowers.

3.1 Mature Flower Structure in Tristylous *Lythrum*, *Oxalis*, and the Pontederiaceae

Mature flower structure in tristylous members of the Lythraceae, Oxalidaceae, and Pontederiaceae are summarized in Table 1 and Fig. 1. The flowers of *Lythrum* (Lythraceae) have six sepals that are united basally into a tube, six petals inserted just below the sinuses of the calyx tube, twelve stamens, and a two-carpellate superior ovary (Fig. 1A; Table 1). Six outgrowths that have been called sepals (Sattler 1973; Cheung and Sattler 1967) or epicalyx appendages (Mayr 1969) develop in the sinuses between calyx lobes after lobe initiation. The 12 stamens are inserted a short distance from the base of the calyx tube (app. 1 mm in *L. salicaria*), and in each morph the six longer stamens, opposite the sepals, alternate with the six shorter stamens, opposite the epicalyx appendages and petals. Our observations on floral organization agree with those of Cheung and Sattler (1967) for *Lythrum salicaria* and Ornduff (1979) for *Lythrum junceum*. In *L. salicaria* the longer stamens (long-level stamens in the M and S morphs, mid-level stamens in the L morph) are inserted slightly higher on the floral tube than the shorter stamens (short-level stamens of the L and M morphs, mid-level stamens of the S morph), but most of the difference in stamen height within morphs results from variation in free filament length. Mature *Lythrum* flowers extend almost horizontally from the inflorescence and are slightly zygomorphic.

Oxalis (Oxalidaceae) flowers have five free sepals and petals, ten stamens that are united basally into a ring, and a five-parted, syncarpous superior ovary (Fig. 1B). The two stamen levels within a flower alternate on the staminal ring, with the five longer stamens opposite the sepals and the five shorter stamens opposite the petals. Differences in anther height depend primarily on differences in filament length.

Table 1. Structural and developmental features of *Lythrum, Oxalis,* and the Pontederiaceae

Characteristic	*Lythrum*	*Oxalis*	Pontederiaceae
Perianth structure	Sepals + petals	Sepals + petals	Tepals
Perianth fusion	Calyx tube, free petals	Free sepals and petals	Tepals fused 1/2 of length
Floral symmetry	Slightly zygomorphic	Radial	Zygomorphic
No. stamen series × no. stamens/series	2 × 6	2 × 5	2 × 3
Insertion of filaments	On calyx tube	On staminal ring on receptacle	On floral tube
Organization of stamen dimorphism	Radial	Radial	Dorsivental
Position of longer stamen level	Opposite sepals	Opposite sepals	Opposite lower tepals
Position of shorter stamen level	Opposite petals	Opposite petals	Opposite upper tepals
Origin of stamen dimorphism within morphs	At initiation	At initiation	Post-initiation
Origin of differences between morphs	?	?	Post-initiation

Flowers of *Eichhornia* and *Pontederia* (Pontederiaceae) have three narrow outer tepals and three broader inner tepals, six stamens, and a tricarpellate, syncarpous ovary, although in *Pontederia* only one carpel develops an ovule (Lowden 1973; Richards and Barrett 1984, 1987; Fig. 1C; Table 1). The six tepals are united into a floral tube for approximately half their length. Mature flowers are oriented horizontally, and the longer and shorter stamens in each morph occur on opposite sides of the flower. The shorter stamen level is on the upper side of the flower, with the central stamen opposite an inner tepal and the two lateral stamens opposite outer tepals. The longer stamen level is on the lower side, with the central stamen opposite an outer tepal and the lateral stamens opposite inner tepals. Anther height depends on both filament length and position of filament insertion on the floral tube. Longer stamens are inserted higher on the floral tube and have longer filaments, while shorter stamens are inserted lower on the floral tube and have shorter filaments.

Comparison of mature flower structure indicates that the two dicotyledonous families resemble each other but are distinct from the monocotyledonous Pontederiaceae in the organization of tristyly. Intraflower stamen dimorphism is organized on a radial pattern with alternating longer and shorter stamens in *Lythrum* and *Oxalis*. In these two genera stamen height within a flower depends primarily on free filament length. In both *Lythrum* and *Oxalis* the relatively long stamen level arises opposite the sepals, whereas the relatively short stamen level is opposite the petals.

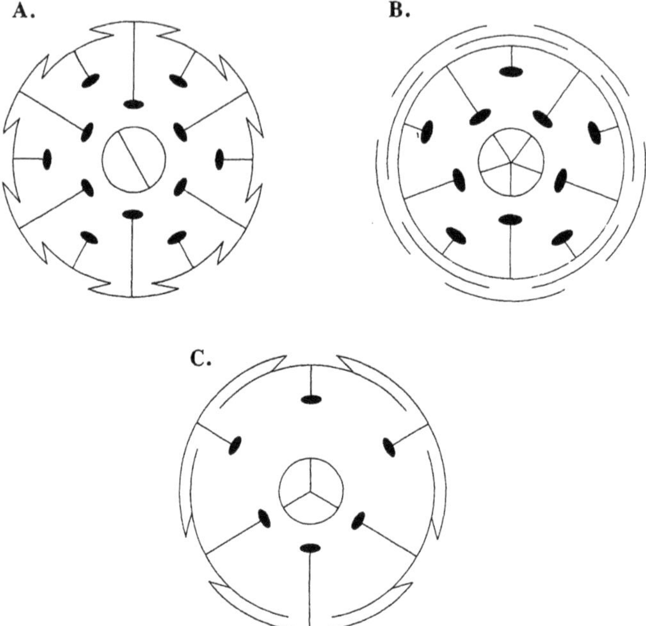

Fig. 1A-C. Floral diagrams of *Lythrum* (**A**), *Oxalis* (**B**), and tristylous Pontederiaceae (**C**). In *Lythrum* stamens are inserted on the floral tube with the longer stamens opposite the sepals and the shorter stamens opposite the petals; in *Oxalis* the stamen bases are united into a ring with the longer stamens opposite the sepals and the shorter stamens opposite the petals; in tristylous Pontederiaceae the stamens are inserted on the floral tube with the shorter stamens on the upper side of the horizontally oriented flowers and the longer stamens on the lower side

Although the perianth in the Pontederiaceae is organized in two series, the within-flower stamen dimorphism is dorsiventral and thus has a fundamentally different architecture from the intraflower stamen dimorphism of the tristylous dicotyledons. Moreover, stamen dimorphism in the Pontederiaceae depends both upon level of insertion on the floral tube and upon free filament length. The differences among families in stamen dimorphism are correlated with perianth structure. The calyx is morphologically distinct from the corolla in the Lythraceae and Oxalidaceae, and each stamen level within a flower is associated with only one perianth series. In the Pontederiaceae, in contrast, distinctions between the inner and outer tepals are less marked, and the within-flower stamen dimorphism is not correlated with tepal series. The lack of differentiation between perianth series in the Pontederiaceae may have provided insufficient developmental cues to the associated stamens to allow for the differentiation of stamen levels from the initial stamen series.

Differences among families in the architecture of tristyly may reflect ancestral floral organizations that provided contrasting developmental contexts for the evolution of tristyly. It is remarkable that the inheritance patterns of the polymorphism are similar, even though the genes controlling tristyly must regulate different developmental processes in these families (see Sect. 5).

3.2 Development of Tristyly in Pontederiaceae

The Pontederiaceae is a small aquatic monocotyledonous family with six to nine genera. Two genera, *Eichhornia* and *Pontederia*, have tristylous species (Eckenwalder and Barrett 1986). These two genera differ in the expression of tristyly. Four out of the five species of *Pontederia* are tristylous, and all species that have been investigated have a strong trimorphic self-incompatibility system. Only three of eight *Eichhornia* species are tristylous, and of these only *E. azurea* has self-incompatible populations (Barrett 1978).

Although flowers of both *Eichhornia* and *Pontederia* have dorsiventral organization when mature (Richards and Barrett 1984, 1987), the floral organs arise in a radial pattern. The six sepals are initiated in two series of three each (Figs. 2, 3), followed by a first stamen series and then a second. The first stamen set is opposite the outer tepal whorl and the second opposite the inner whorl (Figs. 3, 4). Three carpels develop from primordia that arise opposite the outer tepals and first stamen whorl (Figs. 5, 6). Although distinct differences in size between stamen primordia associated with each tepal series are apparent in young flower buds (Figs. 4, 5, 7), these variations are unrelated to differences between stamen levels in mature flowers. The three stamen primordia on the upper half of the flower primordium develop into the shorter stamens within each flower, while the three lower primordia develop into the longer stamens (Fig. 1C, Fig. 7). In both *Pontederia* and *Eichhornia*, size differences between stamen levels become apparent in post-meiotic stamens. Premeiotic differences can be seen, however, in the number and size of pollen mother cells and in time of entry into meiosis (Richards and Barrett 1984 and unpubl. data).

In *Pontederia* and *Eichhornia* differences among morphs in anther and stigma heights arise through differences in both relative growth rates and the duration of growth (Richards and Barrett 1984, 1987, unpubl. data). In *Pontederia* the L and M morphs have similar patterns of growth but differ in the growth rates of the lower, longer filaments and of the styles (Figs. 8, 9). The S morph is qualitatively different from the other two morphs in showing an early inhibition of style and filament expansion, followed by a late acceleration of filament growth (Figs. 8, 9).

Variation in filament length, filament insertion, and style length can arise from differences in cell size and/or cell number. Measurements of *P. cordata* (Richards and Barrett 1987) and *E. paniculata* (Richards and Barrett 1984) indicate that stylar cell length differs among morphs. Long styles have longer cells than mid-length or short styles, and mid-length styles have longer cells than short styles. Cell size differences are not sufficient to account for style length variation, however, so cell numbers must also differ.

Filament cell length also varies among stamen levels. Long filaments have longer cells, short filaments shorter cells, and mid-length filaments have cells of intermediate length. In both *P. cordata* and *E. paniculata*, however, cells of mid-level filaments of the L and S morphs differ in length (Fig. 10). Mid-level filaments of the S morph have longer cells than mid-level filaments of the L morph (Richards and Barrett 1984, 1987, Fig. 10). This is unexpected, since mid-length stamens of the S morph are the upper, shorter stamen level, whereas mid-length stamens of the L morph comprise the lower, longer stamen level. The longer cells of mid-level fila-

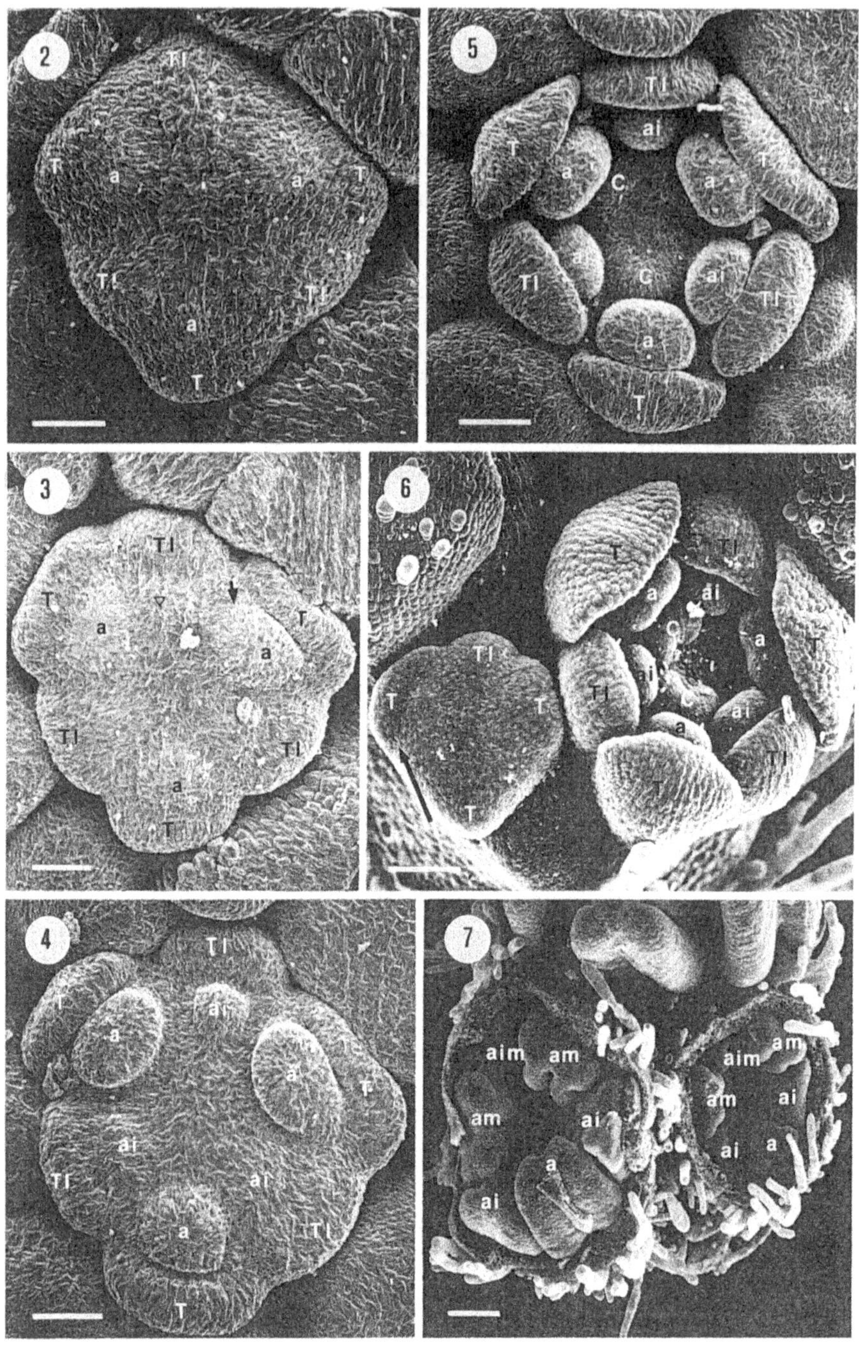
2
Tl
T
a
a
T
Tl
Tl
a
T
5
Tl
ai
T
a
C
a
T
a
Tl
C
ai
Tl
a
T
3
Tl
T
a
a
T
Tl
Tl
a
T
6
T
Tl
a
ai
C
a
T
Tl
T
T
Tl
ai
ai
a
Tl
T
T
4
Tl
T
a
ai
a
T
ai
Tl
ai
Tl
a
T
7
aim
am
aim
am
am
ai
am
ai
ai
a
a
ai

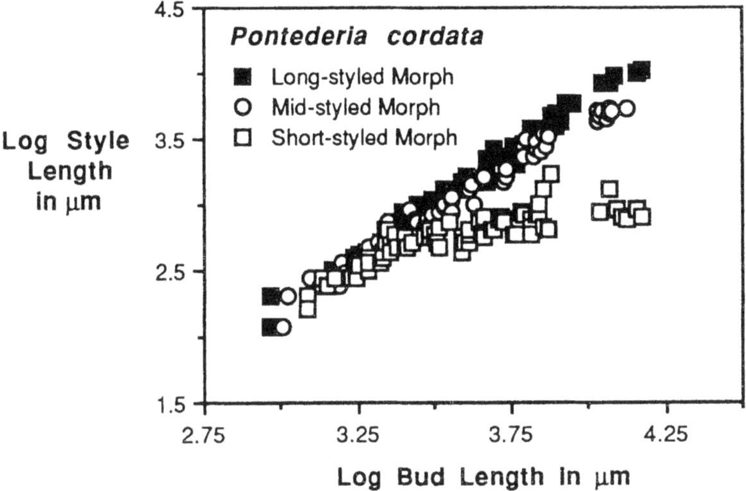

Fig. 8. Log bud length vs. log style length for the three morphs of *Pontederia cordata*. Style relative growth rates are similar in young buds of the three morphs. In later development stylar growth is inhibited in the S morph, while the relative growth rate of the M morph is less than that of the L morph

Figs. 2–7. Scanning electron micrographs illustrating flower development in the S morph of *Pontederia cordata*. *a* stamen primordium in the first-initiated stamen series; *ai* stamen primordium in the second-initiated stamen series; *am* mid-level stamen that develops from a primordium in the first stamen series; *aim* mid-level stamen that develops from a primordium in the second stamen series; *c* carpel; *T* outer tepal; TI inner tepal. **Fig. 2.** Floral primordium that has initiated three outer tepals and stamens and three inner tepals. The two upper stamens, which are more developed than the lower stamen, will belong to the shorter (mid) stamen level. *Bar* 33 µm. **Fig. 3.** Floral primordium with three outer tepals and stamens and three inner tepals. Stamens associated with the inner tepals are just beginning to develop. The upper stamens of both the outer and inner stamen series, which will become the shorter (mid) stamen level, are larger than the lower stamens, which will become the longer (long) stamen level. Within the outer stamen series the upper primordium away from the parent branch (*arrow*) is slightly larger than the other upper stamen primordia, while within the inner stamen series the upper primordium (at *triangle*) is more developed than the two lower stamen primordia. *Bar* 50 µm. **Fig. 4.** A bud that has an exaggerated expression of size differences between the inner and outer stamen series and of earlier development of the upper stamen of the inner stamen series. *Bar* 50 µm. **Fig. 5.** A flower bud in which the tepals have begun to grow around the stamens, and the carpels have been initiated on a radius with the outer tepal and stamen series. The outer stamen series has larger primordia than the inner stamen series. *Bar* 50 µm. **Fig. 6.** Two floral primordia on a branch. In the older primordium the tepals have begun to enclose the bud, while the outer stamen series exceeds the inner in size and has initiated microsporangia. The younger floral primordium is slightly better developed than the bud in Fig. 2. The three outer tepal primordia are distinct, as is the upper, inner tepal primordium. The outer tepal primordium away from the parent branch (*arrow*) is more developed than the other tepal primordia. *Bar* 50 µ. **Fig. 7.** Two older buds with tepals dissected away. In both buds the outer stamen series is larger than the inner stamen series, but in the older bud growth in the long stamen level has begun to exceed that in the mid stamen level, so the lower outer stamen is larger than the upper outer stamens and the lower inner stamens are approximately equal in length to the upper outer stamens. *Bar* 150 µm

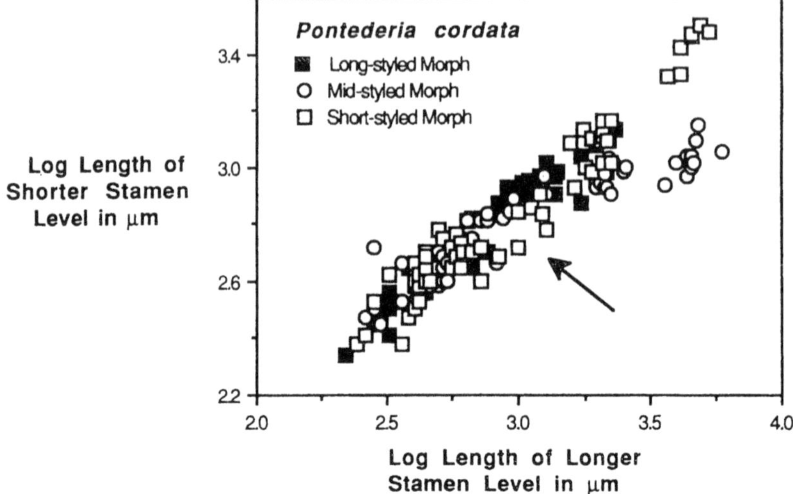

Fig. 9. Log length of the longer vs. the shorter stamen level for the three morphs of *Pontederia cordata*. The L and M morphs have similar patterns of relative growth, while the S morph shows an early inhition (*arrow*) then acceleration in relative stamen growth

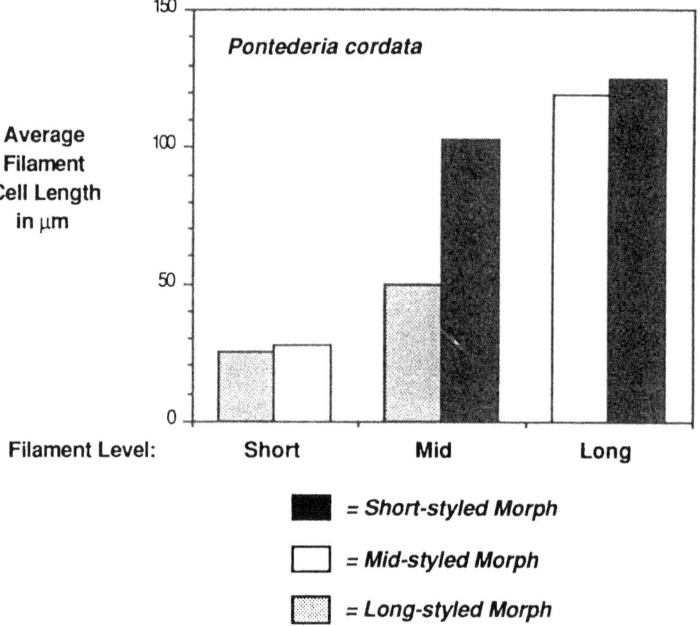

Fig. 10. Average filament cell length for the three morphs of *Pontederia cordata*. Average cell length of long-level filaments is similar, regardless of morph, as is average cell length of short-level filaments. Average cell length of mid-level filaments depends on morph. The S morph, which has mid-level stamens on the upper side of the flower, produces filaments with longer cells than the mid-level filaments of the L morph

ments in the S morph may be a product of the late acceleration in filament relative growth rate observed in this morph (Fig. 9).

3.3 Development of Tristyly in Lythraceae and Oxalidaceae

In common with the Pontederiaceae, the sequence of organ initiation and qualitative aspects of early development are the same in all three morphs of tristylous species of *Lythrum* and *Oxalis*. We have examined development in *Lythrum salicaria* and *Oxalis rubra*.

The floral organs of *Lythrum salicaria* arise in five series (Cheung and Sattler 1967; Sattler 1973; JH Richards pers. observ.). The calyx arises as six sepal primordia that are quickly united into a meristematic ring (Fig. 11). The first stamen whorl is initiated next, opposite the sepal primordia (Figs. 12, 13), followed by origin of the gynoecium in the center of the bud (Figs. 12–14). The second stamen whorl, which alternates with the first set of stamens (Fig. 15), arises after the gynoecium primordium is defined. The epicalyx appendages begin to develop at approximately the same time as the carpel (E in Figs. 13, 14). These appendages arise between the calyx lobes and eventually develop into elongated cylindrical outgrowths (Figs. 15, 16). Relatively late in development the petals arise as individual primordia on the upper edge of the calyx tube between the sepal lobes and thus on a radius with the epicalyx appendages and second stamen whorl.

Lythrum salicaria, therefore, has a temporal separation between initiation of the two stamen whorls. The first-initiated stamens, which become the longer stamens within each flower, are always larger than the second-initiated stamens, which become the shorter stamens (Figs. 15, 25, 26). The first stamens precede the second ones in sporangia and filament initiation (Fig. 15) and exceed them in filament growth (Fig. 16).

In *Oxalis rubra* the five sepals arise in a spiral (Fig. 17), then the floral apex broadens and becomes pentagular with the initiation of the petals, which alternate with the sepals (Fig. 18). Growth of the petals is slow, however, relative to growth of the other floral organs – the petals do not exceed the stamens in length until bud length is ca. 3 mm. The first stamen series is initiated opposite the sepals (Figs. 19, 27), and the second series arises opposite the petals (Figs. 20–22, 28). The petal-opposed stamen series is initiated slightly later than the sepal-opposed series and appears to be lower on the floral apex than the sepal-opposed stamens (Figs. 20–22, 27, 28). Carpel primordia originate on a radius with the petals and second stamen series, forming the five-parted gynoecium (Figs. 23, 28–30).

Thus, as in *L. salicaria*, the two stamen levels within a flower of *O. rubra* differ in time of origin. The differences in early development are accentuated as the anthers grow and initiate sporangia (Figs. 23, 29, 30). Differences in filament length are present prior to significant growth of the stamen ring (Fig. 24).

Stirling (1933, 1936) provides the only published studies of later stages of floral development in tristylous species of *Lythrum* and *Oxalis*. His data from measurements of dissected buds of *L. salicaria* (Stirling 1933) and *O. cernua* (= *O. pes-caprae*) (Stirling 1936), which we have regraphed (Figs. 31–34), suggest that growth patterns in these two species differ from those in the Pontederiaceae. Graphs of petal length

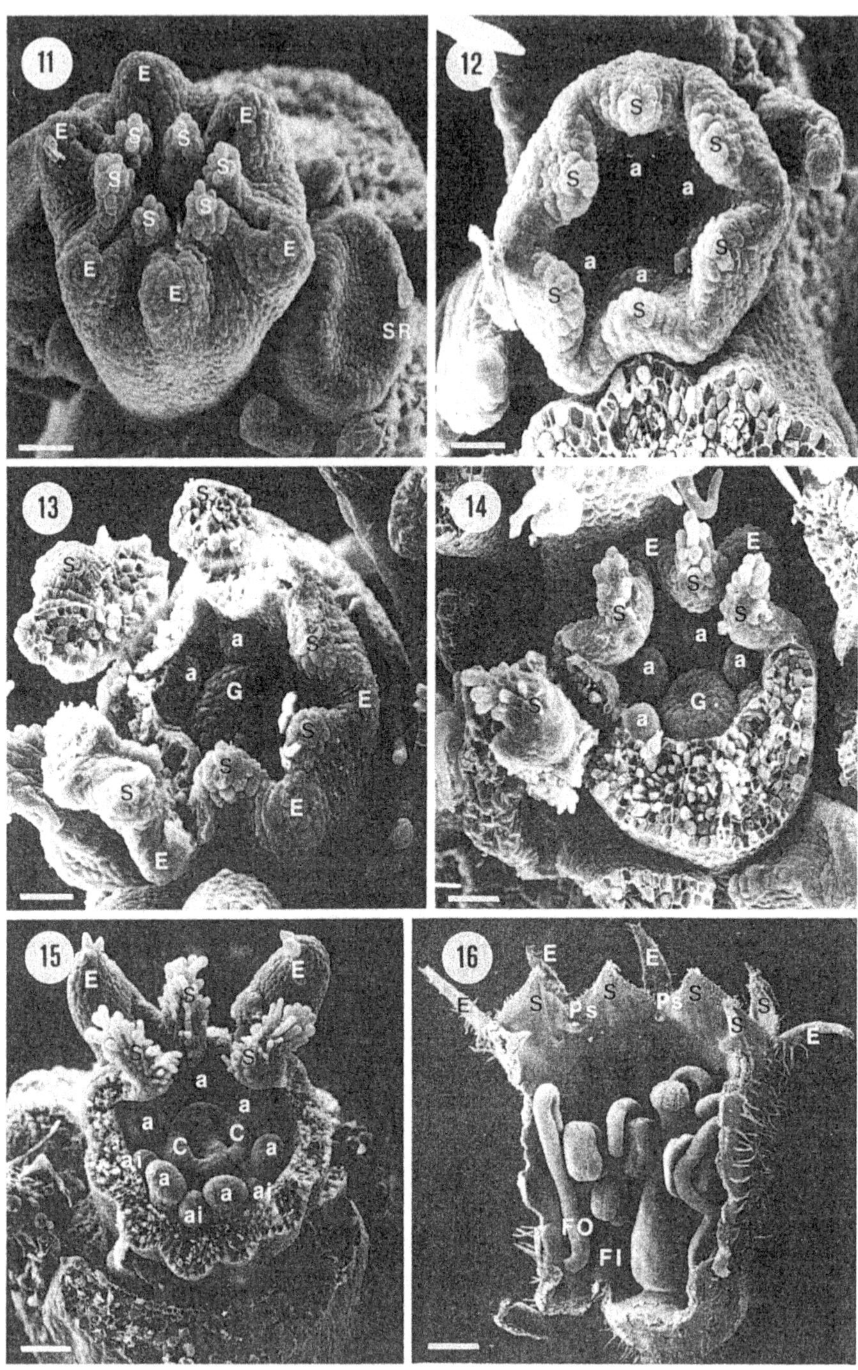
11
E E E S S S S S S S E E E SR
12
S S S a a S a a S S
13
S S S a a G S E S S E S E
14
E E S S S a a a a G a S
15
E E S S S a a a C C a ai a a a ai ai
16
E E E E S Ps S Ps S S S E FO FI

vs. pistil length indicate that in both *L. salicaria* and *O. cernua* the three morphs have different relative growth rates in the later stages of development (Figs. 31, 32). Neither species, however, shows the degree of stylar inhibition that occurs in the S morph of the Pontederiaceae (Fig. 8). Although Stirling provides less information on young buds, the available data on pistil length for both *L. salicaria* and *O. cernua* are consistent with an early similarity in relative growth rates among morphs and an earlier divergence in growth rate of the S morph from the other morphs (Figs. 31, 32).

As in the Pontederiaceae, relative growth of stamens differs among morphs in *Lythrum* and *Oxalis*. In both species graphs of longer vs. shorter filament length show a more divergent pattern in the M morph than in the L and S morphs (Figs. 33, 34). Figures 33 and 34 indicate that in the L and S morphs relative growth rates of stamen levels within a flower are constant and relatively linear. This implies that differences between stamen levels within a morph are established early and maintained throughout development. Development of stamens in the M morph diverges from the other morphs in both *Lythrum* and *Oxalis*, although the pattern of divergence appears to differ in the two species (Fig. 33, 34).

3.4 Comparison of Development Among Tristylous Species

The tristylous species of the Pontederiaceae, Lythraceae, and Oxalidaceae initiate two, temporally separated stamen series. The difference in time of origin establishes a size differential between stamen series that is maintained throughout development. In the Pontederiaceae this difference in time of origin merely produces a subtle variation in height within each stamen level and does not contribute to intraflower stamen dimorphism (Richards and Barrett 1984, 1987). In *Lythrum* and *Oxalis*, in contrast, the first stamen series initiated corresponds to the longer stamen level

Figs. 11–16. Scanning electron micrographs of floral development in the S morph of *Lythrum salicaria*. *a* stamen primordium in the first-initiated stamen series; *ai* stamen primordium in the second-initiated stamen series; *c* carpel; *E* epicalyx appendage; *FO* filament of stamen in first, long-level stamen series; *FI* filament of stamen in second, mid-level stamen series; *G* gynoecium; *Ps* site of petal attachment to calyx tube; *S* calyx lobes; *SR* calyx primordium ring. **Fig. 11.** Two flower buds. In the younger bud the calyx primordia are united into a ring that will form the calyx tube. In the older bud the calyx lobes have grown over to enclose the bud, while the epicalyx appendages are growing between the sepals. *Bar* 50 μm. **Fig. 12.** Bud with first set of stamens initiated opposite sepal lobes. *Bar* 50 μm. **Fig. 13.** Bud with calyx broken away to reveal first stamen set and gynoecial primordium. *Bar* 50 μm. **Fig. 14.** Bud with part of floral tube removed to show first stamen primordia opposite the sepals and the gynoecium primordium, which has begun to form a meristematic ring. *Bar* 50 μm. **Fig. 15.** Older bud broken open to show two sets of stamens and bilocular ovary. Primordia in the first stamen series, opposite the calyx lobes, have initiated anthers and are larger than primordia in the second stamen series. *Bar* 100 μm. **Fig. 16.** Well-developed bud with calyx lobes and epicalyx appendages visible but petals removed. Filaments of both long and mid-level stamens are contorted. The long-level filaments are inserted higher on the floral tube than the mid-level filaments. *Bar* 1 mm

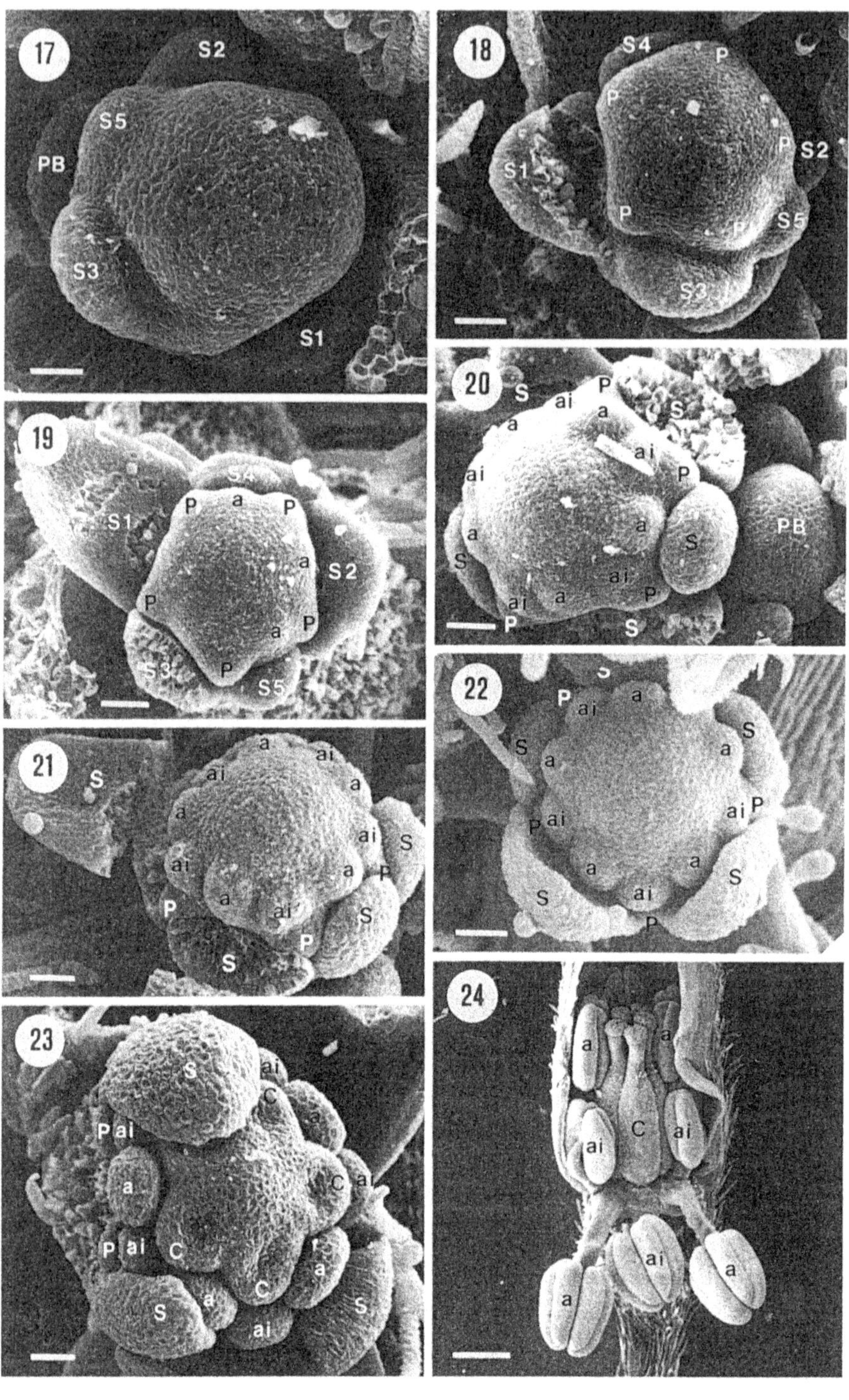
17
S2
S5
PB
S3
S1
18
S4
P
S1
S2
S5
S3
19
S4
S1
P a P
S2
S3
S5
20
S
ai
P
PB
21
S
a
ai
P
22
S
P
ai
a
23
S
ai
C
P ai
a
24
a
ai
C

within each mature flower, while the second stamen series corresponds to the shorter level. The difference between stamen whorls in time of origin is greater for *Lythrum* than *Oxalis*, in which the primordia arise almost simultaneously in some buds.

Our survey of the available data indicates that morph-specific patterns of stamen relative growth differ for *Lythrum* and *Oxalis* as compared to the Pontederiaceae. In both *Lythrum* and *Oxalis* the M morph appears to have the most complex pattern of stamen relative growth and shows relatively late-developing changes in relative growth rates. In the Pontederiaceae, in contrast, the S morph has the most divergent pattern of stamen relative growth when compared to the other morphs. The differences in pattern of stamen relative growth are probably related to the radial architecture of tristyly in the dicotyledonous families vs. the dorsiventral architecture in the Pontederiaceae. Additional developmental data for *Lythrum* and *Oxalis* are needed in order to understand this relationship.

While differences between stamen levels within a flower are present from stamen origin in tristylous species of *Lythrum* and *Oxalis*, these are not necessarily the differences that define stamen levels in tristyly. We do not know whether there are between-level differences in characters other than time of origin. For example, do primordia from different levels have different sizes at origin – e.g., do long-level stamens, which have larger anthers at maturity, have larger primordia than mid- or short-level stamens? This question is important for models of how the tristyly genes affect development. Do the genes regulate some event at primordium origin that results in the tristyly syndrome through developmental amplification or do they act directly on subsequent developmental events? Because many species initiate stamens in two or more whorls but do not develop two stamen levels [e.g., *Silene cucubalus* (Caryophyllaceae) (Sattler 1973) or *Pisum sativum* (Fabaceae) (Tucker 1989)], and because the order of stamen initiation is independent of intraflower stamen dimorphism in the Pontederiaceae, the differences in time of origin between stamen levels in the Lythraceae and Oxalidaceae are not necessarily the first expression of tristyly in these families.

Figs. 17–24. Scanning electron micrographs showing flower development in the S morph of *Oxalis rubra*. *a* stamen primordium in first initiated stamen series, opposite the sepals; *ai* stamen primordium in second-initiated stamen series, opposite the petals; *C* carpel; *P* petal; *PB* prophyll bud; *S* sepal or sepal position if sepal has been removed; *numbers after S* indicate order of initiation. **Fig. 17.** A floral bud that has initiated sepals. *Bar* 20 µm. **Fig. 18.** A bud with five sepals that has initiated petals, forming a pentangular floral apex. *Bar* 23 µm. **Fig. 19.** A bud that has initiated the first stamens opposite the sepals. *Bar* 30 µm. **Fig. 20.** A bud with some sepals removed. The second set of stamens has been initiated opposite the petal primordia. *Bar* 30 µm. **Fig. 21.** A lateral view of a bud slightly older than in Fig. 20 that shows the relation of sepals, petals, and the two stamen sets. The stamens opposite the sepals are slightly larger than those opposite the petals. *Bar* 30 µm. **Fig. 22.** A surface view of a bud similar to that in Fig. 21. The stamens opposite the petals are lower on the apical dome than those opposite the sepals. *Bar* 30 µm. **Fig. 23.** A bud with the five-carpellate ovary initiated. The carpels arise on a radius with the petals and second stamen set. *Bar* 30 µm. **Fig. 24.** An older bud with some sepals removed to show the two stamen levels, which alternate within the flower and have different filament lengths. *Bar* 400 µm

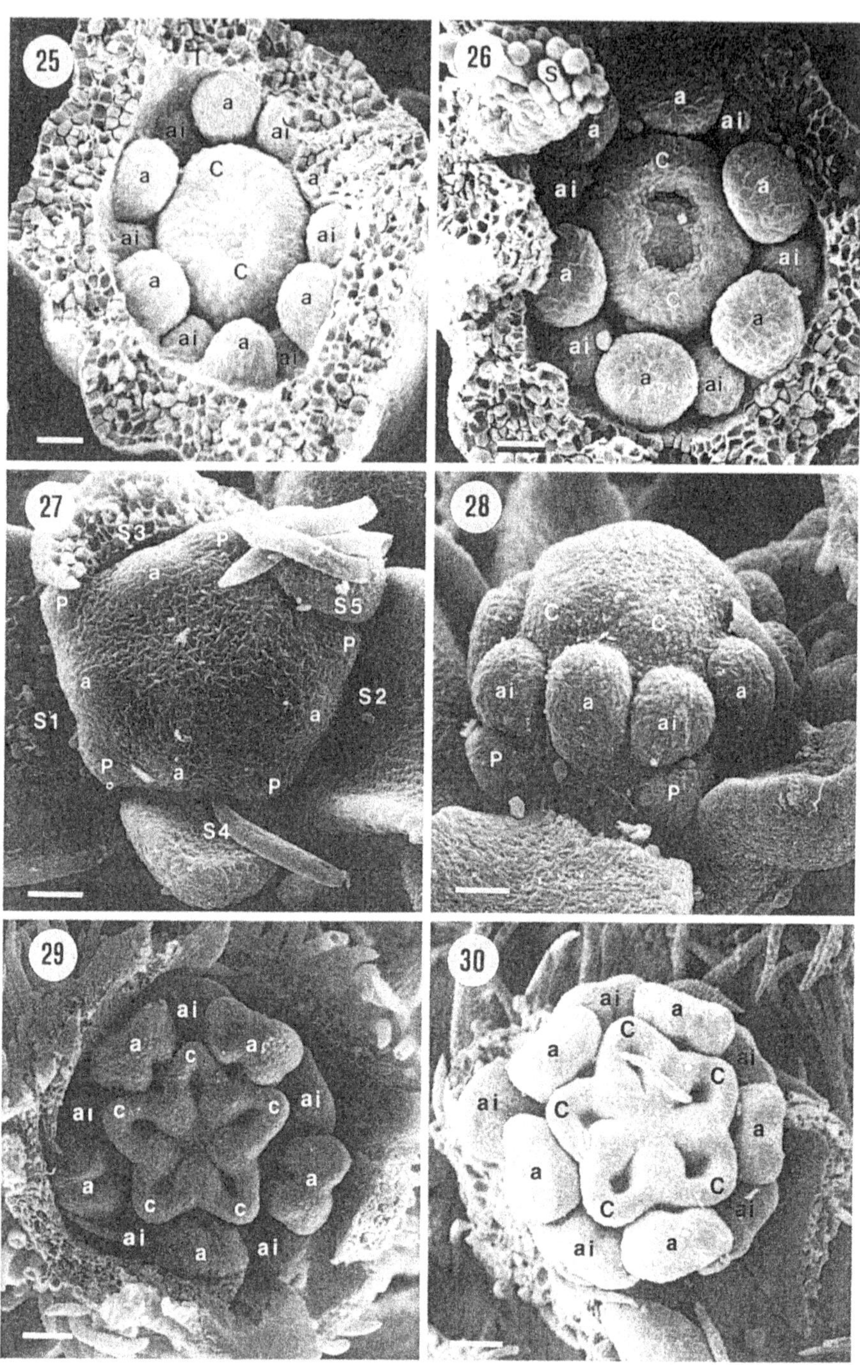
25
a
ai
ai
a
C
a
ai
ai
a
C
a
ai
a
ai
26
S
a
a
ai
ai
C
a
ai
a
C
a
ai
a
ai
27
S3
P
a
S5
P
P
a
S1
S2
a
P
a
P
S4
28
C
C
ai
a
ai
a
P
P
29
ai
a
c
a
ai
c
c
ai
a
a
c
c
ai
a
ai
30
ai
a
C
a
ai
C
ai
C
a
a
C
C
ai
ai
a

In Fig. 35 we illustrate graphically four developmental relationships between stamen levels that could lead to the stamen differences observed in tristylous species. In all the models one factor that contributes to differences between morphs is the absolute amount of stamen growth. For example, the mid- and long-level stamens of the S morph grow more than the short- and mid-level stamens of the L morph. Such differences could result from variations between morphs either in the time over which growth occurs or in absolute growth rates. Differences in amount of growth can be seen by comparing the length of the line describing relative stamen growth in each model. Such differences, however, cannot alone account for the differences between morphs in relative stamen position within a flower. In the descriptions below, we concentrate on the other factors that contribute to morph-specific differences. In Fig. 35A all stamen primordia are equal in size at initiation, and differences among morphs in mature stamen position result from different relative growth rates, as well as differences in amount of growth. In Fig. 35B differences among morphs arise at stamen initiation because of size differences between each stamen series in a flower. These differentials could result from the amount of growth that occurs prior to initiation of the second stamen level or through differences in the size of stamen primordia at origin. Relative growth rates are equal after stamen initiation, but the S morph grows more than the other morphs. This model predicts that the largest difference between primordia within a flower will occur in the M morph. In Fig. 35C stamen primordia are equal-sized at initiation and have similar relative growth rates during early development, but their relative growth rates diverge later in development. Figure 35D incorporates elements of the three previous models: differences are present between stamen levels at time of origin, but the size differential is the same in all three morphs. Relative growth rates are equal among morphs in early development but diverge subsequently. An interesting aspect of this model is that the L and S morphs differ only in the extent of relative growth, whereas the M morph requires a different relative growth rate at some point in development.

Figs. 25–30. Scanning electron micrographs of flower buds of the M (**Fig. 25**) and L (**Fig. 26**) morphs of *Lythrum salicaria*. Labels as in Figs. 11–16. **Fig. 25.** A bud with calyx removed to show the first- and second-initiated stamen series. The stamens from the first series are larger and higher on the floral tube than the stamens from the second series. The gynoecium has formed a meristematic ring and is beginning to be subdivided into two carpels. *Bar* 30 μm. **Fig. 26.** A bud with calyx removed to show the first- and second-initiated stamen series. Differences between stamen series are as in Fig. 25. The two carpels are distinct. *Bar* 30 μm. **Figs. 27–30.** Scanning electron micrographs of flower buds of *Oxalis rubra*. **Figs. 27** and **29** of M morph; **Figs. 28** and **30** of L morph. Labels as in Figs. 17–24. **Fig. 27.** A bud with sepals, petal primordia, and stamen primordia opposite the sepals but not the petals. *Bar* 20 μm. **Fig. 28.** A lateral view of a bud with sepals bent back to reveal petal primordia and two sets of stamens. The stamens opposite the sepals are larger than the stamens opposite the petals. Carpels have been initiated on a radius with the petals and second stamen series. *Bar* 20 μm. **Fig. 29.** An older bud with sepals removed. The stamens that alternate with the carpels are larger than those opposite the carpels. *Bar* 50 μm. **Fig. 30.** An older bud at a developmental stage comparable to the bud in Fig. 29, with similar differences between stamen levels. *Bar* 50 μm

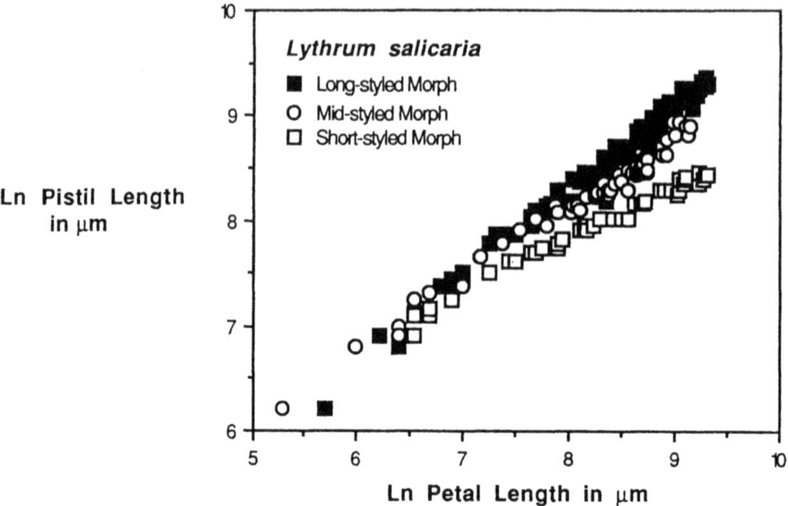

Fig. 31. Ln petal length vs. ln pistil length for the three morphs of *Lythrum salicaria*. The three morphs appear to have similar relative growth rates in early development but diverge in rate subsequently. (Data from Stirling 1933)

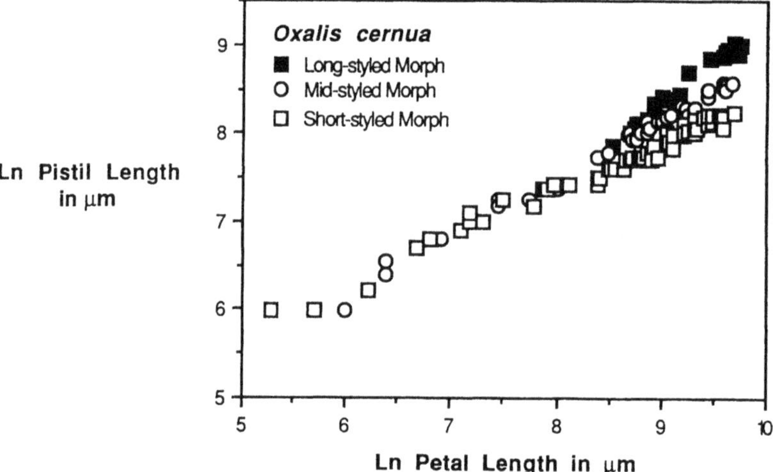

Fig. 32. Ln petal length vs. ln pistil length for the three morphs of *Oxalis cernua* (= *O. pes-caprae*). The data are consistent with a hypothesis of an early similarity but subsequent divergence in relative growth rates. (Data from Stirling 1936)

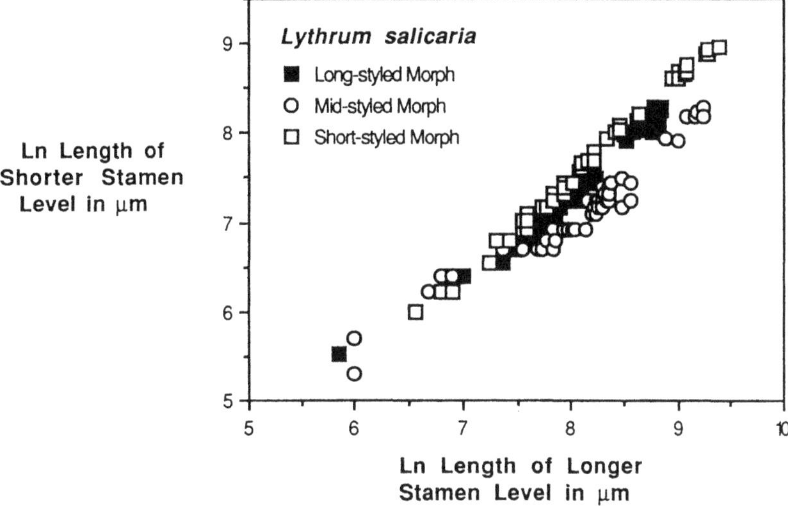

Fig. 33. Ln length of longer stamen level vs. ln length of shorter stamen level for the three morphs of *Lythrum salicaria*. The relative growth rates of all three morphs appear similar in young buds, although more data are needed. The L and S morphs have relatively similar relative growth rates, whereas the M morph diverges distinctly from these two. (Data from Stirling 1933)

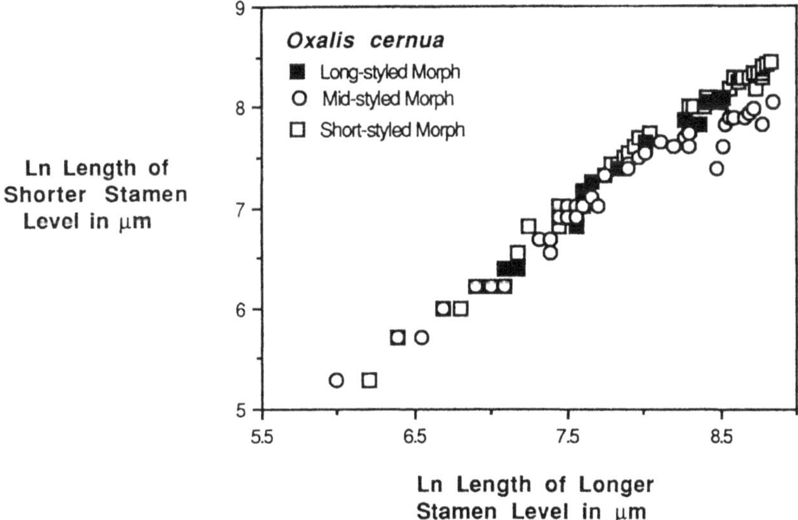

Fig. 34. Ln length of longer stamen level vs. ln length of shorter stamen level for the three morphs of *Oxalis cernua* (= *O. pes-caprae*). The data are consistent with the hypothesis that the three morphs have similar relative growth rates in early stamen development, and the L and S morphs continue this initial rate. The M morph, in contrast, shows a relatively late divergence in growth rate. (Data from Stirling 1936)

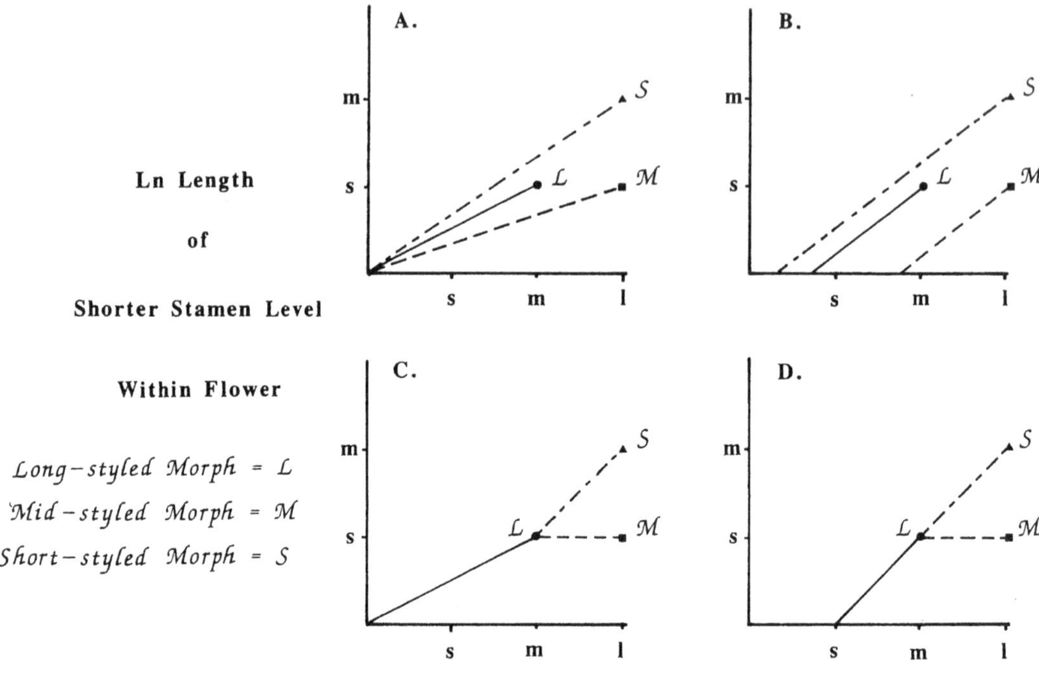

Fig. 35A-D. Four models for relative growth of the longer vs. the shorter stamen level in the three morphs of a tristylous species. **A** Stamen growth in all morphs begins with primordia that are equal in size. The two stamen levels within each flower have different growth rates, however, and the relative difference in growth rates varies among morphs, as does the amount of growth. **B** Stamen primordia from each level within a flower are different in size from initiation, and the degree of difference varies among morphs. Although relative growth rates are similar among morphs, mature organ sizes differ because of the initial differences in primordium size, as well as differences in amount of growth. This model predicts that the primordium size differential is greatest in the M morph. **C** Stamen primordia within a flower are similar sized in all morphs and have the same relative growth rate in early development. Differences among morphs arise because of late-developing changes in relative growth rates, as well as differences in amount of growth. **D** Stamen primordia from each level within a flower differ in size from origin, but the degree of difference is similar among morphs, and early relative growth rates are equal. Differences among morphs in mature stamen height arise because of late-developing changes in relative growth rates. In this case the L and S morphs have similar relative growth rates but differ in the amount of growth, whereas the M morph has a distinctly different relative growth rate in later developmental stages

Although these models are simple and do not include all possibilities, it is clear that none of the tristylous species examined to date follow the pattern found in Fig. 35A. Patterns of primordia initiation and early growth indicate that development of tristyly in the Pontederiaceae is most closely described by some variant of Fig. 35C. The development of tristyly in the Lythraceae and Oxalidaceae, in contrast, is better depicted by Fig. 35B or D. Both figures predict a near or complete overlap of plots for the L and S morphs, as seen in Stirling's data (Fig. 25, 26). These two figures indicate the importance of acquiring quantitative data on early growth in all three morphs to establish when and how divergence occurs.

Similar models, which compare relative growth rates of bud length to gynoecium length, can be constructed for gynoecium development (Fig. 36). Such models are less complex than stamen models, as gynoecium length varies only between morphs and mature flower length is similar in the three morphs. In Fig. 36A gynoecium primordia are similar in size at origin but subsequently differ in growth rates. Gynoecial primordia have similar lengths at origin and the same relative growth rates in Fig. 36B. In this model differences in style length develop because the primordia arise at different times, and thus the extent of growth varies among morphs. In Fig. 36C the primordia have different sizes at origin but subsequently grow at similar rates. Although there is no evidence for major differences in primordium size or time of gynoecium origin in tristylous species, these models could apply to the size or time of differentiation of the gynoecium into style and ovary. Our studies of the Pontederiaceae show no major differences among morphs in style size or bud length at stylar differentiation, but this possibility should be considered for other tristylous species.

In the last three models (Fig. 36D-F) gynoecium or style lengths are similar at origin, and early relative growth rates are equal among morphs. Divergence occurs later in development as a result of inhibition (Fig. 36D) or differences in relative growth rates (Fig. 36E). Many combinations of these latter two variables are possible in modeling the development of tristyly (e.g., Fig. 36F). Although in our models the initial relative growth rate produces long styles, this initial rate could result in the mid or short styles. The other style lengths could then arise from accelerated, as well as inhibited, relative growth rates. The underlying mechanisms for inhibition vs. growth rate reduction may not differ, since a reduction in growth rate is a type of inhibition. Figure 36D, however, implies a difference among morphs in the onset of inhibition, whereas in Fig. 36E changes in relative growth rate occur at the same point in development, but the morphs differ quantitatively either in amount of an inhibitor or in sensitivity of the style to an inhibitor.

The developmental data on style growth of tristylous species are best described by variants on these last three models. Style differences among morphs in the Pontederiaceae arise from differences in time of inhibition (Fig. 8 vs. Fig. 36D). In the Lythraceae and Oxalidaceae styles appear to diverge in growth rates (Fig. 31, 32 vs. Fig. 36E), although additional data on early development in both families are needed to confirm similarity among morphs in initial relative growth rates.

3.5 Development of Distyly

The data on tristyly show that the developmental processes responsible for tristyly differ among families and developmental patterns leading to different stamen levels can even differ within a species. The models of stamen and style relative growth rates (Figs. 35, 36) present additional possibilities. Because distyly occurs in many unrelated taxa and has diverse structural bases, as reviewed in Section 2.2, a similar diversity of developmental pathways should underlie the heteromorphisms of distylous species. The data to evaluate this hypothesis, however, does not currently exist.

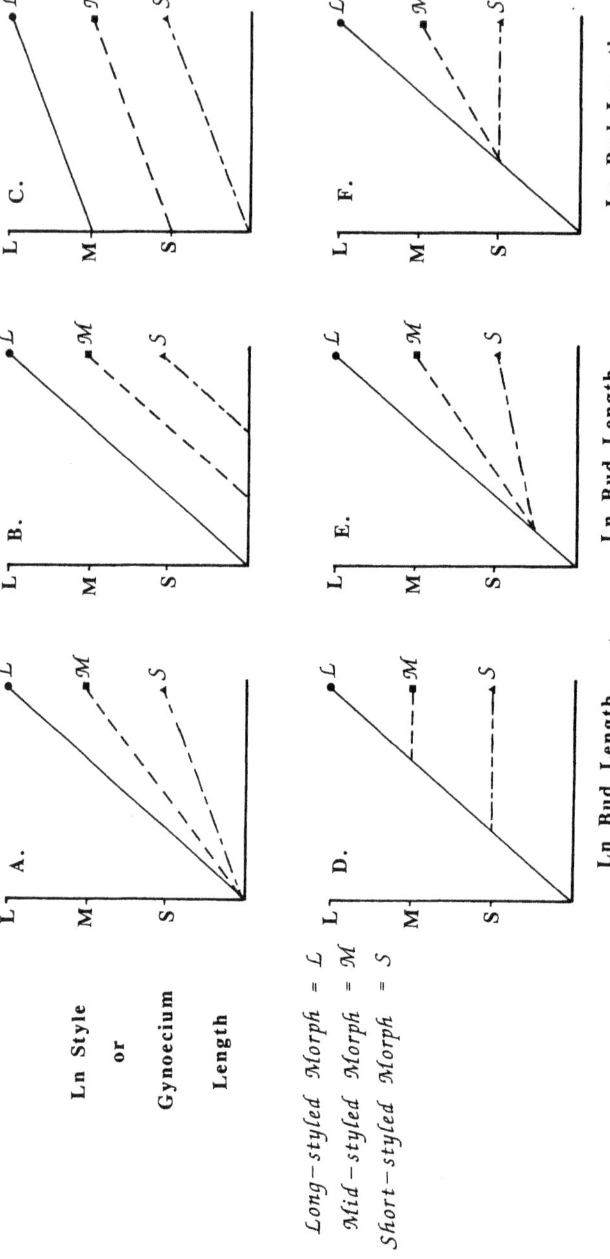

Fig. 36A-F. Six models for relative growth of bud length vs. style or gynoecium length for all morphs of a tristylous species. **A** The style or carpel primordia are equal in size at origin in the three morphs but have different relative growth rates. **B** The style or carpel primordia arise at different times in bud development, first in the L morph, then the M, and finally the S. Although the three morphs have similar gynoecial relative growth rates, the initial timing differences result in different stigma heights. **C** The style or carpel primordia arise at the same time in development in the three morphs, but primordium size at origin differs among morphs. Although gynoecium relative growth rates are the same among morphs, the initial size differences result in different stigma heights in the three morphs. **D** The style or carpel primordia have similar sizes at origin and similar relative growth rates in early development in the three morphs. Stylar variations among morphs result from differences in when style growth stops. **E** The style or carpel primordia are equal sized at initiation and have similar relative growth rates in early development. Relative growth rates diverge subsequently, however, and produce the different stigma heights. **F** This model combines elements of the previous two and provides one example of the many variations on these models that are possible. Style or carpel primordium size are similar among morphs at origin and in early development. Stigma height differences among morphs develop because of early inhibition of growth in the S morph and divergence in relative growth rate in the M morph

Stirling (1932, 1936) studied development of *Hottonia palustris* (Primulaceae), *Menyanthes trifoliata* (Menyanthaceae), and four distylous species of *Primula*. He showed that stamen and pistil growth rates differed between morphs in these taxa, but he did not provide sufficient data to analyze growth patterns. Differences between morphs in stamen or style cell number (Stirling 1932) and cell length (Dulberger 1975, Y. Heslop-Harrison et al. 1981) have been observed in distylous species. In both *Primula sinensis* and *Menyanthes trifoliata* divergence in organ lengths between morphs occurs after meiosis in anthers and ovules (Stirling 1932, 1936), as is found in the development of tristyly in the Pontederiaceae (Richards and Barrett 1984).

We have no evidence to indicate whether stamen and style primordium size or time of initiation differ between distylous morphs (see also the discussion of pollen number dimorphism in Sect. 2.3). If mature flower size is similar in the two morphs, relative growth rates of floral organs can be compared using bud length as a standard. Differences in growth rate, time of origin, size at origin, or time of inhibition (cf. Fig. 36) are possible causes for differences between stamen and style lengths in distylous species. Data on three distylous species of the Rubiaceae show early similarities in style length, followed by subsequent inhibition or reduction in growth rates (Fig. 37, JH Richards unpubl. data). Further comparisons of early development between morphs of distylous species and more complete studies of later stages of growth are needed to understand how the genes that control distyly operate in diverse families. Such data will also help to evaluate evolutionary hypotheses about the origin of the polymorphism (see Sect. 6).

4 Phenotypic Variation in Floral Heteromorphisms

4.1 General Considerations

The flower is usually considered less prone to nongenetic sources of variation than vegetative organs. Many floral traits not only remain relatively constant within individuals but also exhibit a stereotyped plan among populations, species and often families (Stebbins 1951; Berg 1959). Constancy of floral traits is typical of many animal-pollinated plants, where precise positioning of reproductive parts is required for effective pollination. While canalization of floral traits may be especially important in heterostylous species, where the reciprocal arrangement of stamens and styles promotes legitimate pollinations, no explicit comparisons of the patterns of phenotypic variation in heterostylous and nonheterostylous species have been made. We do not know whether floral constancy is, in fact, greater in heterostylous groups. Considerable phenotypic variation in stamen and style length has been reported in distylous *Turnera ulmifolia* (Martin 1965) and *Erythroxylum coca* (Ganders 1979b). How much of this variation is under genetic control and whether it influences legitimate pollination is not known.

Phenotypic variation of floral traits in heterostylous species is often more pronounced where the polymorphism is undergoing evolutionary modification, particularly towards increased self-fertilization through homostyle evolution (see Chap. 10). Two potential sources of phenotypic variation occur in these circumstances.

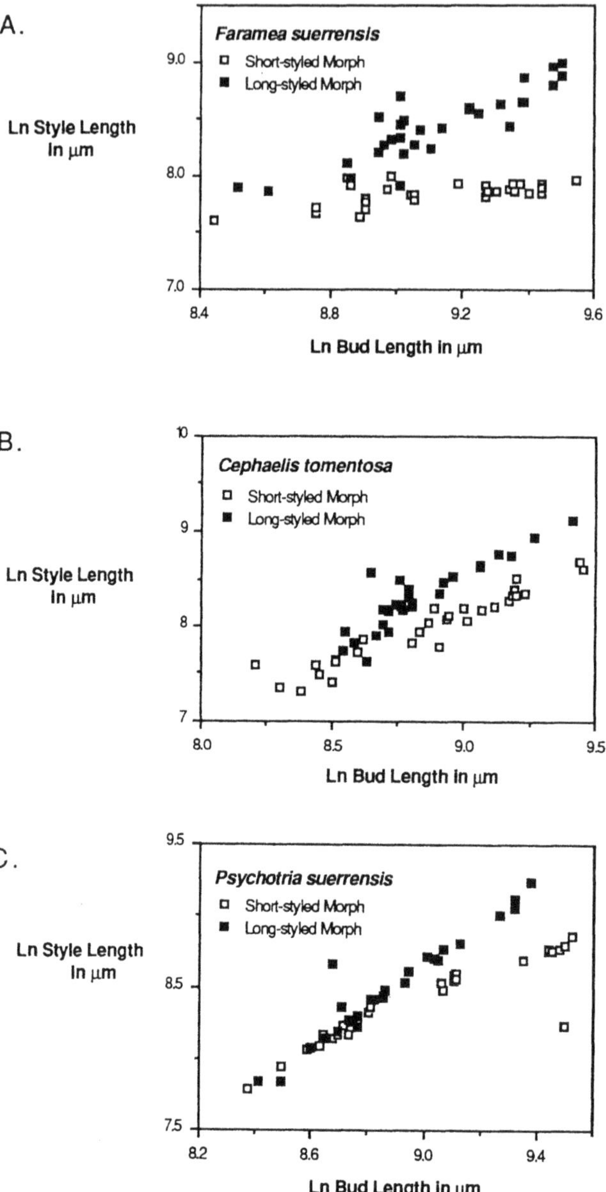

Fig. 37A-C. Ln bud length vs. ln style length for both morphs of three distylous species in the Rubiaceae. Flowers and buds were collected at La Selva, Costa Rica. **A** *Faramea suerrensis;* **B** *Cephaelis tomentosa;* **C** *Psychotria suerrensis*. Growth of the short-morph style of *F. suerrensis* is inhibited when compared to that of the long morph. The morphs of *C. tomentosa* and *P. suerrensis* have similar relative growth rates in early development, but in both species the S morph has a late reduction in stylar growth rate. (JH Richards unpubl. data)

Differences among individuals may result from genetic modifications of floral traits that influence the mating system. In homostylous varieties of *Turnera ulmifolia* variation in stigma-anther separation affects the rate of self-fertilization in natural populations (Barrett and Shore 1987). This variation is polygenically controlled (Shore and Barrett 1990). In contrast, several major genes alter stamen and style length in horticultural varieties of distylous *Primula sinensis*. These genes are nonallelic to the *S* locus and have pleiotropic effects on other aspects of floral phenotype, such as flower color and petal shape (Mather and DeWinton 1941).

A second source of phenotypic variation in floral traits occurs within the individual and is associated with increased inbreeding. Inbred lines of *Primula sinensis*, for example, demonstrate greater intraplant variance in stamen and style length in comparison with their F_1 generations (Mather 1950). A loss of canalization and a reduced ability to buffer against environmental stimuli and accidents of development has been associated with inbreeding and the attendant rise in homozygosity in a wide range of organisms (reviewed in Lerner 1954; Jinks and Mather 1955; Rendel 1959; Levin 1970).

4.2 Floral Variation in *Eichhornia paniculata*

Some of the most striking patterns of floral variation that have been documented in a heterostylous species occur in tristylous *Eichhornia paniculata*, where they are associated with the breakdown of floral trimorphism to monomorphism and the evolution of semi-homostyly. The variation results from both genetic and nongenetic causes and is manifested at a number of levels, including between populations, between genotypes within populations, and between flowers of individual plants. We have attempted to quantify this variation and to determine its genetic, environmental, and developmental basis. The remainder of this section reviews some of our results and discusses their relevance to the ecology and evolution of populations.

4.2.1 Floral Variation Among Populations

The evolutionary breakdown of tristyly in *E. paniculata* involves two important stages: (1) loss of the *S* allele and thus the S morph from populations; and (2) fixation of the *M* allele and hence loss of the L morph from populations. Population structure thus changes from floral trimorphism (L, M, S) through dimorphism (L, M) to monomorphism (M). The decrease in morph diversity is accompanied by the spread and fixation of selfing semi-homostylous variants of the M morph (Barrett 1985a; Barrett et al. 1989). The most significant floral modifications involve alterations in the height of short-level stamens of the M morph, which elongate to the level of the stigma, resulting in automatic self-pollination. Figure 38 illustrates the typical stamen and style configurations of trimorphic, dimorphic, and monomorphic populations of *E. paniculata*. In most dimorphic populations, particularly those in N.E. Brazil, modifications involve elongation of a single outer stamen of the short stamen level. In some monomorphic populations, however, especially in Jamaica, all three stamens of the short-level stamens are elongated to the mid-level position (Fig. 38).

Stamen modifications in the L and S morph of *E. paniculata* have been observed in natural populations. The variants rarely, however, establish successfully in nature to form monomorphic populations, as in the M morph. One exception involves a semi-homostylous L morph in Nicaragua (Barrett 1988). Why the M morph is apparently more susceptible to evolutionary modifications that favor increased self-fertilization is unclear. This situation does not appear to be restricted to *E. paniculata*, since modifications of the M morph are reported in other tristylous species (Stout 1925; Mayura Devi and Hashim 1964; Ornduff 1972; Barrett 1979; Barrett and Anderson 1985). The reason may be that the stigma is located between stamen levels in the M morph. As a result, developmental alterations of stamens or style in either direction will bring anthers and stigmas in proximity with each other. In the L and S morphs reproductively significant alterations can occur in only one direction. The M morph thus has twice the opportunity for such changes. Since alterations of both short-level stamens (*E. paniculata*) and long-level stamens (*E. crassipes*, Barrett 1979) have produced selfing variants of the M morph, both stamen levels are capable of developmental modification.

4.2.2 Floral Variation Within Morphs

An unusual feature of stamen modification in the M morph of *E. paniculata* is the discontinuous nature of the elongation patterns in the short-level stamens. The most common variant observed in N.E. Brazil exhibits a single stamen in the mid-level position (Fig. 39). The change in stamen position results primarily from filament elongation, although the modified stamen is also inserted higher on the floral tube (Fig. 40). The absence of continuous variation in filament length and, instead, the discrete nature of the alteration is consistent with either simple major gene control or polygenic control with a threshold response. Analysis of progeny variation from controlled crosses between modified and unmodified M genotypes from a Brazilian population (B3) are consistent with a model of single recessive gene control for the altered stamen (SCH Barrett unpubl. data). The gene apparently acts relatively late in floral development and appears to have only minimal effects on other facets of floral phenotype. Modification of filament length occurs through cell elongation, not cell division (JH Richards unpubl. data), and is manifested by rapid changes in filament length that occur primarily in the 24 h prior to anthesis (Fig. 41). It seems likely that expansion of filament length in the modified stamen is regulated by hormones, such as gibberellic acid. Hormones have been implicated in the regulation of reproductive organ size in other flowering plants (Greyson and Tepfer 1967; Pharis and King 1985; Koning 1983a,b; Jones and Koning 1986; Koning and Raab 1987).

Further modifications towards semi-homostyly in the M morph of *E. paniculata* involve elongation of the remaining two short-level stamens within a flower to the mid-level position. The genetic basis of these changes is unknown but presumably involves additional modifier genes that regulate positional effects within stamen levels. When a single stamen is modified in *E. paniculata*, it is always the stamen on the side of the flower away from the inflorescence branch (Fig. 39). The modified stamen is inserted on a member of the outer tepal whorl. In addition, although

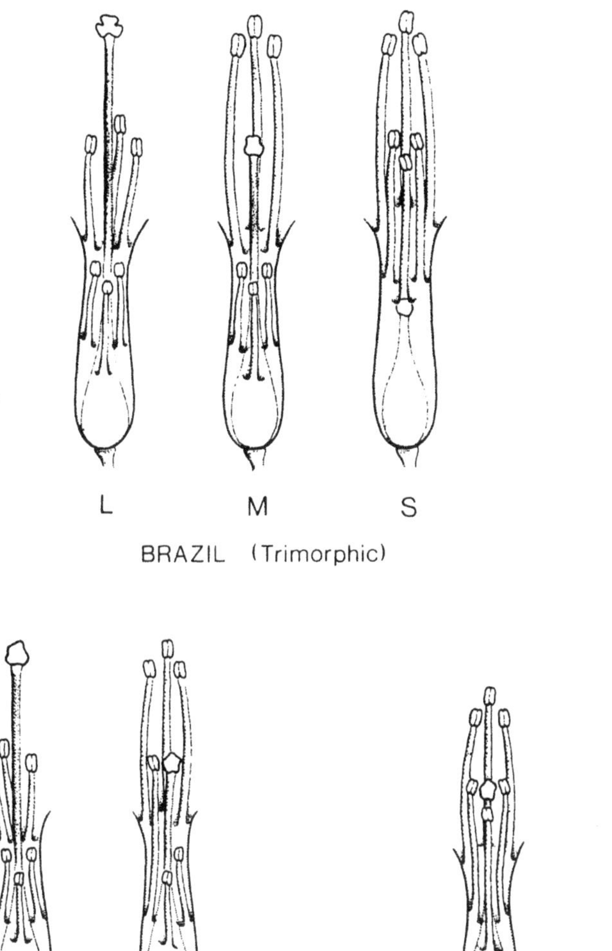

Fig. 38. Stamen and style configurations that accompany the evolutionary breakdown of tristyly to semi-homostyly in *Eichhornia paniculata*. Flowers of the L, M, and S morphs from an outcrossing trimorphic population (B5) are contrasted with two populations from Jamaica (dimorphic J15, monomorphic J3) that have different patterns of stamen modification. In the dimorphic population the L morph is largely unmodified whereas the M morph has a single stamen adjacent to the mid-level stigma. Complete semi-homostyly is evident in the monomorphic population with the three stamens of the "short-level" elongated to the mid position. (For further details see Barrett 1988)

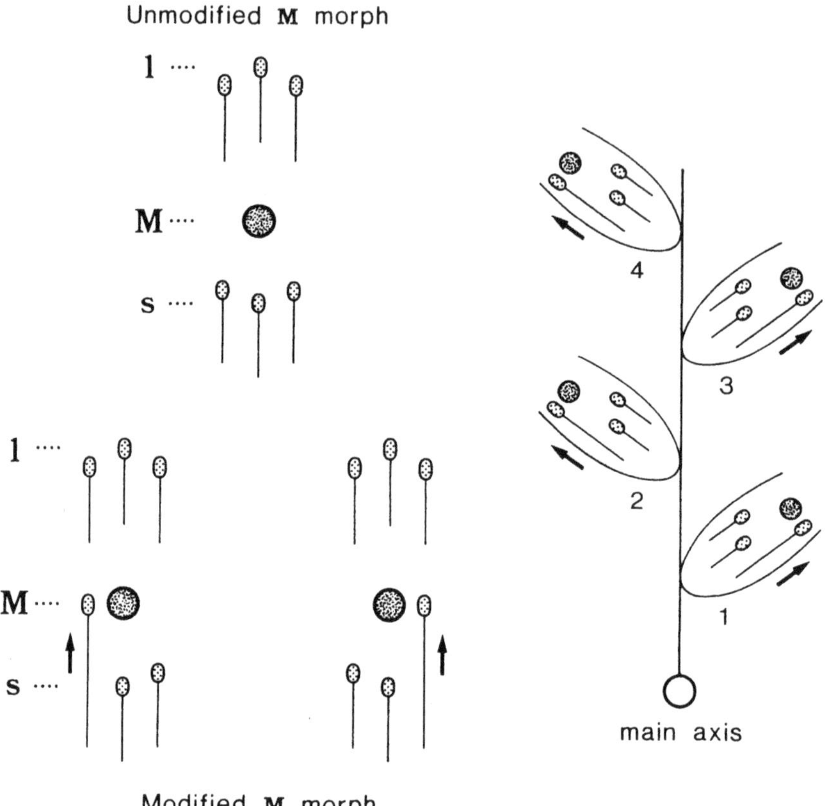

Fig. 39. Patterns of stamen modification in the M morph of *Eichhornia paniculata*. The diagrams illustrate the most common variant observed in Brazilian populations with a single short-level stamen adjacent to the mid-level stigma. Either the right or the left stamen of the short stamen level elongates, depending on the position of the bud on the inflorescence branch

within-series differences in timing of stamen initiation are not apparent in *E. paniculata*, they are present in *Pontederia cordata*, where this outer stamen primordium is the first to develop (Fig. 3). The modifier gene that causes the initial change to the outer stamen acts within this positional/developmental framework. The genes that cause subsequent stamen modification either alter this framework or change the sensitivity of the short-level stamens to these initial positional relationships.

The evolution of complete semi-homostyly in *E. paniculata* (Fig. 38), in which all three stamens are positioned close to the mid-level stigma, is accompanied by manifold changes in other aspects of floral phenotype, including reduction in the size and showiness of perianth parts, weakening of pollen heteromorphism, and decreases in pollen production and ovule number (Barrett 1985b). These changes in floral syndrome are most likely controlled by many additional genes with small effects. The significant point, however, is that the initial change in floral phenotype that precedes the evolution of the selfing syndrome appears to be under relatively

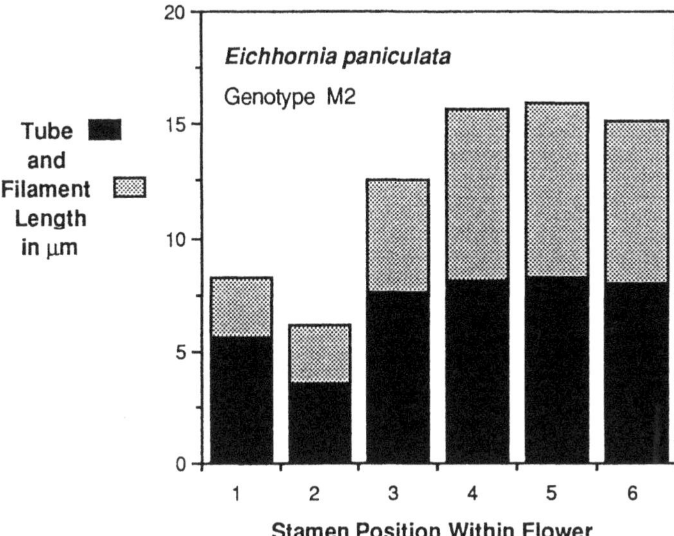

Fig. 40. Length of filament and insertion height (length of floral tube below the filament) in flowers of the M2 genotype (population B3, N.E. Brazil) of *Eichhornia paniculata*. In M2 genotype a single stamen from the short stamen level grows to the level of the mid-length stigma, causing self-pollination. The modified stamen position results primarily from increased free filament length. (JH Richards unpubl. data)

simple genetic control. This illustrates how simple genetic changes that affect floral morphology can alter plant mating systems. Such changes have profound influences on reproductive isolation, character divergence, and speciation (Gottlieb 1984; Barrett 1989).

4.2.3 Intraplant Floral Variation

Until now our discussion of stamen modification in the M morph of *E. paniculata* has involved consideration of phenotypic differences between individuals and populations. A curious feature of the genetic modifications in stamen position in *E. paniculata* is that in some modified plants not all flowers within an inflorescence exhibit the altered phenotype. As a result, inflorescences can be composed of both unmodified and modified flowers. Since the former are incapable but the latter are capable of autonomous self-pollination, genotypes can potentially produce a mixture of selfed and outcrossed seed. Intra-inflorescence variation is particularly evident in populations with both modified and unmodified plants and thus further contributes to a mixed mating system.

A second pattern of intraplant variation is found in primarily selfing populations where all plants possess modified flowers. In this case inflorescences can produce flowers with different numbers (one to three) of modified stamens and, at low

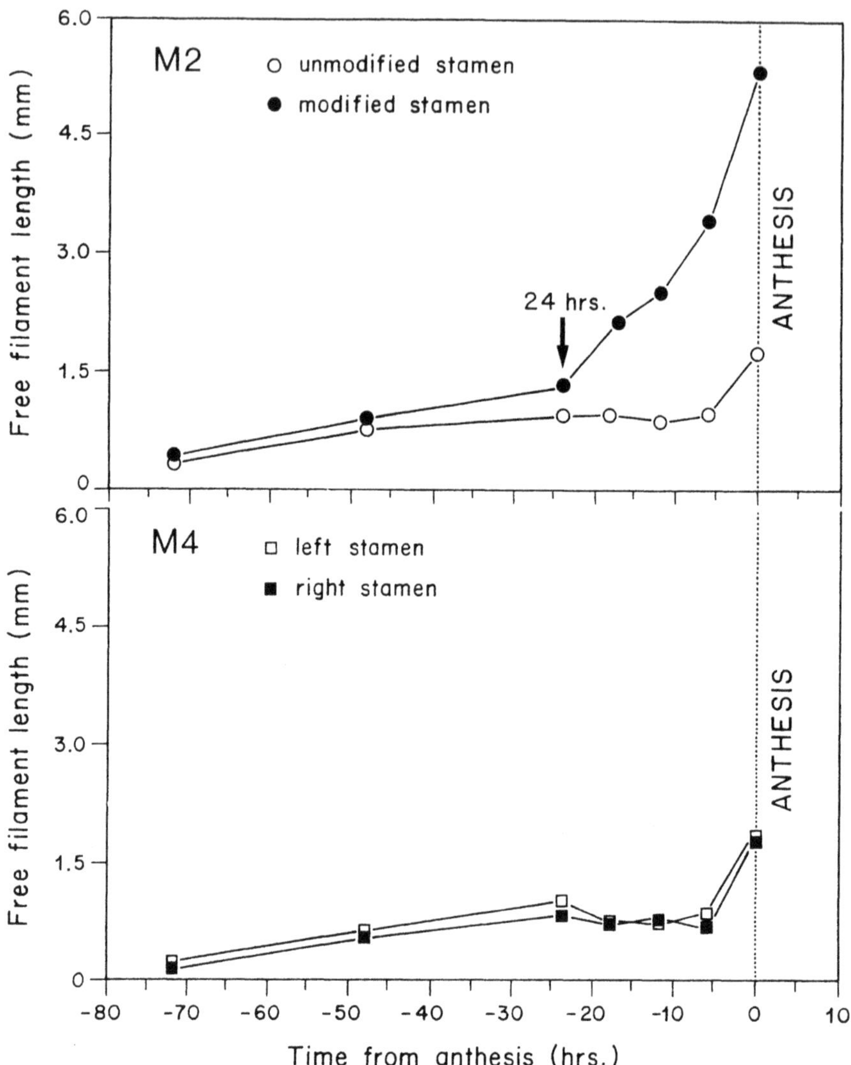

Fig. 41. Growth rate of the free filament of short-level stamens in two genotypes of the M morph of *Eichhornia paniculata* from a population (B3) in N.E. Brazil. Genotype M2 possesses modified stamens that cause autonomous self-pollination of mid-level stigmas. Genotype M4 is unmodified and incapable of autonomous self-pollination. Only one stamen, either the left or right short-level stamen, within each flower of genotype M2 is modified

frequency, unmodified flowers. This pattern of floral instability was first described among Jamaican populations of *E. paniculata* (Barrett 1985b) but is also found in dimorphic and monomorphic populations in N.E. Brazil (Glover and Barrett 1986). The variability in expression of short-stamen modification may result from incomplete penetrance of the modifier gene(s) in different genetic backgrounds, as well as from nongenetic effects associated with development.

Genotypes of *E. paniculata* vary in the frequency of flowers that display stamen modification. In addition, the frequency of modified flowers on successive inflorescences can fluctuate during the blooming period. Figure 42 illustrates these effects for twelve consecutive inflorescences produced by six genotypes of *E. paniculata*. The average frequency of unmodified flowers differs among genotypes (e.g., L9 versus M9) and the degree of instability also varies among genotypes (NA36 versus NA37). No clear pattern is evident with regard to the fluctuations in frequency of modified and unmodified flowers among the six genotypes, suggesting that variation may, in part, result from subtle microenvironmental influences, as well as from random accidents during the course of floral development.

In order to investigate the nature of stamen instability further, Barrett and Harder (1991) used statistical techniques involving logistic regression to examine whether modified flowers are produced at random within inflorescences of *E. paniculata*. The method considers a dichotomous response (unmodified vs. modified flowers) to a group of independent variables (e.g., branch position on an inflorescence, bud position on a branch). The results indicated that modified flowers are more likely to occur on branches positioned towards the distal end of the main inflorescence axis and at bud positions at the proximal end of branches. Flower expansion begins on proximal buds at the base of the inflorescence and proceeds acropetally on the inflorescence and from the base of a branch outwards. Essentially, open flowers occupy a cone that increases in height and then decreases as an inflorescence proceeds through flowering. Flowers most susceptible to modification are thus those that expand at the peak of inflorescence blooming, in the initial phases of fruit development, but when many buds have yet to open. These inflorescence position effects may, therefore, involve complex hormonal and/or nutritional interactions superimposed on a genetic background that allows modification.

The instability of the short stamen position in the M morph of *E. paniculata* does not appear to be the result of genome-wide homozygosity brought about by intense inbreeding. Genotypes NA36 and NA37, which have short-level stamen variation (Fig. 42), result from interpopulation crosses and are heterozygous at a large number of isozyme loci and presumably many other genes. More importantly, the variability in elongation of short-level stamens is not accompanied by increased variation of other floral traits. A multivariate comparison of 14 floral traits in modified and unmodified genotypes from a population in N.E. Brazil failed to detect any increase in intraplant variation associated with short stamen level modification (Seburn et al. 1990). For example, long-level stamens of plants displaying instability in the short-level stamens were as canalized in expression as long-level stamens from unmodified plants. These results suggest that the stamen modifications result from specific changes in the genetic control of short-level stamen variation, as opposed to generalized developmental instability brought about by increased homozygosity.

While developmental instability of the short-level stamens in the M morph of *E. paniculata* does not appear to be correlated with increased variation in other floral

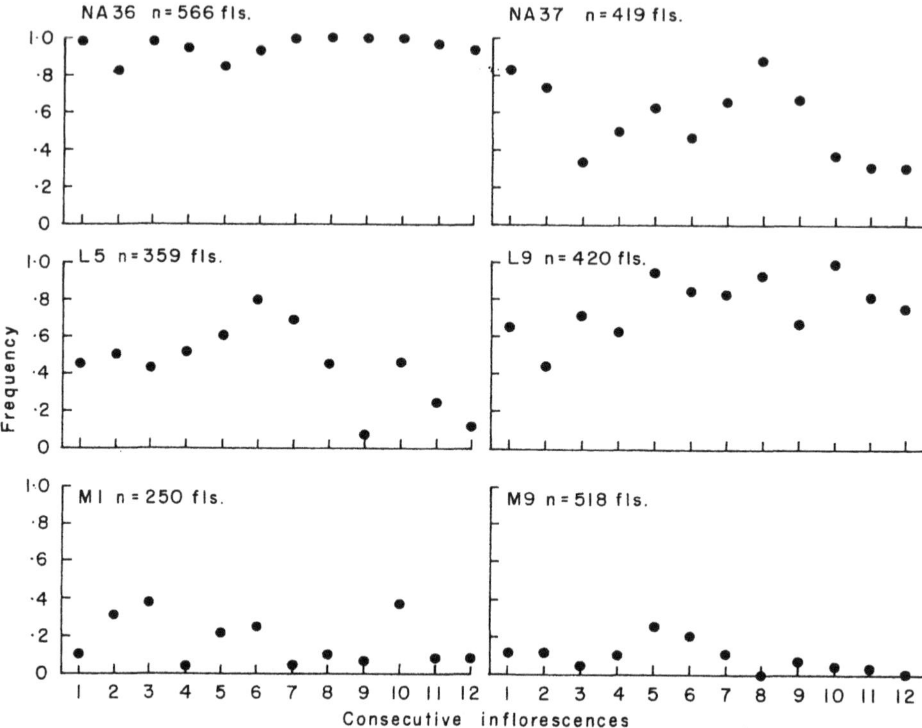

Fig. 42. Comparison of the frequency of unmodified flowers on 12 consecutive inflorescences of six genotypes of the M morph of *Eichhornia paniculata* grown under uniform glasshouse conditions. All flowers on inflorescences were scored as unmodified or modified depending on whether alterations to short-level stamens were evident. Genotypes NA36 and NA37 are F_1 crosses between trimorphic population B1 and monomorphic population J3, genotypes L5 and L9 are from monomorphic population B4, and genotypes M1 and M9 are from dimorphic population B3. See Barrett (1985b) for further details of populations, and Barrett and Harder (1991) for an analysis of developmental instability in the six genotypes

organs, a range of other developmental abnormalities are evident in natural populations with high levels of self-fertilization. These phenodeviants (Lerner 1954) include genotypes with twisted, fasciated, fused, or missing perianth parts, pollen sterility, disturbances in floral pigmentation, and the production of cleistogamous flowers. In most cases the abnormal flowers occur among normally developed flowers within inflorescences and are a relatively stable feature of particular genotypes. More rarely, genotypes are fixed for the abnormality and all flowers are modified. A survey of 16 populations of *E. paniculata* in N.E. Brazil and Jamaica, which had contrasting style morph structure and mating patterns, demonstrated significant differences in occurrence of tepal abnormalities both between regions and between populations within regions (Table 2). In N.E. Brazil populations had relatively few tepal abnormalities, particularly in trimorphic populations. In Jamaica, however, the incidence of tepal instability was significantly higher, and in two populations the majority of flowers

Table 2. A survey of abnormal tepal development in 16 populations of *Eichhornia paniculata* from N.E. Brazil and Jamaica. In each population all flowers open on a single day on one inflorescence of 25 plants were scored for normal or abnormal tepal expression. Abnormalities involved twisted, fused, or missing perianth parts (SCH Barrett, unpubl. data)

Population code	Style morph structure	No. of flowers surveyed	Frequency of abnormality (%)
N.E. Brazil			
B 42	Trimorphic	708	6.2
B 46	Trimorphic	388	20.9
B 56	Trimorphic	204	1.5
B 62	Trimorphic	318	3.8
B 135	Trimorphic	281	3.2
B 65	Dimorphic	169	13.6
B 69	Dimorphic	260	0
B 70	Dimorphic	348	10.0
B 114	Dimorphic	190	5.3
B 63	Monomorphic	200	3.0
			$\bar{x}$ - 6.75
Jamaica			
J 25	Dimorphic	172	17.0
J 17A	Monomorphic	237	72.0
J 17B	Monomorphic	200	82.0
J 18	Monomorphic	96	22.0
J 25	Monomorphic	111	15.0
			$\bar{x}$ - 41.6

exhibited abnormal development. Populations in Jamaica are highly self-fertilizing and abnormal tepal development may be selectively neutral on the island, since pollinators are unlikely to exert much influence on floral form.

5 Genes and the Development of Heterostyly

Heterostyly has evolved repeatedly from an ancestral condition of floral monomorphism (see Chap. 6). Accordingly, the evolution of heterostyly must have occurred through selection of genes that modify development of stamen and style lengths. The reciprocal positioning of stigmas and anthers is hypothesized to be controlled by relatively few loci in both distyly and tristyly. In distyly a single, tightly linked supergene with at least three loci, each with two alleles, segregates as a single gene (Dowrick 1956; Charlesworth and Charlesworth 1979; Muenchow 1981). The supergene controls stigma height, anther height, incompatibility reactions, and various ancillary characters, such as pollen size. In the three-locus supergene model advanced for *Primula* (see Chap. 5) one gene controls stylar morphology and incompatibility, one gene controls pollen size and incompatibility, and a third gene controls anther height (Dowrick 1956). Stamen structure and incompatibility, therefore, are thought to be controlled by separate genes, while style morphology and incompatibility are governed by a single gene.

In tristyly two loci, *S* and *M*, each with two alleles and *S* epistatic to *M*, control the levels of stamens and styles in each flower (Fisher and Mather 1943 for *Lythrum*; Fyfe 1950; Weller 1976b for *Oxalis*; and SCH Barrett unpubl. data for *Eichhornia*). Other loci, however, can affect the expression of the *S* and *M* loci (e.g., Bennett et al. 1986 in *Oxalis*). Charlesworth (1979) proposes that a third locus that creates a gradient of pollen reactions to stigma height is fixed in tristylous species. The existing data, therefore, indicate that relatively simple genetic systems control the inheritance of both distyly and tristyly. While the genetic systems involve a small number of major genes, they are also subject to modifier genes and epigenetic effects that can alter expression of the heterostyly syndrome (cf. Sect. 4).

The structural and developmental data reviewed in this chapter suggest that the genes governing heterostyly control different processes in diverse groups. As we investigate development in heterostylous groups, however, common developmental themes are likely to emerge. Surveys that correlate particular heterostylous developmental patterns with other floral characters, such as presence of an inferior ovary or distinct sepal and petal whorls, would help to clarify the nature of developmental constraints in heterostylous groups. Petal growth in tristylous *Lythrum* and *Oxalis*, for example, is delayed compared to stamen development, whereas in the Pontederiaceae tepal length always exceeds stamen length in early stages of growth. These differences in perianth development may have constrained the organization of tristyly in these families. Comparative studies of heterostylous taxa will provide insights into how ancestral floral structure and developmental patterns have influenced the evolution of distyly and tristyly.

Genetic studies have not yet answered the question of whether the genetic architecture of distyly and tristyly are similar. Although there is evidence for a heterostyly supergene in some distylous species, the genes controlling tristyly may not involve a supergene (Charlesworth 1979; but see Ganders 1979a). Charlesworth (1979) modeled the evolution of tristyly for a species with flowers that already had two anther whorls at different levels. Her model assumes that stigma height automatically determines both anther height and the stigma's incompatibility reaction. Anther height, in turn, determines the incompatibility reaction of the pollen produced. In this floral background tristyly could evolve via a distylous intermediate that has two anther sets. The relation of floral morphology to incompatibility reactions in this model differs from the distyly supergene model, which hypothesizes separate control of anther height and pollen incompatibility.

The contrast between models for the genetic architecture of distyly and tristyly cautions against generalizing from the limited genetic data available. Although the reported inheritance patterns are similar among distylous species, the organization of genes that control the syndrome may differ in these species. Perhaps some distylous species have supergenes, while others have a system that resembles the distylous ancestor in Charlesworth's (1979) tristyly model. Without more data on both distylous and tristylous genetic systems, it is premature to assume that the supergene model describes all distylous species or that the model applies to tristyly.

The available genetic and developmental evidence suggests that distyly is a less complex system than tristyly. The reciprocal herkogamy (Webb and Lloyd 1986) in distylous species can be achieved by quantitative variation of existing processes.

Variation in amount of cell division or elongation or in duration of growth in stamens or styles could lead to the primary differences seen between distylous morphs. The genes that control distyly must regulate these processes. Developmental studies are required to define when and how regulation occurs, as well as to elucidate the diversity of gene action that is likely to occur among the different distylous families.

In contrast to distyly, the development of tristyly requires more than quantitative variation of existing processes in order to develop the three morphs. A dissociation between development of the two stamen levels within a morph had to arise at some point in a tristylous species' evolutionary history. In the radially organized tristylous dicotyledons this dissociation is seen between stamen development in the M morph as compared to the L and S morphs. The relative relation between stamen levels in the S and L morphs is essentially the same, but S morph stamens experience more growth than L morph stamens. The M morph differs from these two in that growth rates of the stamens relative to each other are changed. When considered at the level of the genes controlling tristyly, it can be seen that the *S* locus affects development of both stamen levels, whereas the *M* locus affects growth of a single level.

The dissociation between stamen levels in the dorsiventrally organized Pontederiaceae appears to have occurred prior to the evolution of tristyly. The data on stamen development of the S and L morphs show that the S morph is not a scaled-up version of the L morph but instead has qualitatively different relative growth curves. The *S* locus in the Pontederiaceae primarily affects the upper, shorter stamen level, causing a late-developing elongation, while the *M* locus controls growth of the lower, longer stamen level. Additional evidence for dissociability of growth between stamen levels comes from a study of stamen modification in the mid-styled morph of *Eichhornia paniculata* (Seburn et al. 1990), which shows that variation in length of short-level stamens occurs independently of variation in long-level stamen length.

A precondition for the evolution of tristyly is within-flower stamen dimorphism. This dimorphism can result from the presence of two stamen whorls, as in *Lythrum* and *Oxalis*, or from positional effects related to the horizontal orientation of flowers, as is likely to have characterized the ancestors of tristylous Pontederiaceae. The *S* and *M* loci, which act in this floral background, have distinct effects. The differences between them are most clearly seen in their control of intraflower stamen dimorphism. Our data on tristyly clarify the basic difference between the genes in *Lythrum* and *Oxalis*, which act radially, and those in the Pontederiaceae, which act dorsiventrally. Whether such fundamental differences occur among distylous species remains to be shown.

6 Development and the Evolution of Heterostyly

Models for the evolution of heterostyly differ in (1) what is assumed to be the ancestral condition; and (2) whether the morphological or physiological aspects of the heterostyly syndrome arose first. With the exception of Charlesworth's (1979) model for tristyly, discussed above, these models primarily describe the evolution of distyly. In D Charlesworth and B Charlesworth's (1979) model for the evolution of distyly the ancestral flower was monomorphic with stigma and anthers at the same level within a flower. The model assumes that diallelic self-incompatibility arose first. The morphological characteristics of distyly evolved subsequently through selection pressures to avoid self-fertilization and inbreeding depression. Baker (1966) proposed that distyly evolved in this sequence in the Plumbaginaceae.

Ganders (1979a) criticized the Charlesworths' model for assuming that the ancestral flower was monomorphic with stigma and style at the same level. He argued that most self-compatible, monomorphic species show some degree of stigma-anther separation. Species with no stigma-anther separation regularly self-pollinate and, thus, be unlikely to experience the levels of inbreeding depression required by their models. Ganders (1979a) agreed, however, with the general sequence of events outlined in the Charlesworths' model.

An alternative hypothesis is that the morphological characteristics of the syndrome evolved first and the physiological self-incompatibility arose secondarily. Darwin (1888) first proposed this sequence. Anderson (1973) hypothesized the same pathway to account for the repeated origin of distyly in the Rubiaceae. Lloyd and Webb in Chaps. 6 and 7 propose a more general version for this sequence. They hypothesize that in most cases the ancestral form was a flower with an exserted style (approach herkogamy (Webb and Lloyd 1986)) and that physiological incompatibility arose after the establishment of reciprocal herkogamy.

Developmental studies can be used to test specific hypotheses concerning the ancestral floral form of different heterostylous groups. For example, a form of protandry in which the anthers dehisce prior to stylar elongation and maturation is common in the Rubiaceae (Verdcourt 1958). Anderson (1973) proposed that distyly in the Rubiaceae evolved after mutations preventing stylar elongation arose in a population. The developmental predictions of this proposal are that (1) development of the long- and short-styled morphs of distylous Rubiaceae resemble two stages in the development of protandrous taxa; (2) the long- and short-styled morphs follow similar pathways in early development; and (3) divergence occurs through inhibition of the short style (cf. Fig. 36D,E). These predictions can be evaluated through comparative studies of floral development within distylous taxa and between distylous and protandrous species. Similar studies of distylous species and relatives with approach herkogamy could be used to examine Lloyd and Webb's hypotheses on the origins of heterostyly (see Chaps. 6 and 7).

The interrelationship of reciprocal herkogamy, physiological incompatibility, and the ancillary characters, such as stigma and pollen heteromorphisms, is central to understanding the sequence of events in the evolution of heterostyly. If any of these characters are expressions of the same developmental process, then they are likely to have had a simultaneous origin. Dulberger (1975a) argued that both the morphological and physiological components of heterostyly result from a single phenomenon, which she hypothesized was a difference in growth rates among morphs. The growth rate difference caused pollen size heteromorphisms that, in turn, resulted in differential growth of stamens and styles (Dulberger 1975a).

Arguments have also been made for associations between (1) the functioning of incompatibility and the ancillary characters (Dulberger 1974, 1975b); (2) stamen and style lengths and some ancillary characters (Dulberger 1975a); and (3) incompatibility and stamen and/or style lengths (Mather and DeWinton 1941; D Charlesworth and B Charlesworth 1979). Lloyd and Webb (Chap. 6) hypothesize that incompatibility develops as a response to the different stylar environments provided by the previously evolved reciprocal herkogamy. The interrelationships of these features of the heterostyly syndrome are problematic in part because we lack detailed developmental information on which to build hypotheses. The need for additional evidence is especially important for determining the evolutionary sequence in which reciprocal herkogamy and incompatibility are established. Studies of self-compatible heterostylous groups, such as *Amsinckia* and *Eichhornia*, would be particularly useful to determine whether reciprocal herkogamy imposes functional constraints on pollen-pistil interactions, as implied by the Lloyd and Webb model (see Chaps. 6 and 7).

7 Future Research

Heterostyly provides a unique opportunity for plant biologists to study development as an evolutionary process. This breeding system in polyphyletic, controlled by relatively few major genes, and has a direct effect on the evolution of populations through its influence on mating patterns. In addition, variability in expression of the syndrome occurs both within and between taxa, providing systems in which to study the developmental basis of differences in expression of the syndrome.

In this chapter we have reviewed developmental information on heterostyly and drawn attention to the lack of such data in comparison with the wealth of genetic and ecological work on heterostylous species. We have also formalized models that may help to analyze specific developmental problems in both distylous and tristylous species. The utility of detailed structural and developmental studies in understanding genetic and evolutionary change is seen in our studies of the Pontederiaceae. Research on other heterostylous species that analyzes the developmental correlations among reciprocal herkogamy, heteromorphic incompatibility, and the ancillary characters are needed to understand the functions of these characters.

Distylous species offer a wealth of problems for future developmental research. Quantitative studies of stamen and style growth that establish when and how the morphs diverge are needed to evaluate developmental and genetic relations between morphs and between distylous species and nonheterostylous relatives. Developmental studies will help us to understand what the distyly genes control and how the system has evolved. Studies that establish the developmental basis for ancillary characters, such as pollen size and number dimorphism, are needed to understand their relationship to the primary characters and to evaluate hypotheses about their function. For example, do pollen-size dimorphisms and differences in pollen number arise at the same time in development? Finally, comparisons among distylous species will tell us whether the genes controlling development of distyly operate similarly in different species.

Similar developmental comparisons are needed in tristylous species. Tristyly has the additional developmental problem of the relation of intrafloral stamen dimorphism to the evolution of the heterostyly syndrome. Developmental studies of monomorphic taxa that have two stamen series are needed to establish the effect that time of stamen origin has on developmental events, and thus to understand the development of tristyly in Lythraceae and Oxalidaceae. For example, meiosis occurs at different times in the two anther levels in flowers of *Oxalis* (JH Richards unpubl. data), where intraflower stamen dimorphisms are present at stamen origin. Does timing of meiosis differ between anther series in monomorphic flowers with two stamen series? Does pollen size and/or number vary between anther series of such species in directions that would provide a basis for selection of the particular associations of these characters found in tristylous species? By answering such questions, we will discover the developmental rules (cf. Oster et al. 1988) that underlie evolution of the tristylous syndrome.

Finally, heterostylous groups provide useful experimental tools for integrated studies involving development, genetics and physiology. Quantitative genetic analysis of developmental processes (Cheverud et al. 1983; Atchley 1984) and the determination of phenotypic and genetic correlations of floral traits (Falconer 1981; Shore and Barrett 1990) would help to assess the types of constraints that are involved in the build-up and breakdown of the heterostyly polymorphism. Such data would also shed light on the patterns of developmental integration among components of the syndrome. Work on the effects of growth hormone application on floral organ development, accompanied by determinations of endogenous hormone levels, would help to elucidate the physiological events that lead to differences between mature heterostylous flowers. Comparative studies of the physiology and development of pollen, stamen and style variants, all of which are reported in heterostylous groups, may help to understand the complex processes that underlie the basis for variation in floral form in heterostylous species. Such data may provide more general models for the evolution of floral diversity in the angiosperms.

Acknowledgments. We thank P.K. Diggle for constructive comments on the manuscript, Elizabeth Campolin for drawing several of the figures, and National Science Foundation grant #DCB-8602869 (JHR) and the Natural Sciences & Engineering Research Council of Canada (SCHB) for financial support.

References

Anderson WR (1973) A morphological hypothesis for the origin of heterostyly in the Rubiaceae. Taxon 22:537–542

Atchley WR (1984) Ontogeny, timing of development, and genetic variance-covariance structure. Am Nat 123:519–540

Baker HG (1962) Heterostyly in the Connaraceae with special reference to *Byrsocarpus coccineus*. Bot Gaz 206–211

Baker HG (1966) The evolution, functioning and breakdown of heteromorphic incompatibility systems. I. The Plumbaginaceae. Evolution 20:349–368

Barrett SCH (1978) Floral biology of *Eichhornia azurea* (Swartz) Kunth (Pontederiaceae). Aquat Bot 5:217–228

Barrett SCH (1979) The evolutionary breakdown of tristyly in *Eichhornia crassipes* (Mart.) Solms (water hyacinth). Evolution 33:499–510

Barrett SCH (1985a) Ecological genetics of breakdown in tristyly. In: Haeck J, Woldendorp JW (eds) Structure and Functioning of Plant Populations II: Phenotypic and genotypic variation in plant populations. North-Holland, Amsterdam, The Netherlands, pp 267–275

Barrett SCH (1985b) Floral trimorphism and monomorphism in continental and island populations of *Eichhornia paniculata* (Spreng.) Solms. (Pontederiaceae). Biol J Linn Soc 25:41–60
Barrett SCH (1988) Evolution of breeding systems in *Eichhornia* (Pontederiaceae): a review. Ann MO Bot Gard 75:741–760
Barrett SCH (1989) Mating system evolution and speciation in heterostylous plants. In: Otte D, Endler J (eds) Speciation and its consequences. Sinauer, Sunderland MA, pp 257–283
Barrett SCH, Anderson JM (1985) Variation in expression of trimorphic incompatibility in *Pontederia cordata* L. (Pontederiaceae). Theor Appl Genet 70:355–362
Barrett SCH, Harder LD (1991) Floral variation in *Eichhornia paniculata* (Spreng.) Solms Pontederiaceae II. Effects of development and environment on the formation of selfing flowers. J Evol Biol (in press)
Barrett SCH, Shore JS (1987) Variation and evolution of breeding systems in the *Turnera ulmifolia* L. complex (Turneraceae). Evolution 41:340–354
Barrett SCH, Morgan MT, Husband B (1989) Dissolution of a complex genetic polymorphism: the evolution of self-fertilization in tristylous *Eichhornia paniculata* (Pontederiaceae). Evolution 43:1398–1416
Bawa KS, Beach JH (1983) Self-incompatibility systems in the Rubiaceae of a tropical lowland wet forest. Am J Bot 70:1281–1288
Bennett JH, Leach CR, Goodwins IR (1986) The inheritance of style length in *Oxalis rosea*. Heredity 56:393–396
Berg RL (1959) A general evolutionary principle underlying the origin of developmental homeostasis. Am Nat 93:103–105
Charlesworth D (1979) The evolution and breakdown of tristyly. Evolution 33:486–498
Charlesworth D, Charlesworth B (1979) A model for the evolution of distyly. Am Nat 114:467–498
Cheung M, Sattler R (1967) Early floral development of *Lythrum salicaria*. Can J Bot 45:1609–1618
Cheverud J, Rutledge J, Atchley WR (1983) Quantitative genetics of development: genetic correlations among age-specific trait values and the evolution of ontogeny. Evolution 37:895–905
Darwin C (1888) The different forms of flowers on plants of the same species. Reprinted in 1986 by Univ Chicago Press, Chicago
Dowrick VPJ (1956) Heterostyly and homostyly in *Primula obconica*. Heredity 10:219–236
Dulberger R (1974) Structural dimorphism of stigmatic papillae in distylous *Linum* species. Am J Bot 61:238–243
Dulberger R (1975a) S-gene action and the significance of characters in the heterostylous syndrome. Heredity 35:407–415
Dulberger R (1975b) Intermorph structural differences between stigmatic papillae and pollen grains in relation to incompatibility in Plumbaginaceae. Proc R Soc Lond Ser B 188:257–274
Eckenwalder JE, Barrett SCH (1986) Phylogenetic systematics of Pontederiaceae. Syst Bot 11:373–391
Falconer DS (1981) Introduction to quantitative genetics. Oliver and Boyd, Lond
Feinsinger P, Busby WH (1987) Pollen carryover: experimental comparisons between morphs of *Palicourea lasiorrachis* (Rubiaceae), a distylous, bird-pollinated, tropical treelet. Oecologia 73:231–235
Fisher RA, Mather K (1943) The inheritance of style length in *Lythrum salicaria*. Ann Eugenics 12:1–23
Fyfe VC (1950) The genetics of tristyly in *Oxalis valdiviensis*. Heredity 4:365–371
Ganders FR (1975) Heterostyly, homostyly, and fecundity in *Amsinckia spectabilis* (Boraginaceae). Madroño 23:56–62
Ganders FR (1976) Pollen flow in distylous populations of *Amsinckia* (Boraginaceae). Can J Bot 54:2530–2535
Ganders FR (1979a) The biology of heterostyly. NZ J Bot 17:607–635
Ganders FR (1979b) Heterostyly in *Erythroxylum coca* (Erythroxylaceae). Bot J Linn Soc 78:11–20
Glover DE, Barrett SCH (1986) Variation in the mating system of *Eichhornia paniculata* (Spreng.) Solms. (Pontederiaceae). Evolution 40:1122–1131
Gottlieb LD (1984) Genetics and morphological evolution in plants. Am Nat 123:681–709
Greyson RI, Tepfer SS (1967) Emasculation effects on the stamen filament of *Nigella hispanica* and their partial reversal by gibberellic acid. Am J Bot 54:971–976

Heslop-Harrison Y, Heslop-Harrison J, Shivanna KR (1981) Heterostyly in *Primula*. 1. Fine-structural and cytochemical features of the stigma and style in *Primula vulgaris* Huds. Protoplasma 107:1851–1871

Hufford LD (1988) Roles of early ontogenetic modifications in the evolution of floral form of *Eucnide* (Loasaceae). Bot Jahrb Syst 109:289–333

Jinks JL, Mather K (1955) Stability in development of heterozygotes and homozygotes. Proc R Soc Lond Ser B 143:561–578

Jones LS, Koning RE (1986) Role of growth substances in the filament growth of *Fuchsia hybrida* cv. "Brilliant". Am J Bot 73:1503–1508

Koning RE (1983a) The roles of auxin, ethylene, and acid growth in filament elongation in *Gaillardia grandiflora* (Asteraceae). Am J Bot 70:602–610

Koning RE(1983b) The roles of plant hormones in style and stigma growth in *Gaillardia grandiflora* (Asteraceae). Am J Bot 70:978–986

Koning RE, Raab MM (1987) Parameters of filament elongation in *Ipomoea nil* (Convolvulaceae). Am J Bot 74:510–516

Lerner IM (1954) Genetic homeostasis. Oliver and Boyd, Lond

Levin DA (1970) Developmental instability and evolution in peripheral isolates. Am Nat 104:343–353

Lewis D (1982) Incompatibility, stamen movement and pollen economy in a heterostyled tropical forest tree, *Cratoxylum formosum* (Guttiferae). Proc R Soc Lond Ser B 214:273–283

Lowden R (1973) Revision of the genus *Pontederia* L. Rhododa 75:426–487

Martin FW (1965) Distyly and incompatibility in *Turnera ulmifolia*. Bull Torrey Bot Club 92:185–192

Martin FW (1966) Distyly, self-incompatibility, and evolution in *Melochia*. J Arnold Arbor Harv. Univ 47:60–74

Mather K (1950) The genetical architecture of heterostyly in *Primula sinensis*. Evolution 4:340–352

Mather K, DeWinton D (1941) Adaptation and counter-adaptation of the breeding system in *Primula*. Ann Bot 5:297–311

Mayura Devi P, Hashim M (1964) Homostyly in heterostyled *Biophytum sensitivum* DC. J Genet 59:245–249

Mayr B (1969) Ontogenetische Studien an Myrtales-Blüten. Bot Jahrb 89:210–271

Muenchow G (1981) An S-locus model for the distyly supergene. Am Nat 118:756–760

Opler PA, Baker HG, Frankie GW (1975) Reproductive biology of some Costa Rican *Cordia* species (Boraginaceae). Biotropica 7:234–247

Ornduff R (1964) The breeding system of *Oxalis suksdorfii*. Am J Bot 51:307–314

Ornduff R (1971) The reproductive system of *Jepsonia heterandra*. Evolution 25:300–311

Ornduff R (1972) The breakdown of trimorphic incompatibility in *Oxalis* section Corniculatae. Evolution 26:52–65

Ornduff R (1975) Heterostyly and pollen flow in *Hypericum aegypticum* (Guttiferae). Bot J Linn Soc 71:51–57

Ornduff R (1979) The morphological nature of distyly in *Lythrum* section Euhyssopifolia. Bull Torrey Bot Club 106:4–8

Ornduff R (1980) Heterostyly, population composition and pollen flow in *Hedyotis caerulea*. Am J Bot 67:95–103

Oster GF, Shubin N, Murray JD, Alberch P (1988) Evolution and morphogenetic rules: The shape of the vertebrate limb in ontogeny and phylogeny. Evolution 42:862–884

Pharis RP, King RW (1985) Gibberellins and reproductive development in seed plants. Annu Rev Plant Physiol 36:517–568

Price SD, Barrett SCH (1982) Tristyly in *Pontederia cordata* (Pontederiaceae). Can J Bot 60:897–905

Rendel JM (1959) Canalization of the scute phenotype in *Drosophila*. Evolution 13:425–439

Richards AJ (1986) Plant breeding systems. Allen and Unwin, Lond

Richards JH, Barrett SCH (1984) The developmental basis of tristyly in *Eichhornia paniculata* (Pontederiaceae). Am J Bot 71:1347–1363

Richards JH, Barrett SCH (1987) Development of tristyly in *Pontederia cordata* (Pontederiaceae). I. Mature floral structure and patterns of relative growth of reproductive organs. Am J Bot 74:1831–1841

Sattler R (1973) Organogenesis of flowers. Univ Toronto Press, Toronto

Seburn CNL, Dickinson TD, Barrett SCH (1990) Floral variation in *Eichhornia paniculata* (Spreng.) Solms (Pontederiaceae): I. Instability of stamen position in genotypes from northeast Brazil. J Evol Biol 3:1–21
Shore JS, Barrett SCH (1990) Quantitative genetics of floral characters in homostylous *Turnera ulmifolia* var. *angustifolia* Willd. (Turneraceae). Heredity 64:105–112
Stebbins GL (1951) Natural selection and the differentiation of angiosperm families. Evolution 5:299–324
Stirling J (1932) Studies of flowering in heterostyled and allied species. Part I. The Primulaceae. Publ Hartley Bot Lab 8:3–42
Stirling J (1933) Studies of flowering in heterostyled and allied species. Part II. The Lythraceae: *Lythrum salicaria*. Publ Hartley Bot Lab 10:3–24
Stirling J (1936) Studies of flowering in heterostyled and allied species. Part III. Gentianaceae, Lythraceae, Oxalidaceae. Publ Hartley Bot Lab 15:1–24
Stout AB (1925) Studies on *Lythrum salicaria* II. A new form of flower in the species. Bull Torrey Bot Club 52:81–85
Tucker SC (1989) Overlapping organ initiation and common primordia in flowers of *Pisum sativum* (Leguminosae: Papilionoideae). Am J Bot 76:714–729
Verdcourt B (1958) Remarks on the classification of the Rubiaceae. Bull Jard Bot Etat Brux 28:209–281
Vuilleumier B (1967) The origin and evolutionary development of heterostyly in the angiosperms. Evolution 21:210–226
Webb CJ, Lloyd DG (1986) The avoidance of interference between the presentation of pollen and stigmas in angiosperms. II. Herkogamy. NZ J Bot 24:163–178
Weller SG (1976a) Breeding system polymorphism in a heterostylous species. Evolution 30:442–454
Weller SG (1976b) The genetic control of tristyly in *Oxalis* section Ionoxalis. Heredity 37:387–392
Yeo PF (1975) Some aspects of heterostyly. New Phytol 75:147–153

Chapter 5

The Genetics of Heterostyly

D. LEWIS[1] and D.A. JONES[2]

1 Introduction

Breeding experiments on the genetics of heterostyly began in the 19th century led by Hildebrand (1866) and by Darwin (1877), before Mendel's work was rediscovered in 1900. Bateson and Gregory (1905) introduced their paper on the inheritance of heterostylism in *Primula* with the statement: "In view of the results obtained by Darwin, Hildebrand and others, it seemed likely that the characters long-style and short-style, well known in Primulaceae and other orders, might have a Mendelian inheritance. Our experiments have shown that this is the case in *P. sinensis*, the short style being dominant, the long recessive."

Later, Lady Barlow (1913, 1923), in her papers on tristylic *Lythrum salicaria*, wrote: "In trimorphic species, Darwin's long-styled plants gave only long-styled offspring; I have self-fertilized long-styled plants of both *Oxalis valdiviana* and *Lythrum salicaria* and have always obtained long-styled offspring only. The long-styled form is therefore presumably hypostatic (recessive) and homozygous." She concludes: "The final conclusion of this paper must be the unsatisfactory one that the only scheme so far suggested by no means fits all the facts. The Long = *a a b b*, Mid = *a a B b* and the Short = *A a b b*." In fact we now know that it was very nearly the whole story both for distylic *Primula* and tristylic *Lythrum*. Fisher and Mather (1943), 30 years later, adopting the Barlow (1923) scheme, were able to remove the difficulties by finding that the *B,b* (now *M,m*) factor shows tetrasomic inheritance. The high chromosome number of the species confirmed that it is an ancient autopolyploid.

Later work on distylic and tristylic species in several diverse genera by many workers has substantiated the basic genetic system which is a single factor with two allelomorphs *S,s* in both distyly and tristyly with the addition of a second factor with two allelomorphs *M,m* in tristyly. The dominance of short style over long style has also been confirmed in several species, but the discovery of two exceptions in distyly and one in tristyly where long style is dominant has reduced its heuristic importance.

The main fascination in heterostyly from Darwin to the present day has been its similarity as a breeding system to the separation of the sexes, not only in its genetical consequences, but also in the clean Mendelian segregation en bloc of a whole syndrome of character differences: for there is not only the style length to consider, but anther height, pollen size and legitimacy (incompatibility), and several other

[1]School of Biological Sciences, Queen Mary and Westfield College, University of London, Mile End Road, London E1 4NS, United Kingdom
[2]Department of Botany, University of Florida, Gainesville, FL 32611, USA

Monographs on Theoretical and Applied Genetics 15
Evolution and Function of Heterostyly (ed. by S.C.H. Barrett)

factors that have recently been discovered. The first hints to the solution of this problem of multiple characters came not from theoretical considerations, but from the explanation of exceptions to the two allelomorphic types in distyly and the three in tristyly.

Stout (1925) described the first exception to the two or three types when he found a mid-styled plant in the tristylic *Lythrum salicaria* which was semi-homomorphic in that the long anthers had been replaced by mid anthers. About its cause he wrote: "The appearance of only one plant of the new form among a considerable number of plants together with the fact that this form has not been reported previously suggests a mutation."

Ernst (1928) described four different anomalous types with atypical combinations of characters in *Primula hortensis* and *P. viscosa* and explained their origin by mutation. In an unpublished but official report describing a long homomorphic form in *Primula* × *kewensis*, Pellew (1928) wrote "although the anthers may be high yet the pollen may be of the same average size as in the long-styled plants. It is therefore possible that besides the homo-long-style allelomorphs, a *second pair* affecting the position of the anthers in the tube is involved." Clearly Pellew was not thinking of a mutation of the already discovered *S,s* factor, but the uncovering of another factor with two allelomorphs. Haldane (1933) compared the two possibilities "a series of allelomorphs or very closely linked genes", but left the decision open. With the discovery of more abnormal types and the ever increasing number of characters contained in the system we have gradually accepted and built upon the Pellew explanation of closely linked factors.

2 Definition of Terms

Factor. We have intentionally not used the term gene in the introduction, but confined ourselves exclusively to the Mendelian term "factor", because we were dealing with evidence derived solely from Mendelian analysis without consideration, in detail, of the syndrome of physiological and morphological characters that the factor is controlling.

Gene. The term gene will be used to describe an inherited unit with a single function where the function may be the specification of a length of messenger RNA, functional RNA or a sequence of DNA. At the present level of analysis of the heterostyle system we are unable to describe the genes involved in precise molecular terms and we have to rely on the separation of genetically linked functions (characters) by comparison of evolutionary changes seen between races or species, by mutation, and by recombination arising from crossing-over.

Supergene. The two factors *S* (in distyly and tristyly) and *M* (in tristyly) controlling heterostyly are each composed of a complex of many genes. The actions of these genes are confined to the floral parts of the plant and, if we exclude the Fertile Double and Primrose Eye genes in *Primula sinensis* which are really outside the system, they are confined to the male anthers and pollen and to the female stigma and style parts of the flower. The most complete concept of the complex of genes that we can reasonably postulate is one of mutations of several genes accumulated by selection and assembled by recombination, but in its final form is contained within a length of chromosome which is only rarely disturbed by recombination through crossing-over. The nearest description of this is found in the definition of a supergene (Darlington and Mather 1949) as "a group of genes mechanically held together on a chromosome and usually inherited as a unit. If co-adapted, such genes (which are not necessarily functionally related) may cooperatively produce some adapted characteristic." This does not preclude the possibility that the supergene could contain one or more genes *with* related functions.

Dominance. The natural system of heterostyly is that of a permanent backcross enforced by the genetics of the incompatibility and so is similar to the dimorphism of sex determination, although in the latter one of the breeding pair is hemizygous whereas in heterostyly it is heterozygous. The genetics of heterostyly shows that all the dominant characters are found within one morph. It is, however, not easy to determine dominance in heterostyly in that the system prevents selfing and so is impeding its own resolution. Because a reversal of dominance has been observed within two closely related species in a heterostyled family, we consider the phenomenon of dominance to be an important one and so will describe the evidence for dominance in a later section.

3 Evidence for the Genetic Factors *S,s* and *M,m*

Heterostyly has been described in more than 100 species distributed in approximately 24 families (Ganders 1979). These studies have mainly been concerned with morphological, behavioral, and biological aspects of a breeding system unique to plants and they have demonstrated a remarkable conformity in all these features. Genetical studies have been made in apparently only 23 species in 11 families. These are summarized in Table 1. Because it is directly or indirectly assumed that the genetic system is basically the same in all heterostyled species, it is important to scrutinize the evidence carefully. Table 1 includes species in which a 1:1 segregation of the two types has been obtained by crossing the two legitimate compatible types. This shows that one is heterozygous for the factor and the other is homozygous. Which is the heterozygote carrying the dominant allelomorph can only be determined in one of three ways:

a. Obtain illegitimate selfed or within-type crossed progeny in a diploid. This can only be done in species where the incompatibility is not absolute and a small percentage of seed can be obtained. *Primula vulgaris* and *P. sinensis* are good examples where incompatible pollinations will give enough seed to obtain a result. We would expect to obtain a 3:1 segregation among the progeny of the heterozygote and that all the progeny of the homozygote would be the same as itself.

Table 1. Families and species in which the genetics of distyly and tristyly have been established

Short dominant	Distyly Short	Long	Method	
Primulaceae				
Primula hortensis	*S s*	*s s*	b	Ernst (1955)
P. malacoides	*S s*	*s s*	a	Philp (1929)
P. obconica	*S s*	*s s*	b,c	Dowrick (1956)
P. sinensis	*S s*	*s s*	a,c	Bateson and Gregory (1905), Mather and De Winton (1941)
P. vulgaris	*S s*	*s s*	a,b	Crosby (1949)
P. japonica × *P. burmanica*[a]	*S s*	*s s*	b	Ernst (1950)
Boraginaceae				
Amsinckia spectabilis	*S s*	*s s*	a	Ray and Chisaki (1957)
Pulmonaria officinalis	*S s*	*s s*	a	Darwin (1877)
Loganiaceae				
Gelsemium sempervirens	*S s*	*s s*	a	Ornduff (1980)
Oleaceae				
Forsythia ovata	*S s*	*s s*	a	Sampson (1971)
Polygalaceae				
Fagopyrum esculentum	*S s*	*s s*	a	Garber and Quisenberry (1927)
Turneraceae				
Turnera ulmifolia	*S s*	*s s*	a,b,c	Shore and Barrett (1985)
Long dominant				
Guttifereae				
Hypericum aegypticum	*s s*	*S s*	a	Ornduff (1979)
Plumbaginaceae				
Armeria maritima var. *maritima*	*s s*	*S s*	b	Baker (1966)
	Distyly ex Tristyly			
Short Dominant				
Lythraceae				
Pemphis acidula	*S s*	*s s*	a,c	Lewis and Rao (1971) Lewis (1975 and unpubl.)

	Tristyly				
Short Dominant	Short	Mid	Long		
Lythraceae					
Lythrum salicaria	*Ssmm* *SsMm* *SsMM*	*ssMm* *ssMM*	*ssmm*	a	Barlow (1923) Fisher and Mather (1943)

Table 1. *(Continued)*

	Tristyly				
Short Dominant	Short	Mid	Long		
Oxalidaceae					
Oxalis priceae	*Ssmm* *SsMm* *SsMM*	*ssMm* *ssMM*	*ssmm*	a	Mulcahy (1964)
O. bowiei : *O. hirta* : *O. tragapoda* : *O. valdiviensis* :	*Ssmm* *SsMm* *SsMM*	*ssMm* *ssMM*	*ssmm*	a	Fyfe (1956)
Long dominant					
Oxalis rosea *(articulata)*	*ssmm*	*ssMm* *ssMM*	*SsMM* *SsMm* *Ssmm*	a	von Ubisch (1926) Fyfe (1956)
Pontederiaceae					
Eichhornia paniculata	*SsMM* *SsMm* *Ssmm*	*ssMm* *ssMM*	*ssmm*	a	Barrett et al. (1989 and unpubl.)

[a] *P. japonica* is monomorphic for long style and was crossed with the short-styled form of the distylic *P. burmanica*.

b. In species such as *Primula obconica*, which produces essentially no seed from incompatible pollinations, the dominance relationships can be observed by making compatible crosses with an abnormal homostyle. The analysis of crosses will give a decisive result when crossing the homostyle with both types. If we assume that all the dominant factors are in one type, a presumptive result can be obtained by crossing the homostyle to one type. The best example of this is in *P. obconica*, where a long homostyle crossed to long gave 1:1 long and homostyle, selfed homostyles and homostyle × homostyle crosses gave both homostyle and long in the expected proportions, and homostyle × short gave short, homostyle and long in the expected proportions (Dowrick 1956). We conclude that short is dominant to homostyle and both are dominant to long.
c. The third method is by making autotetraploids of the two types. The duplex *SSss* crossed with the nulliplex (recessive) *ssss* will give a 5:1 ratio of the dominant phenotype to the recessive. So far as we are aware, this has only been done with *Primula sinensis* (Sömme 1930) and with *Turnera ulmifolia* var. *intermedia* (Shore and Barrett 1985). With *P. obconica*, crossing tetraploid homostyles, longs and shorts gave the expected results, confirming the dominance of short, at the same time excluding the possibility of a second factor (supergene *H,h*) for the explantation of the homostyle (Dowrick 1956).

The most striking feature of Table 1 is the uniformity of one diallelic factor *S,s* for distyly and two diallelic factors for tristyly. The dominance of the short-styled form, which was originally thought to be universal, now has three important exceptions. Two of these, *Limonium* (Plumbaginaceae) and *Hypericum* (Guttiferae) may be typical of the families they represent. Baker (1966) has produced evidence that *Armeria* in the Plumbaginaceae may also have dominant long style. The only other detailed report of heterostyly in the Guttiferae is *Cratoxylum formosum* (Lewis 1982), in which no evidence for dominance could be obtained. Unfortunately only a long-term breeding program with a forest tree would reveal dominance! The significance of testing the conformity of dominance within a family is revealed by the third exception which is in *Oxalis articulata*. Here the difference is within a genus which indicates that the difference arose after the basic genetic system had been fully established. The interpretation of the breeding data of Fyfe (1956) has been questioned by Baker [unpubl., quoted and discussed by Mulcahy (1964) and discussed by Ganders (1979)]. Ganders comes to the conclusion "... short is recessive in *Oxalis* "*rosea*" (von Ubisch 1926) and *O.* "*articulata*"." (Fyfe 1956).

We have re-examined Fyfe's data and come to the same conclusion from the statement "Ten longs from the mid × long progeny were crossed to a short. Three of these longs gave only longs, 360 in all, and seven gave 452 longs: 495 shorts". It is difficult to explain this result without assuming long is dominant and short recessive.

Oxalis consists of species with fully developed tristyly and if we accept the exceptional dominant long style in *O. articulata* we must postulate a switch gene which affects the transcription of the whole supergene (factor). The consequences of this change of dominance will be assessed in a later section.

4 Linkage of *S* and *M* Factors at the Chromosomal Level

The linkage studies in *Primula sinensis* by De Winton and Haldane (1931) revealed loose linkage of *S* with blue flower *B* and green stigma *G*, the nearer, *B*, being 7.5% recombination units from *S*. Although both factors affect the flower, there is no heuristic significance in the linkage and they are too far away to act as flanking markers for testing crossing-over within the *S* supergene.

The most important linkage is with the centromere because in the majority of organisms investigated so far recombination close to the centromere is suppressed. The centromeric region is a part of the chromosome where genes could accumulate and be tightly linked although occupying a stretch of chromosome sufficiently large to allow the genes to be physically separated by larger distances than elsewhere in the chromosome.

Evidence for linkage to the centromere comes from "double reduction" in an autotetraploid (Darlington 1929). A triplex plant, i.e., one with three dominant alleles and one recessive *SSSs*, normally produces no *ss* gametes (spores). Occasionally *ss* gametes may be produced at very low frequency. Remembering that centromeres segregate at the first meiotic division and the products of chromosome duplication separate at the second division, in an autotetraploid all the four products of meiosis receive two different centromeres. It is not possible to have a gamete with both halves of the same centromere and so when a gene is completely linked to the centromere *ss* gametes will not occur. On the other hand, when multivalents are formed at zygotene and a single cross-over (or its equivalent) occurs between the locus carrying the recessive allele and its centromere, two different centromeres each carry an *s* allele. When these two centromeres segregate to the same pole at the first meiotic division, one quarter of the gametes at the second division will receive both *s* alleles. The greater the distance between the locus and its centromere the higher the probability of a cross-over between that locus and the centromere. This distance can be estimated by the frequency of double reduction observed from progeny of the cross *SSSs* × *ssss*. Figure 1 shows the situation in three species where double reduction has been measured. The significant conclusions from the data are:

1. *Primula sinensis*. *S* is situated within the cross-over restriction area of the centromere.
2. *Primula obconica*. *S* is situated close to, but outside the cross-over restriction area.
3. *Lythrum salicaria*. If *S* and *M* are unlinked, as all the results show, then they must be on different chromosomes outside the cross-over restriction areas, but at similar distances (10%) from their respective centromeres.

5 Heterostylic Characters

5.1 Comparison Between Species

The number of morphological characters associated with the heterostylic complex is continually increasing as finer analysis is made on well investigated species and as more species with different floral structures, apparently co-evolved with their insect pollinators, are examined. Table 2 summarizes these characters.

Only two of these characters are found in all heteromorphic species and, therefore, can be considered essential. These are (1) self- and intramorph incompatibility and (2) uniformity of dominance of all the characters in one of the two morphs. A third feature that has been found in all the species in which it has been

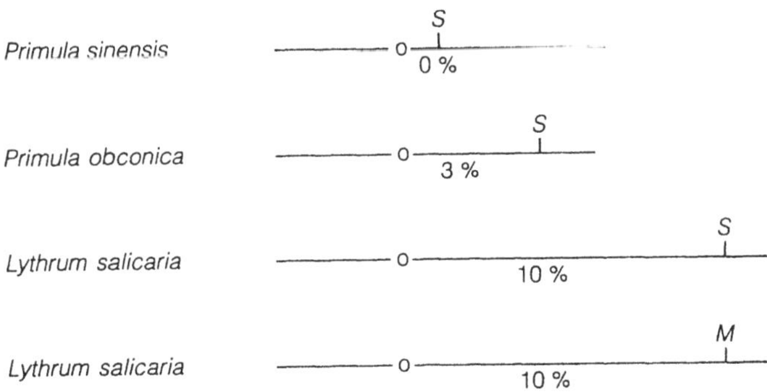

Fig. 1. Map distances of heterostyle supergenes from their centromeres based on evidence of double reduction

Table 2. Heterostylic characters

Female		Male
	Primary	
Uniformity of dominance of all the female and male characters in one morph		
Incompatibility Sporophytic and diallelic		Sporophytic and diallelic
Style length	associated with	Compatible pollen-tube growth rates
	Secondary	
Stigmatic papillae size surface osmotic pressure[b]		Anther filament length point of insertion filament bending[a]
Style Area of conducting tissue		Pollen size sculpturing osmotic pressure[b] carbohydrate[c] color[c]

[a] *Cratoxylum formosum.*
[b] *Linum* spp.
[c] *Lythrum* spp.

sought is (3) style length associated with differential compatible pollen-tube growth. (see Fig. 2 and Table 3)

5.1.1 Self- and Intramorph Incompatibility

Self- and cross-incompatibility have been repeatedly argued to be the driving force for obtaining and maintaining the negative assortative mating necessary for a balanced polymorphism to be maintained. It has also been considered necessary that the controlling *S* gene should exist in two allelic states, one being dominant to the other, and that the pollen should be sporophytically controlled (Lewis 1949). There have been no new facts or theoretical considerations to challenge both of these views.

The incompatibility system in *Anchusa officinalis* (Schou and Philipp 1984) and *A. hybrida* (Dulberger 1970a) is not diallelic and is independent of the heteromorphic characters. The incompatibility recognition site is also exceptional in that it occurs in the ovary and not in the stigma. We consider that it would not be helpful to bring these species into the general discussion without further knowledge of the genetics of the system. It would be interesting to know, however, whether there are compatible pollen-tube growth-rate differences that are adapted to the style length differences in these species as they are in *Primula* (see Sect. 5.1.3).

We must also keep in mind that there are two different recognition processes among heteromorphic species (and see Chap. 3). In the Primulaceae, Guttiferae,

Oxalidaceae, Lythraceae, and Oleaceae the incompatible pollen tubes are inhibited after growing some distance into the style and, in this respect, are similar to the gametophytic multiallelic system in homomorphic species. In the Primulaceae both the incompatible and compatible pollen tubes are affected by temperature in the same characteristic way as the tubes in the gametophytic system (Lewis 1942). In the Linaceae, Plumbaginaceae, and Polygonaceae the incompatible pollen is inhibited on the stigma, which is a striking feature of the sporophytic system of homomorphic species.

These two different recognition systems in heteromorphic species obey the Brewbaker (1959) rule in that inhibition of pollen tube growth in the style is associated with binucleate pollen and pollen germination inhibition with trinucleate pollen. The first group of families has binucleate pollen, the second trinucleate. We must conclude that the *I* gene is quite different in the two groups. This is emphasized by the recent finding of a second gene, which is gametophytic in its operation, in the well-known sporophytic incompatibility system in homomorphic species (Lewis et al. 1988). It would not surprise us if a cryptic second gene in heteromorphic species were to be uncovered by the right techniques. Such a gene might be reduced to homozygosity and provide an essential biochemical part of the incompatibility process.

5.1.2 Uniformity of Dominance of All the Characters in One of the Two Morphs

Dominance of all the alleles of the different character pairs in one morph is essential to have the characters in the right workable combinations and to keep them together by tight linkage. There appears to be no good reason why one particular morph should have become dominant (although see Chap. 6). As discussed earlier, it was thought that dominance was a universal feature of the short style and that it had arisen, possibly arbitarily, in the short style at the outset of the evolution of the system. The finding of dominant long style in the Plumbaginaceae by Baker (1966) and in *Hypericum* in the Guttiferae by Ornduff (1979) does not alter this view because it could be a regular feature of these families. But the occurrence of a species of *Oxalis* with dominant long style when all the rest of the species in this genus have dominant short style is difficult to explain. If we assume that the dominance of all the characters was determined at the time of origin of heterostyly, the dominance could only be changed subsequently if there were a separate dominant operator gene for the whole cluster of genes because they have to be changed en bloc. The dominance change in the fully developed system would then occur by a rare transposition either by crossing-over, mutation, or by a transposon. We realize that this is putting a heavy burden on the one known case of late dominance change, but this is one out of only 11 species in three families that could have revealed such a dominance change. It is obviously desirable to find another example. There is one major complication with this question of dominance change. That is, when the supergene is split up by recombination to form a homostyle the recombined genes retain their original dominance relationships (see Sect. 6.1).

5.1.3 Style Length Associated with Differential Compatible Pollen-Tube Growth

That style length differences in themselves are not essential was shown by Baker (1966), but style length associated with differential compatible pollen-tube growth rate does appear to be of primary importance. This association was first observed by Lewis (1942) in two *Primula* species and has since been recorded for two *Lythrum* species (Esser 1953; Dulberger 1970b), *Linum grandiflorum* (Lewis 1943), *Pemphis acidula* (Lewis and Rao 1971), *Primula veris* (Richards and Ibrahim 1982) and *Cratoxylum formosum* (Lewis 1982) (see Fig. 2 and Table 3).

The important point about this relationship is that the pollen tubes reach the ovary at the same time after pollination whether the style is 7 or 2 mm as in *Lythrum junceum*. The necessity for the pre-fertilization stimulation of the ovary by the auxin

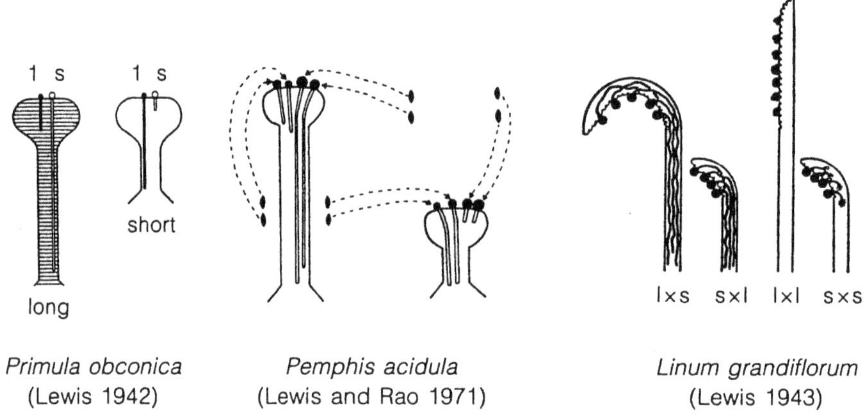

Fig. 2. Diagrams showing the position of legitimate and illegitimate pollen tubes in long and short styles at the time the first tube enters the ovary. Although the optimum time and temperature is different for each species, it is clear that despite the differences in the length of the style legitimate pollen tubes take approximately the same time to reach the ovary

Table 3. Style length and pollen-tube growth rates in *Lythrum*

Lythrum junceum			*Lythrum salicaria*			
Style	length (mm)	Hours to ovary (Dulberger 1970b)	Style	length (mm)	Hours to base of style (Esser 1953)	(Tatebe 1964)
Long	7	6	Long	9	11	11
Mid	4	6	Mid	5	10	10
Short	2	6	Short	2	6	3

IAA produced in the style by the action of the pollen tube growing down that style has been described for *Nicotiana* (Lund 1956). In *Vanilla* there are varying column (style) length differences between species and the ovary is long and thin. When pollen that produces pollen tubes that normally grow down long styles is placed on a short style the pollen tubes enter the ovary, spurn the proximal ovules and travel their normal distance before entering and fertilizing the ovules at the distal end of the ovary (McClelland 1919). These two examples show that both the ovule and the pollen tube have to be suitably mature for entry to take place. It is also clear that the precise timing of the stimulus and the receptivity of the ovules is essential. Pollen tubes can be not only too late in reaching the ovary, but also can be too early. The importance of this relationship is emphasized by the fact that the differential growth rate is correlated in seven of the species with pollen-volume differences, but where there are no pollen-volume differences, e.g., *Linum*, differential growth rate is associated with osmotic pressure differences (Lewis 1943) that are probably the expression of differences in the concentrations of metabolic reserves.

The system offers a positive encouragement to negative assortative fertilization that complements the negative action of incompatibility.

5.2 Possible Developmental Connections

Most of the other characters listed in Table 2 are completely independent developmentally, biochemically, and physiologically, and if we accept that pleiotropy by multiple primary effects of a gene do not exist in this system (cf. Grüneberg 1938), then we must postulate a separate gene with two alleles for each character pair. There is theoretical and experimental evidence for developmental connections which are confined to the male (pollen and anther) or to the female (stigma and style) parts of the system. For example, it is possible to imagine that style length, stigmatic papilla size, and style conducting tissue size might be multiple end results of one developmental process so that one gene would suffice. There is, however, no evidence for developmental connections between the sexual parts. Thus it is probably safer to assume a separate gene for each character pair.

The most comprehensive examination of developmental connections is in the tristylic *Eichhornia paniculata* (Richards and Barrett 1984). This study confirms and greatly extends the observations on distylic *Fagopyrum esculentum* by Stevens (1912), on distylic *Primula*, and on tristylic *Lythrum* spp. and *Oxalis* spp. by Stirling (1936). The main conclusions are that cell size and cell number are differentiated at varying times after flower bud initiation and clearly demonstrate that pollen size and anther height could have a common physiological cause. Similarly, the size of the stigmatic papillae and style length could have a common basis. This connection in development may have been an important factor in the evolution of the fully coordinated system, but we should not necessarily conclude that there is only one gene in control of the morphological characters of the male anther and another one for the female pistil.

We suggest that the most important mutually adjusted connections are not derived from the developmental connections within the two sexual sides of the system, but from natural selection acting on the interaction of the male and female characters which reinforce the negative assortative mating, for example, the speed of pollen tube growth down the style and style length.

6 Homomorphic Variants

6.1 Homomorphic Variants Within Species

The term homomorphic, as applied to the heteromorphic species, is used to describe the condition in abnormal individuals which combine the stylar features of one of the normal types with the anther features of the other. In its simplest form, as in *Primula* species, there are abnormal homostyles which have the long style of the long-style type and the high anthers of the short style type, the so-called long homostyle. The reverse of this combination is the short homostyle.

We can distinguish two situations in which homomorphism occurs. The homomorphic types may be rare variants in wild populations or in cultivation or they may occur as a significant proportion of the individuals in a population. The important point here is that they are either of very recent origin or are of long standing and are interbreeding with the distylic forms from which they have been derived.

When the homomorphic condition has become established as the only type in a population (or subspecies) or in a distinct, but related species, it is described as being monomorphic. In this discussion we will assume that all the monomorphics are secondary, i.e., derived from heteromorphics, as opposed to being primary, i.e., primitive (Ernst 1957).

Homomorphic variants were described as early as 1877 by Darwin. Previously published occurrences were comprehensively classified and reviewed up to 1955 by Dowrick (1956). We will briefly summarize the results of Dowrick's survey together with more recently published work.

Working with *Primula hortensis* and *P. viscosa*, Ernst (1925, 1936) showed that the S supergene could consist of at least three genes $G, A,$ and P. G represented the female gynaecium, A was for male androecium, and P was for pollen size. The long-style form had the genotype $\frac{gap}{gap}$ and the short style $\frac{GAP}{gap}$. The most common homostyle was designated $\frac{gAP}{gap}$ with the gynaecium of the long style and the androecium of the short. Other less common types had appropriate combinations of GAP and gap.

We will not use the order GAP, which is probably only of purely historical origin and was still used by Ernst in 1957, but we follow Lewis (1954) and Dowrick (1956) with the order GPA. With the order GAP, the most frequently occurring recombinants would require a double cross-over in the short styled parent. With the order GPA these are generated by a single cross-over.

The clearest statement about the homostylic genotypes that we have found is by Ernst (1958, p. 108). "Ausser den beiden Allelen P *GAP* und P *gap* sind für *Pr. hortensis* und *viscosa* durch intra- und interspezifische Bestaubungen die weiteren Allele P *gAp* (in *Pr. hortensis*), P *gAP*, P *GaP*, und P *Gap* (in *Pr. viscosa*) nachgewiesen und die Existenz der beiden weiteren Komplexe P *GAp* und P *gaP* wahrscheinlich gemacht worden." On the order *GAP* the genotypes *GaP* and *gAp* require double cross-overs in the same chromosome. The other genotypes, including the two which were not found, require only one cross-over. With the order *GPA* the only genotypes which require a double cross-over are the two that are yet to be found in crosses between normal long and normal short-styled plants. The two missing types were produced in later generations from the four previously observed recombinant genotypes presumably as the result of single cross-overs.

A sumary of the abnormal genotypes obtained by Ernst was tabulated by Dowrick (1956). The table contains five individuals with the *gaP* combination, a combination which would be generated by one cross-over from the triple heterogyote $\frac{GAP}{gap}$ but not from $\frac{GPA}{gap}$ without two cross-overs. But the later papers of Ernst tabulate these *gaP* individuals as arising from a double heterogyote $\frac{GPa}{gpa}$ crossed onto the triple recessive $\frac{gpa}{gpa}$. With the *GPA* order only one cross-over is required but with *GAP* two cross-overs are necessary. The correct origin of the *gPa* recombinants transfers the five individuals from exceptionality to conformity, making the total results, as shown below, decisively in favor of the order *GPA*.

Gene order	*GAP*	GPA	PGA
Single cross-over	67	82	15
Double cross-over	15	0	67

We hope that this partially alleviates the pessimism of the conclusion of Charlesworth and Charlesworth (1979) in their constructive discussion that "there does not seem to be any possibility of ordering the component loci."

In his detailed review of his monumental and life's work on the abnormal heteromorphic forms in *Primula hortensis* and *P. viscosa* and the hybrids between them, Ernst (1957) retabulated all his results. He came to the conclusion that of the 127 occurrences of abnormal types in controlled crosses and selfings 46 could be explained on crossing-over whereas the other 75 could not and were probably the result of mutation. We have examined the tables for *P. hortensis* in detail and find 33 instances of unexpected segregants that are explicable on linkage and crossing-over and 27 which cannot be generated by recombination of the three genes *GPA*. Of these unexpected segregants we can be certain of recombination between the genes stated in 17 cases:

Recombination between *GP* and *A* 11 in 2967 = 0.37%
Recombination between *G* and *PA* 6 in 3220 = 0.19%.

This allows us to make a tentative map

```
        G            P                          A         .
             0.19               0.37
```

The abnormal types unexplained on crossing over are tabulated below:

$a \rightarrow A$	$\frac{GPa}{gpa}$	$\rightarrow$	short-styled	$\frac{GPA}{gpa}$	14 in 477	= 2.93%
$gp \rightarrow GP$	$\frac{gpA}{gpa}$	$\rightarrow$	short-styled	$\frac{GPA}{gpa}$	3 in 599	= 0.50%
$GPA \rightarrow gpa$	$\frac{GPA}{GPA}$ or $\frac{GPA}{gpA}$	$\rightarrow$	long-styled	$\frac{gpa}{gpa}$	6 in 3797	= 0.158%
$a \rightarrow A$	$\frac{GPa}{gpa}$	$\rightarrow$	long homostyle	$\frac{gPA}{gpa}$	4 in 512	= 0.78%.

We give the percentages of these inexplicable anomalies in order to show that they are many orders of magnitude too large to be the result of gene mutation, which varies from 1×10^{-4} to 1×10^{-7}. Furthermore most gene mutation is from a dominant allele to a recessive. Twenty one of the 27 anomalies would be mutations from recessive to dominant ($a \rightarrow A$ and $gp \rightarrow GP$). All but one of these anomalies occur in families raised from selfing or crossing the abnormal cross-over types *GPa* and *gpA*. The one exception is a long-styled plant, $\frac{gpa}{gpa}$, among the 344 progeny of a cross between a normal long $\frac{gpa}{gpa}$ as female and a homozygous short $\frac{GPA}{GPA}$ as male. There are several possible trivial explanations for this; e.g., an unreduced egg nucleus could have been stimulated to develop by the presence of pollen nuclei, but without fertilization. More difficult to explain are the anomalous types arising from *GPa* and *gpA*. We have to emphasize that both the *Primula* species studied by Ernst are ancient tetraploids. His results clearly show disomic inheritance and consequently this implies that only one pair of supergenes of the original heterostylic incompatibility system is functional in normal plants. It may be that a normally quiescent homologous *GPA* pair of supergenes has been activated in these abnormal plants by, for example, a transposon, thereby revealing the dominant genes and so creating the anomalous types. Another possibility is the presence of a switch gene controlling dominance, but the data Ernst has presented could not begin to differentiate between these possible explanations. These results of Ernst provide us with the only solid evidence for an *S* supergene containing linked genes controlling all the heteromorphic characters. His was an immense task spanning more than 30 years, and it is unlikely to be repeated or extended using Mendelian and linkage techniques. At present it is our only guide to the interpretation, on a genetic basis, of anomalies in other heterostyled species.

On the hypothesis of linkage, we would expect to obtain equal numbers of long and short homostyles following crossing-over between *G* and *PA* or between *GP* and *A*. Ernst (1957) has examined the progeny of legitimate reciprocal crosses between normal long- and normal short-styled plants of *Primula hortensis*. We find that he obtained 6 short homostyles among 2887 short styled plants and 16 long homostyles among 2879 long styled plants ($\chi^2_{[1]} = 4.59, 0.05 > P > 0.01$). Thus there is a significant deficiency of short homostyled plants.

When we compare the differences between his reciprocal crosses we find 12 homostyles in 2968 progeny from short female × long male and 10 homostyles in 2798 progeny from long female × short male. It appears that the generation of homostyles does not depend upon which parent is heterozygous.

With the order *GPA* we would expect to obtain homostyles by crossing-over between *G* and *PA* or between *GP* and *A* and we should expect to find the frequency of homostyles in the progeny of legitimate crosses to be approximately the sum of the frequency of cross-overs between *G* and *P* and between *P* and *A*. From the data we have been able to extract from Ernst (1957) we find:

Recombination between	*G* and *PA*	0.19%
Recombination between	*GP* and *A*	0.37%
		0.56%
Homostyles by recombination		0.38%.

These figures have been obtained from different sets of data and the recombinant progeny in the linkage data include normal long- and short-styled plants as well as long homostyles. By the nature of the crosses no short homostyles were to be expected. Thus if we correct for the deficiency of short homostyles, the expected frequency of homostyles is 0.55%. This is a reasonable correspondence with 0.56% and certainly does not negate the hypothesis of linkage.

Although based on small numbers we can have confidence in the proportion 6:16 (0.37:1) because the data in Table 12 of Ernst (1957) also show a deficiency of the short-styled type. In the seven crosses that would be expected to produce short and long homostyles in equal frequency he obtained 670:1678 (0.41:1). This shows that the deficiency of short homostyles in the legitimate crosses is due to low viability of the short homostyle and not to an inequality in the productivity of reciprocal cross-overs. Furthermore, this provides further evidence that conventional crossing-over is the source of the recombined types.

Because we have to rely so much on the *GPA* linkage, with the addition of the *I* genes for incompatibility and other genes for ancillary characters, we must be especially conscious of the limitations of the analytical method. To be absolutely convinced of linkage and crossing-over there should be outside markers to check that a cross-over event has occurred. Such a check might have been possible in a species that was reasonably well mapped, e.g., *Primula sinensis*, but here no homostyles are found probably because of the proximity of the supergenes to the centromere. We have to be satisfied with the consistency of results using single cross-over events and the absence of double cross-overs.

6.2 Comparisons Between Species

Primula vulgaris, P. obconica, P. sinensis, Amsinckia, Turnera. A long homostyle type *gPA* in *Primula vulgaris* has been known since Darwin described it. This homostyle is widely distributed in two regions in England although it occurs as a very rare variant elsewhere. The method by which the homostyle has spread has been the subject of much controversy in population genetics (e.g., Crosby 1959; Bodmer 1960; Piper 1984). It is self-compatible and is in most respects similar to the long homostyle with large pollen of *P. hortensis*.

Primula obconica. This has a long homostyle of the same type, but it has been found only in the autotetraploid (Dowrick 1956). Although it has been extensively studied and grown commercially, *Primula sinensis* has no true homostyle, the types similar to the homostyles of other species being produced by separate genes unlinked both genetically and physiologically with the heteromorphic system (Mather and De Winton 1941).

In his survey of the very large genus *Primula*, Ernst (1950) found short homostyles in three species of the *obconica* group, but he described many species with long homostyles throughout the genus.

The long homostyles of the distylic *Amsinckia spectabilis* are extremely variable and it has been suggested by Ganders (1975) that they arose not by crossing-over, but as the result of modifier genes similar to those found in *P. sinensis*.

The long homostyle in *Turnera ulmifolia* appears to occur only in hexaploid varieties at the margins of the range of the species complex (Shore and Barrett 1985). Again, there is considerable variation in the relative lengths of reproductive organs among populations. Of particular interest is the observation that the three homostylous varieties show similar behavior in crosses with heterostylous populations. Shore and Barrett conclude that these homostyles "have arisen via the same genetic mechanism involving recombination within the distyly supergene."

Plumbaginaceae. The genera *Armeria* and *Limonium* are of particular interest because the long-styled type in *Armeria*, and probably also in *Limonium*, is dominant to the short styled. Baker (1966) has examined both genera for dimorphic and monomorphic (homostyle) species. The situation found is, in principle, the same as in the homostyles of *Primula* with the major exception that the homostyles usually have short styles. Thus of the 16 secondarily monomorphic self-compatible species of *Limonium* listed by Baker (1966), 15 have short styles. The exception was the only annual species in the group. It would be interesting to know whether the long style is dominant in this species or is recessive and an exception in this family as is *Oxalis rosea (articulata)* in the tristylic Oxalidaceae. Baker (1966) also recorded that all three of the monomorphic varieties of *Armeria maritima* are short-styled.

An important point is that where the long-styled form is dominant there is a deficiency of long homostyles and where the short-styled form is dominant there is a deficiency of short homostyles. The dominant genes that are present in the cross-over chromosomes transmitted to the deficient homostyles are, in both cases, the female sets *GSI*s. We suspect that there is a deleterious gene associated with the

GIs part of the complex which is expressed when separated from *IpPA*. We should also note that the male part of the supergene, including the incompatibility gene, has the dominant alleles in both of the predominant types of homostyle in *Primula* and *Limonium*. Thus the genetical explanation for the selective disadvantage of the short homostyles in *Primula*, say, modeled by Charlesworth and Charlesworth (1979), should be equally applicable to long homostyles in *Limonium*. A complication that will need to be considered in further models is the reduced viability of the short homostyle in *Primula hortensis*.

Baker explained that these monomorphic species are derived from dimorphic ancestors and, like most monomorphic species, are self-compatible, as would be expected on the supergene theory. The pollen of the monomorphics has retained its incompatibility reaction; it is incompatible on the long-style (A cob) form and compatible on the short (B papillate) form of dimorphic species. However, the style of the monomorphic species has lost its incompatibility reaction, for it accepts all types of pollen. This sequence in the evolution of self-compatible monomorphic species has a parallel in the sequence of events in homomorphic gametophytic incompatibility where in self-compatible derived species the incompatibility is lost in the style, but partially retained in the pollen (Lewis and Crowe 1958).

Further evidence comes from *Pemphis acidula*, which is a distylic species derived from a tristylic ancestor. In an ancient isolated population in the south of Madagascar all plants seen are long homostyles, self-compatible and with pollen incompatibility, as expected, of the high anthers of short styled plants (Lewis 1975). Later work (in prep.) has shown that the style of the homostyle plants has lost its incompatibility reaction. Evidence from first-generation crosses between longs and homostyles indicates that this is a true loss of incompatibility because it is restored in some, but not all, of the progeny by the addition of the long-style supergene.

Trimorphic Species. The homomorphic anomalies that have been found in tristylic species are classified as semi-homostyles because only one of the two anther levels in a flower has been altered. Most are long semi-homostyles, in which the flower has a long style and long and short anthers, i.e., as in the long homostyle in distylic species, the long anther has the large pollen and the incompatibility reaction which makes it compatible on a long style giving self-compatibility.

The occurrence of homostyles in tristylic species is given below.

Lythrum salicaria	mid semi-homostyle long semi-homostyle	Stout (1925) Esser (1953)
Oxalis dillenii	mid semi-homostyles of two types: 1. with mid and long anthers 2. with mid and short anthers	Ornduff (1972)
Eichhornia paniculata	mid semi-homostyle	Barrett (1985).

There is little or no information on the genetics of these homostyles and the most important conclusion we can draw is that, as in distyly, the various characters of the system can be dissociated and therefore used to support the hypothesis that each part is under the control of a separate gene. In other words, there is nothing against and something in support of the supergene concept of both *S* and *M* in tristyly.

7 No Heteromorphic Chromosomes

We have discussed earlier some of the similarities between sexual dimorphism and heterostyly. In both animals and plants the genes controlling the characters that differentiate the sexes in dioecious species are usually linked together in heteromorphic chromosomes. As tight linkage of many genes is a feature of both dioecy and heterostyly, it is, at first sight, surprising that heteromorphic sex chromosomes, which act as a highly efficient gene linker, are not present in the heterostylic system. The only chromosome difference so far reported is in some early work of Stevens (1912) in which he noted a difference in chromosome size and in nucleolar number in *Fagopyrum*. While admitting that there are few published data, we are sure that were there any differences these would have been observed at the John Innes Horticultural Institution between 1930 and 1950 when work on chromosomes and on heterostyly was in progress, often by the same people. It should be noted, however, that no differences between the karyotypes of long- and short-styled plants of *Byrsocarpus coccineus* were observed by Baker (1962).

The two main differences between sex and heterostyly are (1) that in sexual dimorphism the system is essentially hemizygote × homozygote whereas in heterostyly it is heterozygote × homozygote and (2) which appears to be connected with 1, the differentiation of the sexes is by different genes *M*1, *M*2, *M*3, . . . , *F*1, *F*2, *F*3, etc., but in heterostyly the differentiation is done by alleles of the same genes *G,g P,p I,i* etc.

Although we cannot give an explanation for the frequent occurrence of heteromorphic sex chromosomes we can explain the absence of such chromosomal differentiation in heterostyly in that the genes of the two types are situated allelically on the same piece of chromosome. Any disruption of this situation in heterostyly leads immediately to the breakdown of the complete system. Ornduff (1966) has described an example of current evolution from heterostyly to dioecy in *Nymphoides cordata*. The process cannot have proceeded very far because the male plants retain an ovary with a complete set of fully developed ovules. In these dioecious types that have evolved from distylic species we would not expect to have heteromorphic chromosomes.

8 Hypothesis of the Supergene

Having finished our brief examination of the relevant facts on the genetics of the heteromorphic breeding system, we find it difficult to suggest an alternative hypothesis to the supergene system first adumbrated by Pellew (1928), Mather and De Winton (1941), Mather (1950), and elaborated by Ernst (1936, 1957), Lewis

(1954), and by Dowrick (1956). We can add a few facts and conjectures to fill in some detail to the 1954 model given in Fig. 3. The evidence for the order *GPA* is confirmed by analysis of Ernst's data and rough linkage values between *G* and *P* and between *P* and *A* of 0.19 units and 0.37 units respectively. This indicates that the total supergene is at least 1.0 units long. There is no evidence for the separation of *I*p (incompatibility) from *P* (pollen size) or of *I*s from the stigmatic papillae character in controlled breeding experiments. This is not unexpected because to demonstrate such separation would require the examination of incompatibility relations, preferably by pollen-tube growth tests, in many thousands of plants. We estimate that only a few tens of plants have been examined so far in this way.

In natural populations of homostyled plants, which might contain dissociated types, these have not been found. There is, however, a good reason for this. The pollen grain not only has its incompatibility but also its appropriate compatible pollen-tube growth rate, which is closely connected biochemically with pollen size. Clearly, a pollen grain which was compatible *I*p on a long style would not survive to effect fertilization if its pollen-tube growth rate was one half of that of its competitors

The best positive evidence for the separation of pollen incompatibility (*I*p) and pollen size (*P*) comes from distylous species that are in a transitional or complete stage of evolution from tristyly, for example *Oxalis suksdorfii* (Ornduff 1964) and *Pemphis acidula* (Lewis and Rao 1971; Lewis 1975).

The question of dominance of all the genes in the supergene is a serious problem that cannot be resolved without a complete DNA sequence analysis of the supergene. If we accept the reality of the change of dominance in *Oxalis* species, then we have

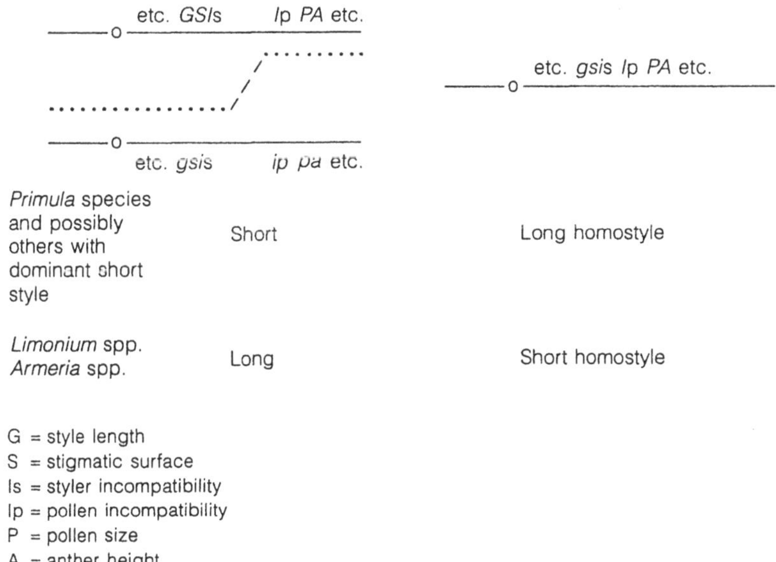

Fig. 3. Hypothetical structure of the supergene controlling heterostyly and the formation of homostyles by crossing-over

to invoke a switch mechanism probably in gene transcription through one or other of the three transcription domains. There would also have to be both positive and negative control to allow for the domain switch of the whole supergene and to survive recombination of the genes in homomorphic variants, particularly $\frac{GpA}{gpa}$ and $\frac{gPa}{gpa}$. It is also extremely likely that the supergenes will include enhancer domains because most of the genes are tissue-specific in their action.

During evolution we consider that the incompatibility came first. Style length and compatible pollen-tube growth rate would then be mutually differentiated by genes for style length and for pollen size which both had a common control system of transcription allowing a gradual lengthening of the style in one morph and increasing the size of the pollen in the other. The reciprocal gradual change could occur by shortening of the style and reducing the size of the pollen grain. With such a system, the delicate balance of style length and compatible pollen-tube growth would be preserved throughout the evolution of the system.

We have to conclude that, apart from circumstantial evidence from correlations, we can only advance by sequence analysis of the DNA involved and the associated molecular biology of the system. The first priority would be the *I* gene itself because it would probably have a considerable output of transcribed RNA and translated protein. We suspect that several of the other characters will have a gene with a distinguishable sequence, but probably with only minute amounts of transcribed RNA. We do not consider that there is any theoretical problem with the assembly of these genes into a supergene. There is enough redundant DNA and a good mechanism for transposition of this DNA to provide a sound genetic system for the evolution of the genes and their assembly.

References

Baker HG (1962) Heterostyly in the Connaraceae with special reference to *Byrsocarpus coccineus*. Bot Gaz 123:206–211

Baker HG (1966) The evolution, functioning and breakdown of heteromorphic incompatibility systems. 1. The Plumbaginaceae. Evolution 20:349–368

Barlow N (1913) Preliminary note on heterostylism in *Oxalis* and *Lythrum*. J Genet 3:53–65

Barlow N (1923) Inheritance of the three forms in trimorphic species. J Genet 13:133–146

Barrett SCH (1985) Floral trimorphism and monomorphism in continental and island populations of *Eichhornia paniculata* (Spreng.) Solms. (Pontederiaceae) Biol J Lin Soc 25:41–60

Barrett SCH, Morgan MT, Husband BC (1989) The dissolution of a complex genetic polymorphism: the evolution of self-fertilization in tristylous *Eichhornia paniculata* (Pontederiaceae). Evolution 43:1398–1416

Bateson W, Gregory RP (1905) On the inheritance of heterostylism in *Primula*. Proc R Soc Ser B 76:581–586

Bodmer WF (1960) The genetics of homostyly in populations of *Primula vulgaris*. Philos Trans R Soc Lond B 242:517–549

Brewbaker JL (1959) Biology of the angiosperm pollen grain. Ind J Genet Plant Breed 19:121–133

Charlesworth B, Charlesworth D (1979) The maintenance and breakdown of distyly. Am Nat 114:499–513

Crosby JL (1949) Selection of an unfavourable gene-complex. Evolution 3:212–230

Crosby JL (1959) Outcrossing on homostyle primroses. Heredity 13:127–131
Darlington CD (1929) The significance of chromosome behaviour in polyploids for the theory of meiosis. John Innes Hortic Inst Conf Polyploids, p 42
Darlington CD, Mather K (1949) The elements of genetics. Allen and Unwin, Lond
Darwin C (1877) The different forms of flowers on plants of the same species. Murray, Lond
De Winton D, Haldane JBS (1931) Linkage in the tetraploid *Primula sinensis*. J Genet 24:121–144
Dowrick VPJ (1956) Heterostyly and homostyly in *Primula obconica*. Heredity 10:219–236
Dulberger R (1970a) Floral dimorphism in *Anchusa hybrida* Ten. Isr J Bot 19:37–41
Dulberger R (1970b) Tristyly in *Lythrum junceum*. New Phytol 69:751–759
Ernst A (1925) Genetische Studien über Heterostylie bei *Primula*. Arch Julius Klaus Stift Vererbungsforsch Sozialanthropol Rassenhyg 1:13–62
Ernst A (1928) Zur Vererbung der morphologischen Heterostylie-Merkmale. Ber Dtsch Bot Ges 46:573–588
Ernst A (1936) Weitere Untersuchungen zur Phänanalyse, zum Fertilitätsproblem und zur Genetik heterostyler Primeln 2. *Primula hortensis* Wettstein. Arch Julius Klaus Stift Vererbungsforsch Sozialanthropol Rassenhyg 11:1–280
Ernst A (1950) Resultate aus Kreuzungen zwischen der tetraploiden, monomorphen *Pr. japonica* und diploiden, mono- und dimorphen Arten der Sektion *Candelabra*. Arch Julius Klaus Stift Vererbungsforsch Sozialanthropol Rassenhyg 25:135–236
Ernst A (1955) Untersuchungen zur Phänanalyse, zum Fertilitätsproblem und zur Genetik heterostyler Primeln 4. Die F2-F5-Nachkommenschaften der Bastarde *Pr. (hortensis × viscosa)*. Erster Teil. Arch Julius Klaus Stift Vererbungsforsch Sozialanthropol Rassenhyg 30:13–137
Ernst A (1957) Austausch und Mutation im Komplex-Gen für Blütenplastik und Inkompatibilität bei *Primula*. Z Indukt Abstammungs Vererbungsl 88:517–599
Ernst A (1958) Untersuchungen zur Phänanalyse, zum Fertilitätsproblem und zur Genetik heterostyler Primeln 4. Die F2-F5-Nachkommenschaften der Bastarde *Pr. (hortensis × viscosa)*. Dritter Teil. Arch Julius Klaus Stift Vererbungsforsch Sozialanthropol Rassenhyg 33:103–251
Esser K (1953) Genomverdopplung und Pollenschlauchwachstum bei Heterostylen. Z Indukt Abstammungs Vererbungsl 85:28–50
Fisher RA, Mather K (1943) The inheritance of style length in *Lythrum salicaria*. Ann Eugenics 12:1–23
Fyfe VC (1956) Two modes of inheritance of the short-styled form in the "genus" *Oxalis*. Nature (Lond) 177:942–943
Ganders F (1975) Heterostyly, homostyly, and fecundity in *Amsinckia spectabilis* (Boraginaceae). Madroño 23:56–62
Ganders F (1979) The biology of heterostyly. N Z J Bot 17:607–635
Garber RJ, Quisenberry KS (1927) The inheritance of length of style in buckwheat. J Agric Res 34:181–183
Grüneberg H (1938) An analysis of the "pleiotropic" effects of a new lethal mutation in the rat (*Mus norvegicus*). Proc R Soc B 125:123–144
Haldane JBS (1933) Two new allelomorphs for heterostylism in *Primula*. Am Nat 67:559–560
Hildebrand F (1866) Über den Trimorphismus in der Gattung *Oxalis*. Monatsber König Preuss Akad Wiss Berl 1866:352–374
Lewis D (1942) The physiology of incompatibility in plants. 1. The effect of temperature. Proc R Soc B 131:13–26
Lewis D (1943) The physiology of incompatibility in plants. 2. *Linum grandiflorum*. Ann Bot 7:115–122
Lewis D (1949) Incompatibility in flowering plants. Biol Rev Camb Philos Soc 24:472–496
Lewis D (1954) Comparative incompatibility in angiosperms and fungi. Adv Genet 6:235–285
Lewis D (1975) Heteromorphic incompatibility system under disruptive selection. Proc R Soc B 188:247–256
Lewis D (1982) Incompatibility, stamen movement and pollen economy in a heterostyled tropical forest tree, *Cratoxylum formosum* (Guttiferae). Proc R Soc B 214:273–283
Lewis D, Crowe LK (1958) Unilateral interspecific incompatibility in flowering plants. Heredity 12:233–256
Lewis D, Rao AN (1971) Evolution of dimorphism and population polymorphism in *Pemphis acidula* Forst. Proc R Soc B 178:79–94

Lewis D, Verma SC, Zuberi MI (1988) Gametophytic-sporophytic incompability in the Cruciferae – *Raphanus sativus*. Heredity 61:355–366

Lund HA (1956) Growth hormones in the styles and ovaries of tobacco responsible for fruit development. Am J Bot 43:562–568

Mather K (1950) The genetical architecture of heterostyly in *Primula sinensis*. Evolution 4:340–352

Mather K, De Winton D (1941) Adaptation and counter-adaptation of the breeding system in *Primula*. Ann Bot 5:297–311

McClelland TB (1919) Influence of foreign pollen on the development of *Vanilla* fruits. J Agric Res 16:245–251

Mulcahy DL (1964) The reproductive biology of *Oxalis priceae*. Am J Bot 51:1045–1050

Ornduff R (1964) The breeding system of *Oxalis suksdorfii*. Am J Bot 51:307–314

Ornduff R (1966) The origin of dioecism from heterostyly in *Nymphoides* (Menyanthaceae). Evolution 20:309–314

Ornduff R (1972) The breakdown of trimorphic incompatibility in *Oxalis* section *Corniculatae*. Evolution 26:52–65

Ornduff R (1979) The genetics of heterostyly in *Hypericum aegypticum*. Heredity 42:271–272

Ornduff R (1980) The probable genetics of distyly in *Gelsemium sempervirens* (Loganiaceae). Can J Genet Cytol 22:303–304

Pellew C (1928) Annual Report John Innes Horticultural Institution for 1928, p 13. Innes Inst, Norwich, England

Philp J (1929) Annual Report John Innes Horticultural Institution for 1929, p 24–25. Innes Inst, Norwich, England

Piper J (1984) Breeding system evolution in *Primula vulgaris*. PhD Thesis, Univ Sussex, England

Ray PM, Chisaki HF (1957) Studies of *Amsinckia*. 1. A synopsis of the genus, with a study of heterostyly in it. Am J Bot 44:529–536

Richards AJ, Ibrahim HB (1982) The breeding system in *Primula veris* L. 2. Pollen tube growth and seed-set. New Phytol 90:305–314

Richards JH, Barrett SCH (1984) The developmental basis of tristyly in *Eichhornia paniculata* (Pontederiaceae). Am J Bot 71:1347–1363

Sampson DR (1971) Mating group ratios in distylic *Forsythia* (Oleaceae). Can J Genet Cytol 13:368–371

Schou O, Philipp M (1984) An unusual heteromorphic incompatibility system. 3. On the genetic control of distyly and self-incompatibility in *Anchusa officinalis L. (Boraginaceae)*. Theor Appl Genet 68:139–144

Shore JS, Barrett SCH (1985) The genetics of distyly and homostyly in *Turnera ulmifolia* L. (Turneraceae). Heredity 55:167–174

Sömme AS (1930) Genetics and cytology of the tetraploid form of *Primula sinensis*. J Genet 23:447–509

Stirling J (1936) Studies of flowering in heterostyled and allied species. 3. Gentianaceae, Lythraceae, Oxalidaceae. Publ Hartley Bot Lab, Univ Liverpool, 15:3–24

Stevens NE (1912) Observations on heterostylous plants. Bot Gaz 53:277–308

Stout AB (1925) Studies on *Lythrum salicaria*. 2. A new form of flower in this species. Bull Torrey Bot Club 52:81–85

Tatebe T (1964) Physiological studies on the fertilization of *Lythrum salicaria* L. 2. On pollen tube growth. J Jpn Soc Hortic Sci 33:155–158

Von Ubisch G (1926) Koppelung von Farbe und Heterostylie bei *Oxalis rosea*. Biol Zentrabl 46:633–645

Chapter 6

The Evolution of Heterostyly

D.G. LLOYD[1] and C.J. WEBB[2]

1 Introduction

Charles Darwin was fascinated by the phenomenon of heterostyly. He described (1862, 1877) how he first thought that pin and thrum plants of *Primula* species represented female and male sexes respectively, but found that they were both functionally hermaphroditic. He demonstrated the infertility of self-pollinations and crosses between plants of the same form, and concluded that the two forms, although hermaphrodites, are "related to each other like males and females... [because plants of each form]... must unite with one of the other form" (Darwin 1862)[3].

Since Darwin drew attention to the remarkable properties of heterostyly, the subject has been intensively investigated as an outbreeding mechanism and as an ideal model system for studying the structure of plant populations. Despite this concentration of effort, the evolutionary origins of heterostyly are still obscure. In this paper, we consider the nature of the flowers of the immediate nonheterostylous ancestors that gave rise to heterostylous descendants and we discuss the evolutionary pathways by which the suites of characters that contribute to heterostyly may have evolved. Our attention is concentrated on the origins of heterostyly, especially distyly, and we do not consider the reversion of heterostyly to monomorphic derivatives (see Chap. 10).

Heterostyly has evolved repeatedly by parallel evolution and occurs in approximately 25 angiosperm families (Ganders 1979, and below). There are a number of morphological and physiological features that commonly differ between the floral morphs in a population. The characters that distinguish the morphs frequently include stigma and anther height, the incompatibility reactions of stigmas and pollen, and morphological differences in the stigmatic papillae and pollen grains (described in Vuilleumier 1967; Ganders 1979; Chap. 3). There are also additional differences between the morphs that are present in only a minority of heterostylous populations.

[1]Department of Plant and Microbial Sciences, University of Canterbury, Private Bag, Christchurch, New Zealand
[2]Botany Division, DSIR, Private Bag, Christchurch, New Zealand

[3]Darwin considered the situation where hermaphrodite individuals are cross-sterile with half the members of their population to be unique (Darwin 1877, pp. 244, 275–7) and comparable to interspecific sterility. Nowadays, we can recognize a near-parallel in heterodichogamous species, in which two morphs mate only with each other because their male and female phases are reciprocally synchronized (see Lloyd and Webb 1986). In these species, however, plants do not function as maternal and paternal parents at the same time.

Monographs on Theoretical and Applied Genetics 15
Evolution and Function of Heterostyly (ed. by S.C.H. Barrett)

These less widespread features include corolla shape, stamen or stigma orientation, the color of the pollen, filaments or the corolla, the possession of hairs, ovule size, and others (Darwin 1877, pp. 245–254; Ganders 1979; Richards 1986). In total, there are over 20 floral characters that sometimes differ between the morphs of heterostylous populations. As yet, no vegetative differences have been found between the floral morphs.

We will not consider all characters individually but choose instead to divide the morph differences into three categories: (1) characters that contribute to the reciprocal herkogamy of anther and stigma positions, principally heights above the base of a flower, (2) the components of self- and intramorph incompatibility, and (3) the rest, which we group together as "ancillary features" of heterostyly. The primary objective of the present chapter is to investigate the nature and evolutionary origins of heterostyly and the sequence in which the three sets of characters have evolved. We concentrate on reciprocal herkogamy and self-incompatibility, and give less attention to the more varied, and probably later-evolved, ancillary features.

Since the differences between the morphs of heterostylous species vary so much, it is necessary to decide which features are essential to define heterostyly. We follow Hildebrand (1866, cited in Ganders 1979) and many subsequent authors in defining heterostyly by the presence of the unique pattern of reciprocal herkogamy. On this criterion, self-incompatibility and any of the ancillary features need not be present for a population to qualify as being heterostylous. The advantages of circumscribing heterostyly by the presence of its reciprocal herkogamy are twofold: reciprocal herkogamy is easier to detect than self-incompatibility, and there are far more species that have reciprocal herkogamy but lack self-incompatibility than have diallelic self-incompatibility but lack reciprocal herkogamy (notably some species of Plumbaginaceae, Baker 1966).

Accordingly, we define heterostyly as a genetically determined polymorphism in which the morphs differ in the sequence of heights at which the anthers and stigmas are presented within their flowers. The definition emphasizes the way the flowers function in relation to their pollinators and incorporates three features of the reciprocal herkogamy of heterostylous species:

1. The anthers and stigmas are separated in every flower by their height above the base of the flower (the flowers are herkogamous). We identify the basic morphological characters of heterostylous flowers as stigma and anther heights, which determine the positions of the pollinating surfaces, rather than stamen and style lengths, which obscure the ecological significance of the features.
2. The relative positions of anthers and stigmas distinguish each morph from the other(s). In distylous populations, the stigmas are above (pin) or below (thrum) the anthers. In tristylous populations, the stigmas are respectively above (long-style), between (mid-style) or below (short-style) the anthers.
3. Both the anther and stigma heights differ between morphs.

The reciprocal herkogamy of heterostylous populations does not require that the heights of the anthers and stigmas coincide exactly in the two or three morphs. There are often considerable disparities in the exact positions of the anthers and stigmas involved in the "legitimate" transfer of pollen from an anther to a stigma at the same level. There may even be appreciable overlap in the absolute positions of anthers (or stigmas) of different morphs, as in *Turnera ulmifolia* (Martin 1965) and *Narcissus fernandesii* (Fernandes 1964) and *Guettarda scabra* (J. Richards and S. Kaptur unpubl.). This variation still preserves the sequence of anther and stigma heights characteristic of each morph, however.

Our definition distinguishes heterostyly from several other types of herkogamy polymorphism. Some species have a stylar polymorphism in which two herkogamous morphs differ in style height but not in anther height (e.g., *Chlorogalum angustifolium*, Jernstedt 1982; *Epacris impressa*, O'Brien and Calder 1989; examples are discussed further below). In these species, the stigmas are above the anthers in one morph and below the anthers in the other. Other polymorphic populations have a herkogamous (long- or short-styled) morph and a nonherkogamous ("homostylous") morph, as in some *Narcissus* species (Henriques 1887; Dulberger 1964). Some enantiomorphous species have individuals with left- and right-handed flowers, in which the anthers and stigmas occur on reverse *sides* in the two kinds of flowers.

A wide range of views has been expressed on the origin of heterostyly. The first hypothesis was presented by Darwin (1877, p. 260 on), who postulated that the reciprocal herkogamy evolved first in response to selective forces increasing the accuracy of pollen transfer from anthers to stigmas at approximately the same height and that the self-incompatibility and ancillary features evolved subsequently. In the 20th century, almost all authors have rejected Darwin's evolutionary scenario, particularly the postulate that self-incompatibility evolved after reciprocal herkogamy. Mather and de Winton (1941) postulated that self-incompatibility and reciprocal herkogamy arose together. There have been many suggestions that self-incompatibility was the first feature of heterostyly to evolve, particularly since Baker (1966) reported the variety of systems in the Plumbaginaceae. This view has been supported by the quantitative model for the evolution of distyly presented by D. Charlesworth and B. Charlesworth (1979). They postulated that diallelic self-incompatibility arose in two steps: (1) a mutation to a new pollen type that is unable to fertilize either itself or individuals of the first type (and is therefore initially male-sterile in effect), followed by (2) a linked stigma mutation that is incompatible with the original pollen type but compatible with the new type.

There are several reasons why recent authors have favored the evolution of self-incompatibility as the initial step in the evolution of heterostyly. The occurrence of self-incompatibility without a herkogamy polymorphism in some species of Plumbaginaceae (Baker 1966) supports this view. Moreover, the prevention of self-fertilization has often been regarded as the selective force primarily responsible for the evolution of heterostyly (e.g., D. Charlesworth and B. Charlesworth 1979). Self-incompatibility by itself appears to prevent selfing more effectively than does reciprocal herkogamy alone. The promotion of cross-fertilization by the transfer of pollen to stigmas at the same height, which Darwin considered the key to heterostyly, has been viewed as a lesser, secondary selective force. The evidence for the promotion of cross-pollination by legitimate transfers has been widely judged to be weak and ambiguous (wrongly in our opinion, as discussed in Chap. 7).

In addition, modern views on the evolution of heterostyly have frequently been directed by genetical considerations. For example, the extent to which legitimate pollen transfer predominates over illegitimate transfer has usually been examined as the degree of disassortative mating, rather than as differences between the relative proficiencies of the two types of pollination, the factor that is most likely to cause any nonrandomness. And in particular, the recognition of the supergene that coordinates variation in the principal features of distyly in *Primula*, and putatively in other genera as well (Ernst 1955; Dowrick 1956; Chap. 5), has dominated much thinking on heterostyly. One consequence of the genetical emphasis has been an implicit assumption that the ancestors of heterostylous species had anthers and stigmas at similar heights (as in homostyly, a result of recombination within heterostyly supergenes) and were at least partly self-fertilized (but see Ganders 1979). This assumption has reinforced the views that the selective advantage of heterostyly lies in preventing self-fertilization and that the self-incompatibility evolved first. It is difficult to argue that Darwin's selective force, the promotion of legitimate pollinations, is responsible for the evolution of heterostyly if one assumes the ancestor already had anthers and stigmas at the same height.

We believe that the ancestors of heterostylous species had anthers and stigmas at different heights and were not "homostylous". The supergene is a specialized genetical mechanism that coordinates variation in a set of derived coadapted characters, by establishing and maintaining close linkage between the genes controlling the component features. The production of homostylous mutants results from a breakdown in linkage by recombination between the components of the supergene. There is little reason why the reshuffled combinations of the specialized supergene should resemble the ancestors of heterostylous species, which would have lacked most of the advanced features of the heterostylous species.

Our arguments in this paper concentrate on floral ecology and on taxonomic, rather than genetical, approaches to the evolutionary origins of heterostyly. To examine what the floral characters of the ancestors of heterostylous species may have been like, we first review the floral morphology and pollination ecology of the *nonheterostylous* members of families that contain heterostylous species. We conclude that the ancestors of heterostylous species were probably herkogamous and in most or all cases possessed "approach herkogamy" (Webb and Lloyd 1986), in which the stigmas protrude beyond the anthers as in the long-styled morphs of heterostylous species. If this conclusion is correct, then the relative plausibility of alternative pathways for the evolution of heterostyly is radically altered, as we discuss below. The feasibility of the selective forces responsible is also affected; we consider that in Chapter 7.

2 Floral Characters of Families in Which Heterostyly Has Evolved

Although it is clear that heterostyly has evolved independently in many families, it is equally clear that heterostyly is not distributed at random among plant families (Darwin 1877; Ganders 1979). In this section, we examine the floral characters of families in which heterostyly has arisen. This topic has received little attention in the heterostyly literature, as most authors have restricted their attention to species that are actually heterostylous. Also, the floral characters that have been studied most are those primary ones that vary between heterostylous morphs and are used to define the syndrome of heterostyly itself. Examining floral characters at the level of plant families may provide clues as to why certain plant groups have a higher probability of evolving heterostyly, as well as give some indication of the functional significance of heterostyly. What we seek to define here is a syndrome of characters that describes the sort of flower that has become heterostylous. For this purpose, we would ideally examine the properties of the sister groups (in the sense of cladistics) of the heterostylous taxa. These are less likely to have reverted from heterostyly than the non-heterostylous representatives within heterostylous taxa. Regrettably, the sister groups of heterostylous taxa are not usually known, so we must make do with an examination of the characters of families with heterostylous representatives.

We have surveyed floral characters of families with distylous and/or tristylous species. The floral characters selected are those that are important in the structural interaction between pollinator and flower, and are generally those that show a predominance of certain character states among families with heterostylous species. Table 1 lists the 18 characters, each defined by the character state usually found in families with heterostyly. Families were assigned to character states mostly according to the family descriptions provided by Cronquist (1981) and Heywood (1978). Information on stigma surface morphology is based on Heslop-Harrison and Shivanna (1977) and Schill et al. (1985). The families are delineated as in Dahlgren (1983).

We have accepted 20 families with distyly only (see Ganders 1979), including the very doubtful Caesalpiniaceae (*Bauhinia*, discussed in Lloyd et al. 1990). We accept five families with tristyly as well as dimorphic populations, namely the Connaraceae (Lemmens 1989) and the Amaryllidaceae (Henriques 1887; Fernandes 1964; Lloyd et al. 1990) as well as the Lythraceae, Oxalidaceae, and Pontederiaceae. In Dahlgren's system, the 25 heterostylous families are referred to 16 of his 33 superorders and to 19 different orders. In only two cases does it seem possible that heterostyly may be the primitive state in related families and therefore may not have arisen independently. These are the Linaceae and Erythroxylaceae (Geraniales) and Gentianaceae and Menyanthaceae (Gentiales). Consequently, heterostyly appears to have evolved at least 23 times and possibly on considerably more occasions if heterostyly has arisen more than once in some families.

2.1 The Flowers of Families with Heterostylous Species

In considering the plants that have become heterostylous, Darwin (1877, p. 259) noted that heterostyly was completely lacking in several large families with flowers adapted for cross-fertilization and which typically have an irregular corolla. He mentioned the Fabaceae (s.l.), Lamiaceae, Scrophulariaceae, and Orchidaceae; one doubtful exception is now known in the Caesalpiniaceae (Fabaceae s.l. – *Bauhinia*, Vogel 1955). He also noted that no heterostylous species is wind-pollinated and that only *Pontederia* has a plainly irregular corolla. (*Eichhornia* and *Lythrum* also have moderately zygomorphic corollas, however). He interpreted these observations by noting that it would be little use for such plants to become heterostylous because they are already well adapted to cross-pollination.

Ganders (1979) discussed a few floral characters other than those that define heterostyly itself. He considered the character states typical of heterostylous taxa

Table 1. Floral character states for the 25 families with known heterostylous species

	Character state	No. of families/ total number	Exceptions[a]
1.	Flower hermaphrodite	24/25	*Clusiaceae, *Santalaceae,
2.	Pollination biotic, predominantly insect	25/25	
3.	Corolla actinomorphic	21.5/25	Acanthaceae, Caesalpiniaceae, *Iridaceae, *Pontederiacae, *Saxifragaceae
4.	Corolla sympetalous	13.5/25	*Amaryllidaceae, Clusiaceae, Connaraceae, Erythroxylaceae, Fabaceae, Linaceae, Lythraceae, Olacaceae, Oxalidaceae, *Polygonaceae, *Santalaceae, Saxifragaceae, Sterculiaceae
5.	Corolla forming a tube (tube, bell, funnel, gullet blossoms)	22/25	Amaryllidaceae, Clusiaceae Polygonaceae
6.	Nectar as reward	23.5/24	*Clusiaceae (no data on Turneraceae)
7.	Reward concealed at base of corolla	23.5/24	*Clusiaceae (no data on Turneraceae)
8.	Anthers and/or stigmas exerted	25/25	
9.	Stamens in one or two whorls	23.5/25	*Acanthaceae, Clusiaceae
10.	Anthers not connate	25/25	
11.	Anther dehiscence by longitudinal slits	25/25	
12.	Pollen presentation peripheral	25/25	
13.	Style elongated	21.5/25	Gentianaceae, *Menyanthaceae, *Olacaceae, *Polygonaceae, Saxifragaceae
14.	Stigmas relatively localized and presented centrally	25/25	
15.	Stigma surface papillate	20.5/23	*Acanthaceae, Polygonaceae, *Sterculiaceae, *Rubiaceae (no data on Connaraceae, Olacaceae)
16.	Ovary superior	21/25	Amaryllidaceae, Iridaceae, Rubiaceae, Santalaceae
17.	Ovary syncarpelous	24/25	Connaraceae
18.	Carpels 1–5	24.5/25	*Sterculiaceae

[a] Families are assigned by most frequent character state; where states are reported as equally frequent, families are included in both the Exception column and marked * and the Number of families column and scored 0.5.

rather than their families, but for most of the characters he discussed this makes little difference. He noted that pollination of heterostylous species was mostly by insects, and that hummingbird pollination is known but there are no wind- or bat-pollinated flowers. He also observed that heterostylous flowers typically have a small, definite number of stamens and with the exception of *Pontederia* have actinomorphic corollas. He suggested that most heterostylous species are sympetalous, and that in those that are apopetalous there is nevertheless formation of a floral tube. He concluded that tube formation is advantageous in positioning insect mouth parts for efficient pollination. Ganders pointed out that heterostyly is lacking in both the most primitive and most advanced families of angiosperms. He also observed that on the whole flowers of heterostylous species are rather small.

An additional characteristic of families in which tristyly occurs is that all have flowers with two stamen whorls. This has been considered a necessary precondition for the evolution of anthers at two distinct heights in the same flower (Yeo 1975), but that view has been discredited (Richards and Barrett 1987).

Our analysis of the character states that predominate in families with heterostylous species (Table 1) is basically in agreement with the observations of Darwin and Ganders but adds further characters and indicates that there are surprisingly severe constraints on the sort of flower that has become heterostylous.

The flowers of families with heterostylous members are hermaphroditic and, as pointed out by Darwin and Ganders, are biotically pollinated, usually by insects. Only three of the 25 known heterostylous families are in the monocotyledons. The structure of the flower is relatively simple, with the open corolla actinomorphic or slightly zygomorphic, and forming a variously shaped floral tube that is probed by the visitor. The type of flower that has become heterostylous appears to have been one with an intermediate level of specialization. Flower shapes are generally in the tube, or less often funnel classes of Faegri and van der Pijl (1979), rather than in the dish-bowl, flag, gullet, or brush classes. Nectar is generally present and concealed at the base of the floral tube. The floral tube may be formed in various ways, as noted by Ganders (1979), but contrary to Gander's statement, the petals are frequently not fused. Because flower size varies greatly within families, we do not include this character in our analysis.

It is clear why some character states restrict or prevent the evolution of heterostyly. In particular, unisexual flowers cannot become heterostylous, although they may be derived from heterostyly (Beach and Bawa 1980). Abiotically pollinated flowers, and less specialized flowers with brush blossoms or with open dish-shaped corollas and exposed nectar, lack the precision required for reciprocal pollen transfer and generally have not become heterostylous. Also, many primitive families have forms of complete dichogamy which preclude the evolution of heterostyly. Equally, many taxa with specialized flowers are excluded; for example, those with strongly zygomorphic gullet and flag blossoms, or closed or triggered flowers. Such flowers generally have precise placement of stigmas and pollen. They are often horizontally oriented and have zygomorphic corollas accompanied by a dichogamy pattern that brings the pollinating surfaces into the same position sequentially. Thus as noted by Darwin, heterostyly is absent in the Lamiaceae, Scrophulariaceae and Orchidaceae, and is rare (or absent?) in the Fabaceae s.l. It is also absent in the Asteraceae and related families with reduced flowers aggregated into blossoms.

Androecial characters reveal a similar pattern of intermediate advancement. Most families with heterostylous species have a small, definite number of stamens in one or two whorls. The anthers are always free, although the stamens are often epipetalous, and dehisce by longitudinal slits. Families characterized by numerous stamens in many whorls, which present pollen over a broad area, generally have no heterostylous representatives. At the other extreme, families with precise presentation mechanisms involving fused anthers, pollinia, or secondary pollen presentation have no heterostylous species.

The ovary is usually superior and comprises few, usually united carpels. No families with numerous free carpels have given rise to heterostyly, and few families with inferior ovaries have heterostyly species.

Stamens and/or styles are usually exerted from the floral tube with the pollen presented peripherally and the stigmas relatively localized and central. Families with numerous, dispersed stigmas have no heterostylous species. At the other extreme, families in which the stigmatic surface is eccentric within the flower and the pollen is presented over a narrowly restricted area generally have no heterostylous species.

As suggested by the morphological features, the taxonomic distribution of heterostyly is strongly nonrandom. Heterostyly is completely absent in the first four superorders of dicotyledons in Dahlgren's system, including the Magnoliiflorae, and in the first three superorders of the monocotyledons, including the Alismatiflorae. Conversely, heterostyly is rare or absent in the advanced superorders such as the Lamiiflorae and Zingiberiflorae. Heterostyly is relatively more common in the dicotyledons; whereas the monocotyledons make up 22% (92/461) of Dahlgren's families, they provide only 12% (3/25) of the heterostylous families.

2.2 Families with Exceptional Character States

For six of the 18 characters listed in Table 1, all of the 25 heterostylous families have the same predominant character state. For the remaining characters, however, the predominant character state, either in some whole families or in parts of families, differs from that prevailing in most heterostylous families. In many of the exceptional cases, heterostylous species actually conform to the general syndrome because they belong to subfamilies, tribes, or genera that differ from most of their family in having the character state that prevails in other heterostylous families. In other exceptions, the character is modified in some way so that the flower nevertheless fits the general prerequisites for the evolution of heterostyly.

Unisexual flowers are found in the Clusiaceae and to a less extent in the Sterculiaceae, but not in the subfamilies in which distyly has arisen. In the Santalaceae, unisexuality is not present in the only heterostylous genus, *Arjona*, or apparently in related genera.

In five heterostylous families the corolla is usually or often zygomorphic; in the Acanthaceae, Iridaceae, and Pontederiaceae, genera with heterostylous species do not have strongly zygomorphic corollas, while the only heterostylous genus in the Saxifragaceae, *Jepsonia*, is actinomorphic. *Bauhinia* is the only genus reported to have heterostyly in the Caesalpiniaceae, a family with zygomorphic corollas and complex pollination mechanisms which make it an unusual source of heterostyly (and in which heterostyly is unconfirmed). The questionably distylous species *B. esculenta* (Vogel 1955) has a zygomorphic corolla but not the typical flag blossom of many legumes. It belongs to a group of species seen by Wunderlin et al. (1981) as an early offshoot with some primitive characters. There is insufficient evidence of heterostyly in *B. esculenta* for it to be considered confirmed (Lloyd et al. 1990).

As mentioned above, many heterostylous families do not have sympetalous corollas, but in all families with the exception of the Clusiaceae and Polygonaceae there is usually a floral tube, even if the blossom shape is primarily dish- or bowl-shaped. The floral tube may in part be formed by the calyx, filaments, or appendages of the petals as noted by Ganders (1979). In *Arjona* and related genera of the Santalaceae, the corolla is sympetalous.

The Clusiaceae is an exceptional family in many characters. Typically, the corolla is a dish, there are many stamens, the flowers are nectarless and the pollen reward is well exposed. Of the four distylous genera, three have been studied in detail. Ornduff (1975) noted that in gross morphology the flowers of *Hypericum aegypticum* are similar to those of some distylous species in other families. He illustrated the numerous stamens aggregated into three distinct bundles, and suggested that the bundles may function like the fewer stamens characteristic of distylous flowers. Robson (1972) considered that *H. aegypticum* and its allies form a derived group within *Hypericum*, and noted that the stiff petals of these species form the petals into a floral tube. Furthermore, he noted that these species are unusual in *Hypericum* in having stamens united to above the middle and petal appendages with nectariferous tissue at their bases. He described the same association between formation of a floral tube, union of the stamens and nectar production in *Eliaea* and *Cratoxylum* section *Tridesmos*, both of which have distylous species, and suggested that this group of character states had arisen in these two genera independently from *Hypericum*. Thus the heterostylous representatives of the Clusiaceae and their close relatives are not anomalous in their floral syndrome.

The Polygonaceae are also atypical in several features. They frequently have dish-shaped corollas and this is true of the only heterostylous genus, *Fagopyrum*. They also often have short styles associated with their dish blossoms, as in *Fagopyrum* (Darwin 1877, Figure on p. 112). *Fagopyrum* must be considered unusual among heterostylous plants in these characters. Nevertheless, the flowers of this and related genera may function in a way similar to flowers with a well-defined tube and exerted styles because approach herkogamy has been described for dish blossoms (Webb and Lloyd 1986). Observations of how insects approach and operate these flowers are needed.

In addition to the Polygonaceae, four other families have relatively short styles. In all these cases, the families or at least the heterostylous genera have superior ovaries elongated below the short style (e.g., many Gentianaceae) or the whole flower is small (e.g., many Saxifragaceae).

In many Acanthaceae, there is less than a complete set of stamens. This is true of the only heterostylous genus, *Oplonia*, which has only two stamens. The reduction from four to two stamens seems to be associated with the development of zygomorphy, not very pronounced in *Oplonia* (Stearn 1971). It is not associated with a pollination mechanism of the degree of complexity that precludes the evolution of heterostyly.

Smooth stigma surfaces occur in several families, but as the data for this character are so sparse the exceptions are difficult to interpret. Information for stigma surfaces is based on the relatively few species for which data is available in each family; nevertheless, the presence of papillate stigmas may be a prerequisite for the evolution of the self-incompatibility system usually associated with heterostyly.

Inferior ovaries predominate in four families. The reason for the association of heterostyly with superior ovaries is unclear, but in all Rubiaceae, most Iridaceae, and *Arjona*, the only heterostylous genus in the Santalaceae, the corolla is sympetalous and usually forms a long floral tube, while in the advanced Amaryllidaceae (including *Narcissus*) there is a tubular corolla. This suggests that in apopetalous families a superior ovary may increase the probability that the flower will have a floral tube.

Only one of the families, Connaraceae, has free carpels. However, they number only one to five and may be fused at the base, and in the heterostylous genera the stigmas are relatively localized and central within the flowers (Hemsley 1956; Baker 1962). Fused carpels may predominate in families with heterostylous species because they are associated with long styles and a relatively localized stigma position.

2.3 Conclusion

The character states that predominate in families with heterostylous species describe flowers that the probe of a pollinator descends into to seek nectar. Probed flowers without a particular bilateral orientation of pollinators were described as stereomorphic by Leppik (1972). The pollinating surfaces contact the probe in the course of its passage into, or out of, the flower. The contacts with the stigmas and anthers are kept separate by their low numbers and by their segregation at different heights and into central versus peripheral positions. The important feature of the passing contacts of these "depth-probed" flowers is that the pollinating surfaces are contacted *in succession* as the probe enters or leaves the flower. The contacts are not in a random sequence, as in the unordered herkogamy of flowers that lack depth, nor are they with only one of the two pollination surfaces, as in flowers that are completely dichogamous or unisexual.

In many depth-probed flowers, the stigmas project above the anthers, a condition described as approach herkogamy (Lloyd and Webb 1986). The path the pollinator takes when probing such flowers in search of nectar at the base of the floral tube brings it into contact first with the exerted central stigmas, and second with the peripherally presented pollen. Flowers with approach herkogamy are common in moderately advanced families such as those that have evolved heterostylous species. We postulate that the syndrome of characters in families with heterostylous species is that of depth-probed flowers with approach herkogamy.

Unfortunately, we cannot verify from the literature that the nonheterostylous species of these families actually have approach herkogamy, because the relative positions of stigmas and anthers are not usually indicated in taxonomic descriptions. The data required to confirm or contradict our hypothesis are not generally available. Nevertheless, it is suggestive that most of the character states listed in Table 1 are consistent with the syndrome of approach herkogamy. The corolla characters are associated with the path to concealed nectar, and the few unspecialized anthers are suited with contact with a passing probe. That the stigmas are generally exerted from the floral tube explains the predominance of elongated styles, and the central, localized stigmatic area is more often achieved when styles, or at least carpels, are fused.

We conclude that the ancestors of heterostylous species were most likely to have approach herkogamy associated with a depth-probed flower, and were therefore at least partly outcrossed. This is consistent with the strongly nonrandom taxonomic distribution of families with heterostylous species, concentrated among dicotyledons of intermediate advancement. The syndrome that is often suggested or implied as the ancestor of heterostyly, "homostyly", facilitates self-pollination and occurs in many angiosperm families. It is therefore difficult to reconcile with the limited taxonomic distribution of heterostyly, and it is more likely to be a derivative condition associated with self-fertilization than the predecessor of heterostyly.

3 Postulates About Evolutionary Events

The evolution of heterostyly is a complex matter, requiring a series of events for each of the three principal components – reciprocal herkogamy, self-incompatibility, and ancillary characters. In this section we will discuss the major features of the evolution of heterostyly, with particular emphasis on distyly. Consideration of ecological, morphological, and genetical aspects of heterostyly has led us to four primary postulates concerning the evolution of distyly:

1. The immediate ancestors of heterostylous species had separated anthers and stigmas, and in particular they possessed approach herkogamy.
2. The evolution of reciprocal herkogamy preceded that of self-incompatibility and the ancillary features.
3. The initial step in the evolution of reciprocal herkogamy was the evolution of a stigma-height polymorphism.
4. Heteromorphic, diallelic self-incompatibility is qualitatively distinct from homomorphic, multiallelic self-incompatibility.

None of these postulates has been widely advocated in the past, and current orthodoxy is diametrically opposed to the second one especially. We present our arguments for each postulate in turn.

3.1 Ancestors of Heterostylous Species Had Approach Herkogamy

Heterostyly frequently reverts to a monomorphic homostylous condition in which the anthers and stigmas are held at approximately the same height (see Ganders 1979; Charlesworth 1979; Chap. 10). Such homostyly is the result of recombination within the supergene that coordinates variation in the components of the full heterostyly syndrome (Chap. 5). Most authors who have considered the evolution of heterostyly have explicitly or implicitly assumed that the ancestor resembles the reversion product. To our knowledge, only Ganders (1979) has previously postulated a herkogamous ancestry of heterostylous species. His reasoning was that an outbreeding advantage (also widely held to be the selective advantage of heterostyly) was unlikely to be sufficiently strong to account for the evolution of heterostyly unless the ancestor was already partly outcrossing.

There are several additional reasons for postulating that the immediate ancestors of heterostylous groups were herkogamous. The first point, as discussed in the introduction, is that it is unlikely that the primitive condition would resemble homostyly, the breakdown product of a highly evolved and specialized supergene. Second, a herkogamous ancestor simplifies the evolution of reciprocal herkogamy and therefore provides the most parsimonious explanation. The evolution of a herkogamous polymorphism in stigma height requires only one evolutionary step from a herkogamous ancestor – although the initial products do not have exactly complementary positions of stigmas and anthers (see below and Fig. 1). Two steps are required to generate a pair of herkogamous morphs from a homostylous ancestor.

A third reason for postulating that the ancestors of heterostylous taxa were herkogamous is the common occurrence among flowering plants of herkogamy and in particular of one form, approach herkogamy, in which the stigma protrudes beyond the anthers (Webb and Lloyd 1986) as in the pin morph of distylous species. We postulate that the immediate ancestors of heterostylous taxa were most likely to have had approach herkogamy. Although it has been little studied, approach herkogamy is widespread in outcrossing angiosperms. It probably occurs in at least 15% of all flowering plants, as our conservative estimate, and is found in a wide variety of dish, tube, gullet, and trumpet blossoms, based on the classification of Faegri and van der Pijl (1979). Approach herkogamy is actually much more common than heterostyly. Most heterostylous families have blossoms of these types, as discussed above. Moreover, there are a number of cases known where heterostylous taxa within these families have close relatives (other races or species) with approach herkogamy (e.g., *Narcissus bulbocodium*, Bateman 1954a, Fernandes 1966; *Nivenia stokoei*, Ornduff 1974; *Palicourea fendleri*, Sobrevila et al. 1983; *Turnera ulmifolia*, Barrett and Shore 1987). The monomorphic varieties of *T. ulmifolia* are derived from heterostylous ancestors, and this is also possible for the other monomorphic herkogamous populations.

Reverse herkogamy, which resembles the thrum morph of distylous species in presenting anthers beyond the stigmas, is far less common than approach herkogamy (Webb and Lloyd 1986) but it could be the ancestor of some heterostylous taxa.

At present, our hypothesis of a derivation of heterostyly from approach herkogamy is plausible but unconfirmed (as are the alternatives). Cladistic studies of heterostylous groups and their relatives, similar to that recently conducted by Eckenwalder and Barrett (1986), are required to determine whether the ancestors of heterostylous groups had approach herkogamy. It will become clear below and in Chapter 7 that an assumption of an ancestral herkogamous condition radically alters the possibilities for the evolutionary pathways and selective forces by which heterostyly has evolved.

3.2 Reciprocal Herkogamy Preceded Self-Incompatibility and the Ancillary Features

Darwin (1877, p. 267) proposed that the self-incompatibility of heterostylous species evolved after the reciprocal herkogamy but "almost simultaneously", as an incidental byproduct of the coadaptation of the different types of pollen to the appropriate style lengths of legitimate pollinations. Mather and de Winton (1941) suggested that the patterns of herkogamy and self-incompatibility arose together. Almost all other writers on heterostyly have assumed that the self-incompatibility has preceded the reciprocal herkogamy (but see Anderson 1973). As Lewis (1982) wrote: "It has been argued, calculated and generally accepted that the self-incompatibility arose

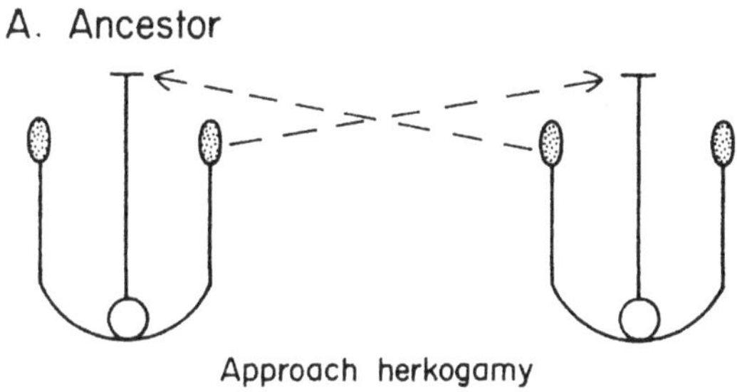

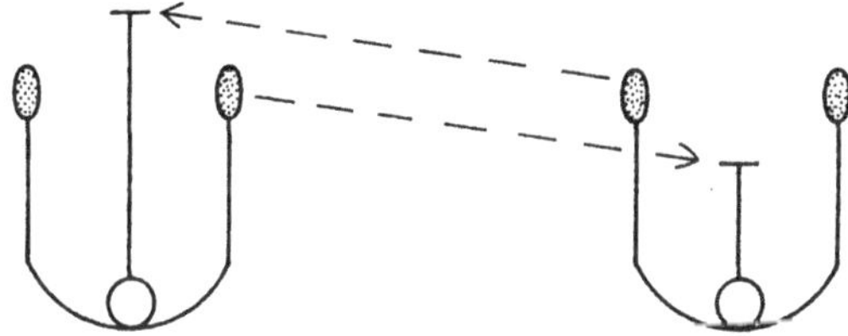

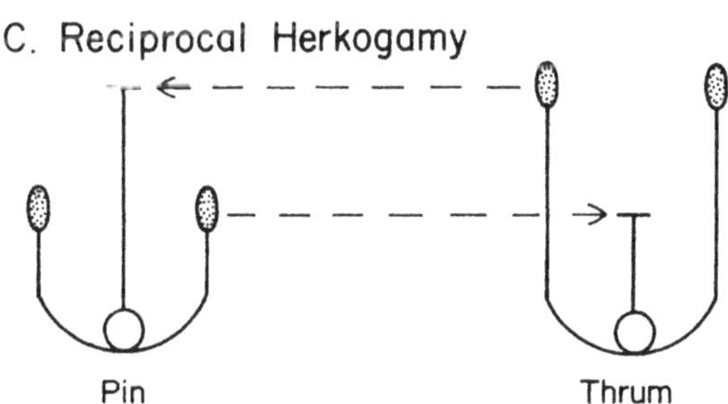

Fig. 1A-C. The postulated principal stages in the evolution of reciprocal herkogamy. **A** The ancestral condition, approach herkogamy. **B** A stigma-height polymorphism. **C** Reciprocal herkogamy. The *arrows* show the directions of most proficient pollen transfer

first. . . ." In contrast, we revive Darwin's proposal that the reciprocal herkogamy of heterostylous taxa preceded self-incompatibility and the ancillary features.

The most critical evidence bearing on this issue is the nature of intermediates possessing only either heteromorphic self-incompatibility or a herkogamous polymorphism. Where reciprocal herkogamy has evolved before self-incompatibility, we may find relict intermediate populations with a herkogamy polymorphism but without diallelic self-incompatibility. On the other hand, where the first step in the evolution of heterostyly has been the introduction of diallelic self-incompatibility, we may find extant intermediates that are self-sterile but do not have a herkogamy polymorphism. The search for intermediate conditions is complicated by the secondary loss of some features of heterostyly in many species (B. Charlesworth and D. Charlesworth 1979; Chap. 10). Hence the occurrence of species with only either diallelic self-incompatibility systems or a herkogamous polymorphism *outside* taxa that contain heterostylous species provides the most convincing evidence for the order of evolution of the herkogamy and self-incompatibility polymorphisms, because reversion from dimorphic to monomorphic conditions is unlikely.

In nonheterostylous groups, herkogamous stigma-height polymorphisms are known in three genera (two *Rhododendron* species, Philipson and Philipson 1975; *Epacris impressa*, Brown and Crowden 1984, O'Brien and Calder 1989; *Chlorogalum angustifolium*, Jernstedt 1982). *Epacris impressa* has a self-incompatibility system that is not associated with the stigma-height polymorphism. [*Veronica gentianoides* is sometimes cited as a fourth example, but the two style lengths were found in plants from different populations (Correns 1924).] These stigma-height polymorphisms are best interpreted as the first stage towards the evolution of reciprocal herkogamy. In contrast, we know of no confirmed reports of diallelic self-incompatibility systems in groups that lack heterostyly. Early reports of a two-locus, diallelic system in Brassicaceae preceded the elucidation of the one-locus, multiallelic system in the family (Bateman 1954b) and can be discounted. Even allowing for the paucity of detailed genetic analyses of self-incompatibility systems, the occurrence of only two or three compatibility groups ought to be possible to detect from a high frequency of cross-incompatibility. The absence of such reports provides some evidence against the evolution of diallelic self-incompatibility mechanisms before the herkogamy polymorphisms of heterostylous species.

Examining now the diversity of conditions *within* taxa containing heterostyly (genera or part-families, which may or may not be monophyletic clades), there are a considerable number of reports of heterostylous, self-compatible populations or species (see Ganders 1979), but only one known series of species that has a diallelic self-incompatibility system but lacks a herkogamy polymorphism (Plumbaginaceae:Staticeae, Baker 1966). This comparison must be treated cautiously, however, as it is not clear in the absence of secure phylogenies how many of the self-compatible, heterostylous species or nonheterostylous species with diallelic incompatibility are primitively so. Stronger evidence is provided by the existence in at least two heterostylous groups of species with multiallelic self-incompatibility systems combined with heterostyly or other herkogamy polymorphisms. Among the self-incompatible species, *Anchusa undulata* subsp. *hybrida* and *A. officinalis* are distylous (Dulberger 1970; Schou and Philipp 1984), while *Narcissus triandrus* is tristylous (Henriques 1887, 1888; Bateman 1952; Fernandes 1964; Lloyd et al. 1990) and *Narcissus tazetta* and several other species have a stigma-height dimorphism only (Dulberger 1964; Barrett et al. unpubl. data). In both genera, the self-incompatibility is not linked with the herkogamy locus. *Villarsia parnassiifolia* (Menyanthaceae) is another distylous species that may have a multiallelic self-incompatibility system (Ornduff 1988). Moreover, in the Boraginaceae, *Narcissus*, and *Villarsia*, homomorphic self-incompatibility is found beyond the heterostylous taxa. The most parsimonious hypothesis for both *Narcissus* and *Anchusa* is that the heterostyly evolved in species that already had multiallelic self-incompatibility systems – not the two- or three-group self-incompatibility systems characteristic of heterostylous species. The phylogeny of events in *Villarsia* is uncertain (Ornduff 1988).

The principal reason why previous authors have favored the evolution of the self-incompatibility mechanisms before reciprocal herkogamy, apart from the primacy often given to genetical considerations over ecological factors in recent discussions of heterostyly, seems to be a widespread conviction that reciprocal herkogamy could not establish on its own. Starting from a "homostylous" ancestor, the usual assumption, this may well be so. D. Charlesworth and B. Charlesworth (1979) pointed out that reciprocal herkogamy cannot readily evolve from a homostyle in response to selection improving the proficiency of cross-pollinations between anthers and stigmas at the same height, the selective force that Darwin (1862, 1877) postulated. A homostylous ancestor would already have anthers and stigmas at the same height, and a change to a pin or thrum form would be likely to *decrease* the amount of crossing. The Charlesworths assumed a homostylous ancestor and proposed that the primary selection force responsible for both the self-incompatibility and herkogamy of heterostylous species was the restriction of self-fertilization.

When a herkogamous ancestry for heterostyly is considered, the relative workability of an outbreeding advantage and the promotion of legitimate cross-pollinations as selective forces responsible for the initial events in the evolution of heterostyly is reversed. The ancestral herkogamy in a monomorphic species is likely to restrict self-fertilization and interference between the androecia and gynoecia (Webb and Lloyd 1986), but it requires cross-pollination between anthers and stigmas at different levels. The introduction of a second herkogamous morph may therefore promote cross-pollination (see Chap. 7), but it need not restrict self-fertilization or self-interference any more than occurs in the original monomorphic ancestor (cf. Ganders 1979). The realization that reciprocal herkogamy is not likely to bring about a decrease in self-fertilization when it evolves from monomorphic herkogamy led Ganders (1979) to suggest that reciprocal herkogamy is unlikely to evolve in the absence of self-incompatibility. Ganders, however, did not distinguish between the (passive) prevention of self-fertilization and an (active) improvement in the proficiency of cross-fertilization. Both "ensure cross-fertilization" (Darwin 1877, p. 258), but they do so in different ways. When the distinction is made firmly (Lloyd and Yates 1982), the difficulty that Ganders saw in the evolution of heterostyly from a herkogamous ancestor without the support of a preexisting diallelic self-incompatibility system disappears.

3.3 The Initial Step in the Evolution of Heterostyly is a Stigma-Height Polymorphism

We postulate that the initial step in the evolution of reciprocal herkogamy, and subsequently other features of heterostyly, is the introduction into a herkogamous population of a second herkogamous morph with a stigma at a different height. Assuming that the monomorphic ancestor had approach herkogamy, the invasion of a mutant with reverse herkogamy would establish a stigma-height polymorphism (Fig. 1).

The primary reason for favoring an initial polymorphism in stigma height rather than anther height is that a number of stigma-height polymorphisms are known. Some of these occur in genera with heterostylous species (e.g., *Narcissus cyclamineus* and *N. minor*, Henriques 1887; *N. tazetta*, Dulberger 1964; *Villarsia parnassiifolia*, Ornduff 1986), where they could be primitive or secondarily derived from heterostyly. Other stigma-height polymorphisms occur in taxa without heterostylous representatives (*Rhododendron* species, Philipson and Philipson 1975; *Chlorogalum angustifolium*, Jernstedt 1982; *Epacris impressa*, Brown and Crowden 1984, O'Brien and Calder 1989) and have presumably evolved directly from monomorphic ancestors. [The polymorphism in *Epacris impressa* may have evolved from reverse herkogamy, which is common in the Epacridaceae (Webb and Lloyd 1986)]. There are other species in which the style is highly variable among plants, but the distribution of stigma heights is not bimodal or has not been documented (e.g., *Gesneria pendulina*, Darwin 1877, p. 261; *Mirabilis froebelii*, Baker 1964). In contrast, we know of no instances of the reverse kind, an anther-height polymorphism without a stigma-height polymorphism. In species with epipetalous stamens, as in many heterostylous groups (above), a sudden discrete change in stamen length may be precluded by the adherence of filaments to the corolla.

Moreover, in a number of distylous species the dimorphism in the style is much more pronounced than that of the stamens. The difference between morphs in stigma height exceeds the difference in anther height by more than 50% in *Hedyotis nigricans* (Bir Bahadur 1970), *Anchusa officinalis* (Philipp and Schou 1981), *Quinchamalium chilense* (Riveros et al. 1987), *Menyanthes trifoliata* (Olesen 1987), and *Cephaelis tomentosa* and *Psychotria suerrensis* (J.H. Richards, in Barrett and Richards 1990). We know of only one species in which the difference in anther height exceeds that in stigma-height by a comparable margin. *Linum grandiflorum* was described as having only a stigma height polymorphism by Darwin (1864, 1877), but it actually has a much greater difference between morphs in anther height (V. Stevens unpubl., B. Murray pers. commun.) and therefore (just) qualifies as being heterostylous[4]. The greater separation of styles in several species suggests that there is stronger selection to segregate the stigma positions than that to separate the anther positions. This is consistent with our postulate that the polymorphism in stigma height evolved before that in anther height.

---

[4]The discrepancy between the various observations of *Linum grandiflorum* is curious. Is it caused by parallel differences between the morphs in anther height and the length of the petal claws, so the anthers are at similar positions *relative to the outspread petal limbs*? This could lead to the differences appearing to be in style length, rather than filament length, as Darwin suggested.

In enantiomorphy, a left- and right-handed heteromorphism that parallels heterostyly, the initial change may also involve the style rather than the stamens, since *Saintpaulia* species have a heteromorphism in the direction of the style but the anthers are always in the same position (Willis 1973, D.G.L. pers. observ.). A similar kind of enantiomorphous arrangement of the gynoecium only may occur in *Exacum* (Willis 1973).

The postulated change in stigma height involves a mutation which would shorten the style. This would be straightforward developmentally, since it does not require any new morphogenetic capability.

The hypothesis that the first step in the evolution of distyly is the invasion of a reverse herkogamy mutant into a population of plants with approach herkogamy may also explain one of the puzzling features of the genetics of distyly. The inheritance of distyly has been determined in 16 genera belonging to 11 families (Ornduff 1979, 1988; Schou and Philipp 1984; Shore and Barrett 1985; Chap. 5) and in *Narcissus tazetta*, a species with a stigma-height polymorphism only (Dulberger 1964). Short styles are dominant and the short- and long-styled morphs are *Ss* and *ss* respectively in all except *Hypericum aegypticum* (and possibly *Limonium vulgare*, see Ornduff 1979, and Chap. 5). [*Armeria maritima* is sometimes considered another exception (Baker 1966), but it lacks a herkogamy polymorphism.] Similarly, among tristylous taxa, short- and mid-styles are dominant to long styles in *Lythrum salicaria* and five species of *Oxalis*, (Chap. 5) and *Eichhornia paniculata* (S.C.H. Barrett, pers. commun.). Another species of *Oxalis* has the reverse dominance, presumably the result of a secondary change within the established system of tristyly.

An advantageous mutant can be favored by selection as soon as it arises if it is caused by a dominant allele and is therefore expressed in every individual that carries it. A recessive allele, however, must become homozygous before it can be subject to selection, and it is therefore more likely to become extinct by chance before it can become established (Fisher 1922; Haldane 1927; Crow and Kimura 1970). If this factor has favored the establishment of dominant stylar mutants, it would lead to the observed pattern of dominant short-styled forms *only* if the form with reverse herkogamy (subsequently the thrum or short-styled morph) is the second form to appear, as we have postulated. Similarly for tristylous species, the fact that the alleles for both short- and mid-styles are dominant over those for long styles at their separate loci suggests that long styles were ancestral here too and that the mid-style and short-style forms evolved subsequently.

3.4 Heteromorphic Self-Incompatibility is Qualitatively Distinct from Multiallelic Self-Incompatibility

In its gross cytological features, the inhibition of incompatible pollen tubes in species with multiallelic self-incompatibility is broadly similar to the inhibition of intramorph pollen in heterostylous species. Both phenomena are often collected together as "incompatibility systems." Several authors have proposed that the diallelic, one- or two-locus systems of heterostylous species are derived from multiallelic systems by a reduction in the number of alleles per locus (Crowe 1964; Muenchow 1982; Wyatt 1983). Others have pointed out a number of differences in the anti-selfing mechanisms of nonheterostylous and heterostylous species and suggested that the two systems have evolved independently (D. Charlesworth and B. Charlesworth 1979; Charlesworth 1982; Gibbs 1986). We support the latter view, and further propose that the differences between multiallelic self-incompatibility and the one- or two-locus, diallelic systems of heterostylous species are fundamental to the extent that they operate in quite different ways and are qualitatively distinct. The distinctions concern four aspects:

Number of Alleles. Multiallelic systems typically have many alleles (dozens) in one population. This is indicated by detailed population surveys in a number of species (see Lewis 1949; Sampson 1967; Gibbs 1986; Stevens and Kay 1989), and is supported by the high levels of cross-compatibility characteristic of incompatibility systems in many species. To our knowledge, no extensive survey of any homomorphic species has revealed a low number (below ten) of alleles. Moreover, a few-allele incompatibility system that is not associated with morphological differences would be vulnerable to (1) invasion by new alleles, which would experience a strong frequency-dependent advantage when they first appear (Wright 1939), or (2) extinction of the population (Imrie et al. 1972), or (3) the evolution of self-compatibility if the level of cross-pollination falls (Nagylaki 1976; Lloyd 1979, and Chap. 7). The model proposed by Muenchow (1982) to describe how a four-allele sporophytic incompatibility system might evolve into a diallelic heterostylous system starts with a number of alleles that is already closer to the heterostylous systems than to multiallelic systems and therefore it does not bridge the large gap between the systems.

Genetic and Cytological Natures of the Systems. There are a number of genetic and cytological differences between the two systems. First, sporophytic multiallelic self-incompatibility is characterized by a diversity of genetic interactions. Different S allele combinations in the same species show dominance, independence, or competition. The allelic interactions are often different in pollen and styles. Dominance in both pollen and styles, as in the self-incompatibility of heterostylous species, is uncommon. Second, the two families that definitely possess sporophytic multiallelic self-incompatibility (Brassicaceae, Asteraceae) have trinucleate pollen grains and simultaneous cytokinesis, whereas many of the heterostylous families have binucleate pollen grains and/or successive cytokinesis (Brewbaker 1959; Pandey 1960; Gibbs 1986; Chap. 5). The possibility of the evolution of heteromorphic self-incompatibility from multiallelic incompatibility systems is diminished even further by the total absence of heterostyly in the Brassicaceae and Asteraceae. One heterostylous species, *Anchusa officinalis* (Boraginaceae), may have a sporophytic multiallelic system, but reciprocal differences indicate different reactions in the pollen and style (Schou and Philipp 1984), unlike diallelic self-incompatibility. Third, as pointed out to us by Brian Murray, in multiallelic sporophytic systems the site of interaction is the stigma surface, since the sporophytically derived recognition factors are on the outside of the pollen grain and the recognition event takes place on the stigma surface. In many diallelic systems, however, the site of inhibition is in the style and therefore pollen wall "proteins" cannot be involved in the recognition reaction. Fourth, the two systems experience different types of mutations. Diallelic systems can generate mutant self-incompatibility alleles that have one allele reaction in the pollen and another in the style. These types are unknown in multiallelic systems, which, however, show mutants with a lack of reaction in either the pollen or the style, which in turn is unknown in diallelic systems (D. Charlesworth and B. Charlesworth 1979; Charlesworth 1982).

The most dramatic demonstration of the genetical distinctions between multiallelic and diallelic self-incompatibility systems comes from the operation of tristyly. As Darwin (1865) first noted, each of the long-, mid- and short-styled morphs produces pollen at two levels with different compatibility reactions. Since two types of pollen are produced by the same genotype, the difference between their incompatibility reactions must be determined by the developmental milieu in each type of anther – that is, by environmental factors in the broad sense. This environmental control is quite different from the genetically determined molecular specificities that are universally considered to control multiallelic incompatibility systems. Moreover, in two families with tristylous species, there are self-incompatible *distylous* species or populations that have evolved secondarily from tristylous ancestors (Oxalidaceae, Ornduff 1972; Weller 1976, 1979; Lythraceae, Lewis and Rao 1971; Ornduff 1978; see also Charlesworth 1979). The diallelic self-incompatibility mechanisms in these distylous species, at least, have origins quite different from that of the multiallelic incompatibility systems.

Molecular Natures of the Systems. Multiallelic self-incompatibility reactions depend on the recognition of molecular specificities expressed by *S*-gene products in the pollen and styles (Lewis 1960; Nettancourt 1977; Gibbs 1986). Virtually all molecular explanations of multiallelic self-incompatibility, including the widely accepted model of Lewis (1960), include two basic postulates: (a) There is one molecular mechanism in a population (in which the specificities at each locus are controlled by a single polypeptide coded by a functional gene). Different incompatibility alleles and phenotypes are governed by variant specificities of the *same* molecule. (b) Pollen and style reactions are governed by a *shared* specificity. These systems operate by the mutual recognition of specificities of one molecular species that are shared in incompatible pollen and pistils. The data pertaining to these points are discussed in Nettancourt (1977) and Richards (1986). There is accumulating evidence that pollen grains and styles of a self-incompatible plant produce an antigen common to both. Molecular biologists are currently searching for a gene product whose variation can explain the many *S*-gene specificities present in a population. An alternative hypothesis, that different self-incompatibility alleles have arisen as separate recessive lethals that have become secondarily linked (Mulcahy and Mulcahy 1983) has been severely criticized (Lawrence et al. 1985).

Heterostylous populations have only two or three pollen reactions and the same number of stigma/style behaviors. For such simple systems, it is not necessary to have matching specificities in pollen and pistils, whose mutual recognition causes intramorph incompatibility. All that is required is that two or three kinds of pollen, whose differences are controlled environmentally or genetically, must react differently in two or three pistil environments. Over 100 years ago, Darwin (1877) suggested for distylous species that the sexual elements secondarily became "co-adapted" for legitimate pollinations and "ill-adapted" for illegitimate pollinations in populations that had already evolved reciprocal herkogamy. On this scenario, the two or three intra-morph incompatibility reactions evolve separately and need not share a common molecular basis.

There is some evidence that the self-incompatibility of heterostylous species *operates through the failure of each type of pollen tube to grow under particular conditions, rather than from the recognition of shared pollen and style specificities.* The strength of intramorph incompatibility differs among the morphs in many heterostylous species (Darwin 1877; D. Charlesworth and B. Charlesworth 1979). As Darwin (1877, p. 266) discovered, the infertility of illegitimate unions increases as the distance between the appropriate stigmas and anthers increases in tristylous species of *Lythrum* and *Oxalis* and in some, but not all, distylous species. Moreover, in a number of distylous groups the site of the inhibition of illegitimate pollen is different in pin and thrum styles (Lewis 1943; Ghosh and Shivanna 1980; Stevens and Murray 1982; Shivanna et al. 1981; Bawa and Beach 1983; Anderson and Barrett 1986; Richards 1986; Dulberger 1987; Wedderburn and Richards 1990). In a review of the mechanisms of diallelic systems, Gibbs (1986) concluded that "a general picture has emerged that the SI reaction may be manifested differently in the two morphs".

The best evidence against the participation in heterostylous species of a recognition system involving shared pollen and style specificities again comes from tristylous families. Consider the evolution of tristyly to distyly, which as mentioned above has occurred in the Lythraceae and Oxalidaceae. In cases where, for example, the mid-style form of tristylous populations is lost, pollen from the mid-level anthers of the long-styled morphs becomes "compatible" with the low stigmas of the short-styled morph, whereas pollen from the mid-level anthers of the short-styled form is able to penetrate the high stigmas of the long-styled morph (e.g., Weller 1979). That is, the illegitimacy reaction of mid-level anthers changes in *different* directions in the long- and short-styled morphs. It would be virtually impossible for this to occur if the intramorph incompatibility reactions of these tristylous and distylous populations depended on the recognition of shared pollen and style specificities.

On the other hand, diallelic self-incompatibility systems may involve an immunological type of recognition for at least part of its operation in some species. In *Primula* species, stylar extracts from pin plants inhibit the growth of pin pollen tubes more than that of thrum pollen tubes while thrum stylar extracts inhibit the growth of thrum tubes specifically (Golynskaya et al. 1976; Shivanna et al. 1981). There are lectin-binding sites in the stigma surface of *Primula obconica* (B. Murray and V. Stevens, pers. commun.) and *Linum grandiflorum* (Ghosh and Shivanna 1980).

Associated Morphological Features. The most striking distinction between the diallelic and multiallelic systems of self-incompatibility is the association between diallelic specificities and morphological characters of the pollen and stigmatic papillae. Many authors have been impressed by the close association between the diallelic self-incompatibility reactions and the secondary characters of the pollen and stigmas, and to a less extent the style and filament lengths. Even in the dimorphic species of the Staticeae (Plumbaginaceae) that lack reciprocal herkogamy but have diallelic self-incompatibility (Baker 1966), the pollen grains and in most species also the stigmatic papillae differ between the incompatibility morphs. These associations between the incompatibility reactions and morphological features hold even in

Chap. 5). The close associations led Mather and de Winton (1941) to suppose that the incompatibility reactions and the reciprocal herkogamy arose at the same time. On the other hand, Dulberger (1975a,b and Chap. 3) postulated that the ancillary dimorphisms of stigmatic papillae and the shapes and sizes of pollen exines participate in self-incompatibility. We concur with Dulberger's view, but believe this is only one side of the explanation and postulate that the morphological features have arisen in part to restrict the success of self-pollinations and in part to facilitate legitimate cross-pollinations. This dual hypothesis combines the suggestion of Darwin (1877) that the morphological features of heterostyly promote legitimate cross-pollination and that of Dulberger (1975a,b) that the secondary features participate in self-incompatibility.

We postulate that heteromorphic self-incompatibility is due to a combination of intramorph failures arising incidentally from specialization for legitimate pollinations (as Darwin postulated) and active selection restricting self-fertilization. Insofar as these factors depend on preexisting differences in pollen and pistil physiology, they can only have arisen after reciprocal herkogamy has evolved to provide the different pollen and pistil environments. The molecular nature of the self-incompatibility mechanism may differ between morphs, as Anderson and Barrett (1986) have also suggested, and also among heterostylous species. Moreover, there could be more than one physiological factor contributing to the failure of pollinations within any one morph, as suggested by Shivanna et al. (1981).

The property that pollen grains become ill-adapted for growth in one type of style as a result of adaptation to another type of style in the same species was likened by Darwin (1862, 1877) to the failure of pollinations between species. There is evidence for such divergence of capabilities in the pollen reactions of self-incompatible tristylous species, in which the pollen grains from different anthers of the *same* morph have a different potential to grow in styles of different length. The potential for such divergence is also shown by the behavior of cleistogamous and chasmogamous flowers of nonheterostylous species. Cleistogamous flowers are solely pollinated by their own pollen, and chasmogamous flowers are pollinated only by pollen from chasmogamous flowers. In *Collomia grandiflora*, both pollen grains and stigmatic papillae differ markedly between the two floral forms (Lord and Eckard 1986). Artificial cross-pollinations between the two floral forms are unsuccessful (Lord and Eckard 1984). The loss of capability in unaccustomed combinations is parallel to the postulated process in heterostylous species, although in cleistogamous species the pattern is reversed in that pollen is adapted to grow in styles of the same form rather than the complementary form. Another striking parallel to the inhibition of illegitimate pollen has been observed in crosses between species of *Rhododendron*. Crosses with species that have styles that are markedly shorter or longer than those of the pollen parent are usually unsuccessful (Williams and Rouse 1988).

Structural differences in pollen and stigmas that reduce the growth of intramorph pollen in heterostylous species may be considered part of the self-incompatibility mechanism, whereas differences that increase the adhesion, germination, or growth of intermorph pollen are better considered to be adaptations promoting cross-pollination. Some ancillary features of heterostyly may function in part by reducing self-fertilization or self-interference, and in part they may actively promote cross-pollination. The importance of the two selective forces may vary between pollen and stigma characters, among heterostylous groups, and even for the same character between morphs of the same species. Yeo (1975), one of the minority of writers on heterostyly who have stressed pollination ecology, also concluded that the functional significance of each character should be considered separately, and could differ between morphs.

Where the illegitimate reactions are based on differential pollen responses to particular pistil environments, there need not be any morph-specific stylar reaction (and a gene responsible for it) other than the gene or genes responsible for the morphological differences between the morphs. This may explain the "lack of recombination" in *Primula* between a postulated, but unidentified, gene controlling the stylar illegitimacy reaction and the gene controlling style length and stigmatic papillae (Ernst 1955; Dowrick 1956; Charlesworth 1979; Chap. 5). The former gene may not exist as a separate entity. Similarly, the absence of recombinants in *Primula* between pollen size and incompatibility reactions may be because the two characters are controlled by the same gene. There are, however, several lines of evidence that indicate that genes controlling stigma and anther heights in particular are distinct from those controlling the self-incompatibility reactions, in some genera at least. In *Primula*, anther height is controlled by a separate gene from that controlling the pollen size and the pollen incompatibility reaction. The two genes have been resolved by recombination (Ernst 1955). In *Mitchella repens*, style length and the stylar incompatibility reaction are separated in a self-incompatible long-homostyle which has long styles with the stylar reaction normally associated with short styles (Ganders 1979). This shows that style length and the stylar reaction can be disassociated, but it does not prove that the two characters are controlled naturally by different genes, as D. Charlesworth and B. Charlesworth (1979) pointed out. In the Staticeae (Plumbaginaceae), morph differences in the pollen and stigma reactions and structural characters are accompanied by differences in stigma and anther heights in some species, but not others (Baker 1966).

It may well be that the suite of androecial (and separately gynoecial) characters normally associated with distyly morphs is controlled in part by separate major genes linked into the supergene and in part by a miscellany of modifier genes whose operation hinges upon the developmental milieu associated with particular stigma and anther heights, but which are not necessarily linked to the genes for these heights. The operation of these genes could be coordinated with the activities of the supergene by their being "morph-limited" (by analogy with sex-limited genes) rather than morph-linked. The mix of morph-linked and morph-limited genes may vary among the heterostylous groups, even for the standard features of heterostyly, and it may also vary between the androecial and gynoecial components of the supergene in any one species.

4 Discussion

The evolutionary pathway we have postulated involves the evolution of reciprocal herkogamy from a herkogamous ancestor via a stigma-height polymorphism, followed by the evolution of self-incompatibility (probably separately in each morph) lowed by the evolution of self-incompatibility (probably separately in each morph) and ancillary features. This pathway differs from almost all others proposed this century in its origin, the sequence of appearance of the components, and the nature of the self-incompatibility. It resembles more closely the first hypothesis presented by Darwin (1877), particularly in postulating that reciprocal herkogamy preceded self-incompatibility. Darwin did not explicitly discuss the nature of the flowers that gave rise to heterostyly, but he appears to have assumed (e.g., 1877, pp. 261, 265) that the ancestor was not herkogamous but had anthers and stigmas at approximately the same height. In line with his firmly held belief in the gradual nature of evolution in general, Darwin postulated that the divergence of heterostylous morphs occurred gradually from an initial "great variability in the length of the pistils and stamens, or of the pistil alone." In contrast, we postulate that the initial event was the invasion of a mutant with a large change in style length. Our scenario is also more detailed than Darwin's in postulating the sequence of gynoecial and androecial events during the evolution of reciprocal herkogamy.

Our proposals on the evolution of heterostyly differ widely from the pathway postulated by D. Charlesworth and B. Charlesworth (1979). Their model assumes a "homostylous ancestor and postulates a sequence of a pollen mutation followed by a stigma mutation, providing self-incompatibility before reciprocal herkogamy evolves. This model and ours are the two most detailed suggestions on the evolution of heterostyly. Both are backed up by mathematical models of the selective forces responsible for the evolutionary changes (see Chap. 7).

The plausibility of the alternative scenarios for the evolution of heterostyly hinges on the ancestral condition that is postulated. The postulate of ancestral homostyly in the Charlesworths' model makes an initial selection of reciprocal herkogamy unlikely. Ancestral herkogamy, as postulated in our pathway, allows reciprocal herkogamy to evolve much more readily without support from an existing self-incompatibility mechanism. When one herkogamous morph is already present, a herkogamy polymorphism can evolve in one step, rather than the two steps required from a nonherkogamous ancestor. The introduction of the polymorphism in stigma and anther height in turn makes the evolution of self-incompatibility and ancillary features easier developmentally, since the further physiological and anatomical differences can be built onto the existing differences in the stylar and anther environments of the herkogamy morphs. In Chapter 7, we argue that the appearance of a herkogamy polymorphism not only provides a foundation on which the developmental mechanisms for self-incompatibility and the ancillary features can be based, but it also fosters their selection.

The suggestion that the first step in the evolution of heterostyly, the herkogamy polymorphism, makes the subsequent steps easier may explain why there are so few examples of species which have achieved only the first step in this complex pathway. Where heterostyly can be selected, intermediate conditions may be passed through quickly, leaving few extant examples. Heterostylous families and part-families contain many species or populations with only some of the dimorphic features of relatives with more complex heterostyly syndromes. But in many cases, close relationships between populations with less and more complex syndromes suggest that the less complex syndromes are probably of secondary origin and are not ancestral (see Ganders 1979; B. Charlesworth and D. Charlesworth 1979; Chap. 10).

The postulates of the scenario we have suggested have been based primarily on evidence from comparative morphology and pollination ecology rather than the genetical nature of heterostylous systems. It is therefore particularly satisfying that our postulates can readily explain three distinct aspects of the genetics of heterostylous species that have hitherto been unexplained. These are: (1) the dominance of thrums in almost all cases of distyly that have been investigated genetically (Ornduff 1979, discussed above), (2) The nonseparation of genes for stylar incompatibility and stigma height, and for pollen incompatibility and pollen size in the *Primula* supergene, as discussed above, and (3) the existence of multiallelic self-incompatibility systems (or at least systems with intramorph cross-compatible groups, presumptive evidence for multiple alleles) in some species with various herkogamy polymorphisms. Such putatively multiallelic systems occur in distylous species of *Anchusa* (Boraginaceae) (Dulberger 1970; Schou and Philipp 1984) and in species of *Narcissus* that are tristylous (Fernandes 1964; Lloyd et al. 1990, for the self-incompatibility system see Bateman 1952) or have a herkogamy-homostyly polymorphism (Dulberger 1964). The occurrence of these species is a major anomaly if diallelic self-incompatibility is postulated to evolve before heterostyly, but it is easily accommodated by our hypothesis that heterostyly arose first. The ability of our postulates to explain the three separate genetical phenomena provides considerable support for them.

The term heterostyly covers a multiplicity of variations on a theme – a central (and we believe primary) element of reciprocal herkogamy usually associated with diallelic self-incompatibility and a variable accompaniment of ancillary features. Even allowing that much of this variety results from the secondary breakdown of complex heterostylous systems, the diversity and repeated evolution of heterostylous systems should caution us that there is likely to be more than one evolutionary pathway by which heterostyly has evolved. (The Boraginaceae and the Staticeae of the Plumbaginaceae are the most obvious candidates for divergent routes to heterostyly.) Richards and Barrett (Chap. 4) emphasize that tristyly has been achieved by different developmental means within the three well-known tristylous families. We have earlier recognized some possible variations from our postulates (particularly reverse herkogamy as the ancestral condition and an initial anther height polymorphism rather than one in stigma height). The postulates about the evolutionary precedence of a herkogamy polymorphism and the nature of the associated self-incompatibility systems are, in our opinion, less likely to be violated. But heterostylous species are so diverse that it would be unwise to rule any alternative out completely.

All attempts to date to uncover the evolutionary origins of heterostyly have been hindered by ignorance of the exact phylogenetic relationships among the species in taxa with heterostylous representatives. The floral characters of the ancestors of heterostylous species cannot be confidently established until we know the sister groups of the heterostylous taxa. Similarly, we need a solid knowledge of the cladistic relationships of heterostylous populations and their relatives before we can determine whether any species with less complex heterostyly syndromes represent intermediate steps towards the more complex dimorphisms in their relatives. We will not have a sound understanding of the evolution of heterostyly until we can reliably distinguish intermediate stages in the evolution of complex heterostyly polymorphisms from secondary reversions to simpler systems.

Acknowledgments. We are grateful to Brenda Casper, Deborah Charlesworth, Fred Ganders, Brian Murray, and Bob Ornduff for their helpful comments on a draft of the manuscript, and especially to Spencer Barrett for giving us the benefit of his extensive knowledge of heterostyly through his numerous and invaluable suggestions.

References

Anderson WR (1973) A morphological hypothesis for the origin of heterostyly in the Rubiaceae. Taxon 22:537–542

Anderson JM, Barrett SCH (1986) Pollen tube growth in tristylous *Pontederia cordata* (Pontederiaceae). Can J Bot 64:2602–2607

Baker HG (1962) Heterostyly in the Connaraceae, with special reference by *Byrsocarpus coccineus*. Bot Gaz Chicago 123:206–211

Baker HG (1964) Variation in style length in relation to outbreeding in *Mirabilis* (Nyctaginaceae). Evolution 18:507–509

Baker HG (1966) The evolution, functioning and breakdown of heteromorphic incompatibility systems. I. The Plumbaginaceae. Evolution 20:349–368

Barrett SCH, Richards JH (1990) Heterostyly in tropical plants. In: Gottsberger G, Prance GT (eds) Reproductive biology and evolution of tropical woody angiosperms. Mem N Y Bot Gard 55:35–61

Barrett SCH, Shore JS (1987) Variation and evolution of breeding systems in the *Turnera ulmifolia* L. complex (Turneraceae). Evolution 41:340–354

Bateman AJ (1952) Trimorphism and self-incompatibility in *Narcissus*. Nature (Lond) 170:496–497

Bateman AJ (1954a) The genetics of *Narcissus*. 1 – sterility. Daffodil Tulip Year Book 1954:23–29

Bateman AJ (1954b) Self-incompatibility systems in angiosperms. II. *Iberis amara*. Heredity 8:305–332

Bawa KS, Beach JH (1983) Self-incompatibility systems in the Rubiaceae of a tropical lowland wet forest. Am J Bot 70:1281–1288

Beach JH, Bawa KS (1980) Role of pollinators in the evolution of dioecy from distyly. Evolution 34:1138–1142

Bir Bahadur (1970) Heterostyly in *Hedyotis nigricans* (Lam.) Fosb. J Genet 60:175–177

Brewbaker JL (1959) Biology of the angiosperm pollen grain. Indian J Genet Plant Breed 19:121–133

Brown N, Crowden RK (1984) Evidence of heterostyly in *Epacris impressa* Labill. (Epacridaceae). In: Williams EG, Knox RB (eds) Pollination 1984: Proceedings of a Symposium held at the Plant Cell Biology Research Centre, University of Melbourne. Melbourne Univ Press, Melbourne, pp 187–193

Charlesworth D (1979) The evolution and breakdown of tristyly. Evolution 33:486–498

Charlesworth D (1982) On the nature of the self-incompatibility locus in homomorphic and heteromorphic systems. Am Nat 119:732–735

Charlesworth B, Charlesworth D (1979) The maintenance and breakdown of distyly. Am Nat 114:499–513

Charlesworth D, Charlesworth B (1979) A model for the evolution of distyly. Am Nat 114:467–498

Correns C (1924) Lang- und kurzgrifflige Sippen bei *Veronica gentianoides*. Biol Zentralbl 42:610–630

Cronquist A (1981) An integrated system of classification of flowering plants. Columbia Univ Press, New York

Crow JF, Kimura M (1970) An introduction to population genetics theory. Harper and Row, New York

Crowe LK (1964) The evolution of outbreeding in plants. I. The angiosperms. Heredity 19:435–457

Dahlgren R (1983) General aspects of angiosperm evolution and macrosystematics. Nord J Bot 3:119–149

Darwin C (1862) On the two forms, or dimorphic conditions in the species of *Primula*, and on their remarkable sexual relations. J Linn Soc Lond Bot 6:77–96

Darwin C (1864) On the two forms, or dimorphic conditions in the species of Primula, and on their remarkable sexual relations. J Linn Soc Lond Bot 6:77–96
Darwin C (1865) On the sexual relations of the three forms of *Lythrum salicaria*. J Linn Soc Lond Bot 8:169–196
Darwin C (1877) The different forms of flowers on plants of the same species. Murray, Lond
Dowrick VPJ (1956) Heterostyly and homostyly in *Primula obconica*. Heredity 10:219–236
Dulberger R (1964) Flower dimorphism and self-incompatibility in *Narcissus tazetta* L. Evolution 18:361–363
Dulberger R (1970) Floral dimorphism in *Anchusa hybrida* Ten. Isr J Bot 19:37–41
Dulberger R (1975a) *S*-gene action and the significance of characters in the heterostylous syndrome. Heredity 35:407–415
Dulberger R (1975b) Intermorph structural differences between stigmatic papillae and pollen grains in relation to incompatibility in Plumbaginaceae. Proc R Soc Lond Ser B 188:257–274
Dulberger R (1987) Fine structure and cytochemistry of the stigma surface and incompatibility in some distylous *Linum* species. Ann Bot 59:203–217
Eckenwalder JE, Barrett SCH (1986) Phylogenetic systematics of Pontederiaceae. Syst Bot 11:373–391
Ernst A (1955) Self-fertility in monomorphic primulas. Genetica 27:391–448
Faegri K, van der Pijl L (1979) The principles of pollination ecology, 3rd edn. Pergamon, Oxford New York
Fernandes A (1964) Contribution à la connaissance de la génétique de l'hétérostylie chez le genre *Narcissus* L. I. Resultats de quelques croisements. Bol Soc Broteriana 38:81–96
Fernandes A (1966) Contribution à la connaissance de la génétique de l'hétérostylie chez le genre *Narcissus* L. II. L'hétérostylie chez quelques populations de *N. triandrus* var. *cernuus* et *N. triandrus* var. *concolor*. Genet Iber 17:215–239
Fisher RA (1922) On the dominance ratio. Proc R Soc Edinb 52:321–341
Ganders FR (1979) The biology of heterostyly. NZ J Bot 17:607–635
Ghosh S, Shivanna KR (1980) Pollen-pistil interaction in *Linum grandiflorum*. Planta 149:257–261
Gibbs PE (1986) Do homomorphic and heteromorphic self-incompatibility systems have the same sporophytic mechanism? Plant Syst Evol 154:285–323
Golynskaya EL, Bashkirova NV, Tomchuk NN (1976) Phytohemagglutins of the pistil in *Primula* as possible proteins of generative incompatibility. Sov Plant Physiol 23:69–76
Haldane JBS (1927) A mathematical theory of natural and artificial selection. Part V. Selection and mutation. Proc Camb Philos Soc 23:838–844
Hemsley JH (1956) Connaraceae. In: Turrill WB, Milne-Redhead E (eds) Flora of Tropical East Africa. Crown Agents for Overseas Governments and Administrations, Lond
Henriques JA (1887) Amaryllideas de Portugal. Bol Soc Broteriana 5:159–174
Henriques JA (1888) Additamento ao catalogo das Amaryllideas de Portugal. Bol Soc Broteriana 6:45–47
Heslop-Harrison Y, Shivanna KR (1977) The receptive surface of the angiosperm stigma. Ann Bot 41:1233–1258
Heywood VH (ed) (1978) Flowering plants of the world. Oxford Univ Press, Oxford Lond
Hildebrand F (1866) Ueber den Trimorphisms in der Gattung *Oxalis*. Monatsber König Preuss Akad Wiss Berl 1866:352–374
Imrie BC, Kirkman CJ, Ross DR (1972) Computer simulation of a sporophytic self-incompatibility breeding system. Aust J Biol Sci 25:343–349
Jernstedt JA (1982) Floral variation in *Chlorogalum angustifolium*. Madroño 29:87–94
Lawrence MJ, Marshall DF, Curtis VE, Fearon CH (1985) Gametophytic self-incompatibility re-examined: a reply. Heredity 54:131–138
Lemmens RHMJ (1989) Heterostyly in Connaraceae. Acta Bot Neerl (Abstr) 38:224–225
Leppik EE (1972) Origin and evolution of bilateral symmetry in flowers. Evol Biol 5:49–72
Lewis D (1943) The physiology of incompatibility in plants. II. *Linum grandiflorum*. Ann Bot 7:115–122
Lewis D (1949) Incompatibility in flowering plants. Biol Rev Camb Philos Soc 24:472–496
Lewis D (1960) Genetic control of specificity and activity of the *S* antigen in plants. Proc R Soc Lond Ser B 151:468–477

Lewis D (1949) Incompatibility in flowering plants. Biol Rev Camb Philos Soc 24:472–496
Lewis D (1960) Genetic control of specificity and activity of the *S* antigen in plants. Proc R Soc Lond Ser B 151:468–477
Lewis D (1982) Incompatibility, stamen movement and pollen economy in a heterostyled tropical forest tree, *Cratoxylum formosum* (Guttiferae). Proc R Soc Lond Ser B 214:273–283
Lewis D, Rao AN (1971) Evolution of dimorphim and population polymorphism in *Pemphis acidula* Forst. Proc R Soc Lond Ser B 118:247–256
Lloyd DG (1979) Some reproductive factors affecting the selection of self-fertilisation in plants. Am Nat 113:67–79
Lloyd DG, Webb CJ (1986) The avoidance of interference between the presentation of pollen and stigmas in angiosperms. I. Dichogamy. NZ J Bot 24:135–162
Lloyd DG, Yates JMA (1982) Intrasexual selection and the segregation of pollen and stigmas in hermaphroditic plants, exemplified by *Wahlenbergia albomarginata* (Campanulaceae). Evolution 36:903–913
Lloyd DG, Webb CJ, Dulberger R (1990) Heterostyly in species of *Narcissus* (Amaryllidaceae), *Hugonia* (Linaceae) and other disputed cases. Plant Syst Evol 172:215–227
Lord EM, Eckard KJ (1984) Incompatibility between the dimorphic flowers of *Collomia grandiflora*, a cleistogamous species. Science 223:695–696
Lord EM, Eckard KJ (1986) Ultrastructure of the dimorphic pollen and stigmas of the cleistogamous species, *Collomia grandiflora*. Protoplasma 132:12–22
Martin FW (1965) Distyly and incompatibility in *Turnera ulmifolia*. Bull Torrey Bot Club 92:185–192
Mather K, de Winton D (1941) Adaptation and counter-adaptation of the breeding system in *Primula*. Ann Bot 5:297–311
Muenchow G (1982) A loss-of-alleles model for the evolution of distyly. Heredity 49:81–93
Mulcahy DL, Mulcahy GB (1983) Gametophytic self-incompatibility reexamined. Science 220:1247–1251
Nagylaki T (1976) A model for the evolution of self-fertilisation and vegetation reproduction. J Theor Biol 58:55–58
Nettancourt D de (1977) Incompatibility in angiosperms. Springer, Berlin Heidelberg New York
O'Brien SP, Calder DM (1989) The breeding biology of *Epacris impressa*. Is this species heterostylous? Aust J Bot 37:43–54
Olesen JM (1987) Heterostyly, homostyly, and long-distance dispersal of *Menyanthes trifoliata* to Greenland. Can J Bot 65:1509–1513
Ornduff R (1972) The breakdown of trimorphic incompatibility in *Oxalis* section *Corniculatae*. Evolution 26:52–65
Ornduff R (1974) Heterostyly in South African flowering plants: a conspectus. J S Afr Bot 40:169–187
Ornduff R (1975) Heterostyly and pollen flow in *Hypericum aegypticum* (Guttiferae). J Lin Soc Lond Bot 71:51–57
Ornduff R (1978) Features of pollen flow in dimorphic species of *Lythrum* section *Euhyssopifolia*. Am J Bot 65:1077–1083
Ornduff R (1979) The genetics of heterostyly in *Hypericum aegypticum*. Heredity 42:271–272
Ornduff R (1986) Comparative fecundity and population composition of heterostylous and non-heterostylous species of *Villarsia* (Menyanthaceae) in Western Australia. Am J Bot 73:282–286
Ornduff R (1988) Distyly and monomorphism in *Villarsia* (Menyanthaceae): some evolutionary considerations. Ann MO Bot Gard 75:761–767
Pandey KK (1960) Evolution of gametophytic and sporophytic systems of self-incompatibility in angiosperms. Evolution 14:98–115
Philipp M, Schou O (1981) An unusual heteromorphic incompatibility system. Distyly, self-incompatibility, pollen load and fecundity in *Anchusa officinalis* (Boraginaceae). New Phytol 89:693–703
Philipson MN, Philipson WR (1975) A revision of *Rhododendron* section *Lapponicum*. Notes R Bot Gard Edinb 34:1–72
Richards AJ (1986) Plant breeding systems. Allen and Unwin, Lond
Richards JH, Barrett SCH (1987) Development of tristyly in *Pontederia cordata* (Pontederiaceae). I. Mature floral structure and patterns of relative growth of reproductive organs. Am J Bot 74:1831–1841

Robson NKB (1972) Evolutionary recall in *Hypericum* (Guttiferae)? Trans Proc Bot Soc Edinb 41:365–383

Sampson DR (1967) Frequency and distribution of self-incompatibility alleles in *Raphanus raphanistrum*. Genetics 56:241–251

Schill R, Baumm A, Wolter M (1985) Vergleichende Mikromorphologie der Narbenoberflachen bei den Angiospermen; Zusammenhange mit Pollenoberflachen bei heterostylen Sippen. Plant Syst Evol 148:185–214

Schou O, Philipp M (1984) An unusual heteromorphic incompatibility system. III. On the genetic control of distyly and self-incompatibility in *Anchusa officinalis* L. (Boraginaceae). Theor Appl Genet 68:139–144

Shivanna KR, Heslop-Harrison J, Heslop-Harrison Y (1981) Heterostyly in Primula. 2. Sites of pollen inhibition, and effects of pistil constituents on compatible and incompatible pollen-tube growth. Protoplasma 107:319–337

Shivanna KR, Heslop-Harrison Y, Heslop-Harrison J (1983) Heterostyly in Primula. 3. Pollen water economy as a factor in the intramorph-incompatibility response. Protoplasma 117:175–184

Shore JS, Barrett SCH (1985) The genetics of distyly and homostyly in *Turnera ulmifolia* L. (Turneraceae). Heredity 55:167–174

Sobrevila C, Ramirez N, Xena de Enrech N (1983) Reproductive biology of *Palicourea fendleri* and *P. petiolaris* (Rubiaceae), heterostylous shrubs of a tropical cloud forest in Venezuela. Biotropica 15:161–169

Stearn WT (1971) A survey of the tropical genera *Oplonia* and *Psilanthele* (Acanthaceae). Bull Br Mus Nat Hist Bot 4:259–323

Stevens JP, Kay QON (1989) The number, dominance relationships and frequencies of self-incompatibility alleles in a natural population of *Sinapis arvensis* L. in south Wales. Heredity 62:199–206

Stevens VAM, Murray BG (1982) Studies on heteromorphic self-incompatibility systems: physiological aspects of the incompatibility system of *Primula obconica*. Theor Appl Genet 61:245–256

Vogel S (1955) Über den Blütendimorphismus einiger südafrikanischer Pflanzen. Oesterr Bot Z 102:486–500

Vuilleumier BS (1967) The origin and evolutionary development of heterostyly in the angiosperms. Evolution 21:210–226

Webb CJ, Lloyd DG (1986) The avoidance of interference between the presentation of pollen and stigmas in angiosperms. II. Herkogamy. NZ J Bot 24:163–178

Wedderburn F, Richards AJ (1990) Variation in within-morph incompatibility inhibition sites in heteromorphic *Primula* L. New Phytol 116:149–162

Weller SG (1976) Breeding system polymorphism in a heterostylous species. Evolution 30:442–454

Weller SG (1979) Variation in heterostylous reproductive systems among populations of *Oxalis alpina* in southeastern Arizona. Syst Bot 4:57–71

Williams EG, Rouse JL (1988) Disparate style lengths contribute to isolation of species in *Rhododendron*. Aust J Bot 36:183–191

Willis JC (1973) A dictionary of the flowering plants and ferns, 8th edn (revised by Airy Shaw HK). Cambridge Univ Press, Cambridge

Wright S (1939) The distribution of self-sterility alleles in populations. Genetics 24:538–552

Wunderlin RP, Larsen K, Larsen SS (1981) Tribe 3. Cercideae Bronn (1822). In: Polhill RM, Raven PH (eds) Advances in legume systematics, vol 1. R Bot Gard Kew, pp 107–116

Wyatt R (1983) Pollinator-plant interactions and the evolution of breeding systems. In: Real L (ed) Pollination biology. Academic Press, Orlando, pp 51–95

Yeo PF (1975) Some aspects of heterostyly. New Phytol 75:147–153

Chapter 7

The Selection of Heterostyly

D.G. LLOYD[1] and C.J. WEBB[2]

1 Introduction

Since heterostyly became well known over a century ago, botanists have consistently proposed that its adaptive significance lies in the encouragement it provides for outcrossing. Darwin (1877, p. 258) for example, wrote that "We may feel sure that plants have been rendered heterostyled to ensure cross-fertilization, for we know that a cross between the distinct individuals of the same species is highly important for the vigour and fertility of the offspring." There has, however, been a variety of opinions expressed as to precisely how heterostyly ensures cross-fertilization.

Several different selection forces must be carefully distinguished. Darwin (1877) proposed that the reciprocal herkogamy of heterostylous species evolved before the self-incompatibility and ancillary features. He postulated that the complementary placement of the anthers and stigmas of different morphs facilitated the transfer of pollen in "legitimate" cross-pollinations between anthers and stigmas at the same height, because the two pollinating surfaces of legitimate crosses come into contact with the same parts of a pollinator's body. This "cross-promotion" hypothesis postulates that there has been active selection improving the proficiency of legitimate cross-pollination.

Alternatively, cross-pollination may be passively encouraged by preventing or discouraging self-pollination, thereby allowing more opportunities for crossing without actually altering the process of cross-pollination, Such "anti-selfing" hypotheses have been widely invoked in the modern era to explain the selection of heterostyly and many other breeding systems. Remarkably, Darwin did not use this hypothesis to explain either the reciprocal herkogamy of heterostylous species, which he attributed to the promotion of crossing, or their self-incompatibility. He saw that the incompatibility mechanism that inhibits self-fertilization in heterostylous species also curtails cross-fertilization with other individuals of the same morph – half the population on average in the case of distyly. He argued pointedly that a mechanism that resulted in each plant being cross-sterile with "half its brethren" (1877, p. 263) could scarcely favor cross-fertilization. Darwin regarded this as an obstacle to the direct selection of self-incompatibility, and he was forced to regard the evolution of self-incompatibility as an "incidental and purposeless" (1877, p. 265) byproduct of the coadaptation of the two or three kinds of pollen tubes to the legitimate styles in

[1]Department of Plant and Microbial Sciences, University of Canterbury, Private Bag, Christchurch, New Zealand
[2]Botany Division, DSIR, Private Bag, Christchurch, New Zealand

Monographs on Theoretical and Applied Genetics 15
Evolution and Function of Heterostyly (ed. by S.C.H. Barrett)

which they predominantly grow. We will argue below that Darwin's obstacle does indeed hinder the evolution of self-incompatibility in heterostylous species, but the barrier can be overcome more readily in populations in which reciprocal herkogamy already exists and reduces the frequency of illegitimate cross-pollinations.

Features of heterostyly may encourage cross-fertilization in a third way, as "anti-interference" mechanisms (Lloyd and Yates 1982; Webb and Lloyd 1986) by which the mutual interference between stamens and carpels in the same flowers is reduced. Baker (1954, 1964, see Sect. 6, Historical Appendix) recognized that pollen deposited illegitimately on stigmas achieves no reproductive fitness and is wasted. He offered an alternative "pollen wastage" hypothesis for the adaptive significance of reciprocal herkogamy, suggesting that it reduces the wastage of pollen from redeposition on stigmas of the form that produced it. Some of the ancillary morphological features of the pollen and stigmas of heterostylous species may also have been selected because they reduce pollen wastage on illegitimate stigmas. Moreover, interference with pollen fitness could operate at pollen pickup if styles and stigmas obstruct the access of pollinators to pollen. In addition, some of the features of heterostyly may reduce interference on maternal fitness. Yeo (1975) suggested that the herkogamy of thrum flowers (with pollen above the stigmas) and dimorphism in pollen and stigma characters may reduce the amount of illegitimate pollen that is deposited on stigmas. Such "stigma clogging" could limit the access of legitimate pollen or reduce its growth. The experimental application of illegitimate pollen onto the same stigmas as legitimate pollen has been found to adversely affect seed set in *Lythrum salicaria* (Nicholls 1987), but had little or no effect in *Pontederia cordata* (Barrett and Glover 1985) and *Turnera ulmifolia* (Shore and Barrett 1984).

In this paper, we compare the ability of the cross-promotion, anti-selfing, and anti-interference hypotheses to explain the principal features of heterostyly. There is no necessity for the three major components of heterostyly – reciprocal herkogamy, self-incompatibility and the various ancillary features – to have been selected in the same way, so each feature must be considered in its own context. We first examine the evidence that heterostyly promotes legitimate pollinations and reduces pollen wastage, as Darwin and Baker respectively suggested. Then we present phenotypic selection models that compare the ability of selection against selfing and selection favoring the proficiency of legitimate crosses to cause the evolution of reciprocal herkogamy and self-incompatibility. In considering both the evidence and the models, we examine the movement of pollen between and within morphs in terms of the proficiencies of the transfer of pollen in different directions rather than as disassortative mating. Thus we look directly at the transfer processes themselves instead of their genetical consequences.

2 A Reevaluation of Evidence on Darwin's Hypothesis that Heterostyly Promotes Legitimate Cross-Fertilization

Darwin (1877, p. 261) postulated that the reciprocal herkogamy of heterostylous species evolved before the self-incompatibility, by a process that we would now describe as disruptive selection from a monomorphic ancestor with anthers and stigmas at the same height. He proposed that the advantage of reciprocal herkogamy is that it increases the amount of pollen transferred between anthers and stigmas at the same height, which he called legitimate crosses. Although Darwin predominantly argued that heterostyly was selected because it enhances the proficiency of legitimate crosses, he sometimes expressed his argument ambiguously in stating that heterostyly may "favour the intercrossing of distinct individuals" (1862b) or "ensure cross-fertilisation" (1877, p. 258). Darwin did not firmly distinguish between selection against self-fertilization (favoring crossing passively) and selection actively favoring more proficient cross-fertilization, although he postulated both factors on various occasions.

Four kinds of evidence may be sought in support of Darwin's hypothesis of the promotion of legitimate cross-pollinations:

2.1 Examination of Pollinators

If more pollen is transferred in legitimate pollinations than in illegitimate combinations, this is likely to be because the pollen of different morphs is picked up and redeposited from different parts of a pollinator's body. A number of studies have confirmed this supposition, usually using the distinctive size of the pollen from different heights as a means of identifying the pollen source. Darwin (1862b, 1877) pushed various objects (the proboscises of dead bees, bristles, and needles) into the corolla tubes of *Primula* flowers, first of one form, then of the other. He found that pollen from thrum flowers (presented at a high level) was deposited predominantly around the base of the probes. In contrast, the pollen of pin flowers (presented at a lower level) was deposited a little above the tip of the probes, but there was some intermingling of the two types of pollen. Darwin (1865, 1877) also observed that after bees had visited the tristylous species *Lythrum salicaria*, green pollen from high-level anthers was deposited predominantly on the inner sides of the hind legs and abdomen, whereas yellow pollen from the mid- and low-level anthers was deposited primarily on the underside of the thorax.

Other workers have examined the location of thrum and pin pollen on insects captured while foraging on heterostylous species. There is a tendency for the two types of pollen to be found on different parts of insect bodies in distylous species of *Fagopyrum* (Rosov and Screbtsova 1958, cited in Ganders 1979), *Pulmonaria* (Olesen 1979) and *Cratoxylum* (Lewis 1982). There has been no report of pin and thrum pollen having similar distributions. By inserting dead bees into the flowers of the tristylous species, *Pontederia cordata*, Barrett and Wolfe (1986) similarly found that pollen from the three heights was deposited most abundantly on different parts of the bees. In a subsequent study of three taxa of bees collected from a natural population of *Pontederia cordata*, Wolfe and Barrett (1989) found some redistribution of short-level pollen on the bees' bodies, apparently as a result of grooming activities in flight. But presumably the short-level pollen that is deposited on short-level stigmas is that which has not been redistributed on the bees' bodies, and all three types of stigmas will still receive a predominance of legitimate pollen.

These observations of pollen loads on insects support Darwin's hypothesis by showing that pollen pickup from the anthers of different morphs is segregated on insect bodies. But by their nature they cannot show that the redeposition of pollen on stigmas is similarly segregated, since they are subject to the reservation that the pollen observed on foraging insects is that which has been picked up and not redeposited. Hence they provide an incomplete demonstration of the full process of segregated transfer.

2.2 Pollen Loads on Stigmas

During the last 20 years, many workers have sought more direct evidence by examining the composition of the pollen loads on the stigmas of the morphs of heterostylous populations. The rationale has been that a more proficient transfer of pollen in legitimate combinations would be indicated by a higher proportion of legitimate pollen on each type of stigma than would be expected from random mating. Evidence for "disassortative pollination" has been sought by statistically comparing the relative production of the pollen types in a population as a whole with the proportions observed on the various stigma types.

In a review of the 13 distylous species examined to that time, Ganders (1979) observed that on thrum stigmas there was "excess" pin (legitimate) pollen on four species as expected by Darwin's hypothesis, the proportions were random in three species, and three species showed "excess" thrum (illegitimate) pollen. No species showed an "excess" of thrum (legitimate) pollen on pin stigmas. These results and comparable patterns obtained since (Ornduff 1980a,b; Lewis 1982; Schou 1983; Nicholls 1985; Price and Barrett 1984; Barrett and Glover 1985; Glover and Barrett 1986a) have been widely interpreted as offering only limited support for Darwin's hypothesis. As Ganders (1974, 1979) pointed out, however, intraflower selfing would cause more illegitimate pollen to be deposited than expected from random mating. Consequently, pollen loads on intact stigmas "indicate nothing about levels of disassortative pollination in the interflower component of the pollen loads" (Ganders 1979). Moreover, geitonogamous (interflower) self-pollinations exceeding random expectations occur whenever pollinators move frequently between flowers of the same inflorescence, or clonal growth or habitat segregation of morphs cause flowers of the same morph to be clumped together in a population. The statistical comparisons of observed pollen loads on intact stigmas with those expected from random pollination are therefore inappropriate.

The effects of autogamous and geitonogamous self-pollinations can be isolated by comparing pollen loads on intact and emasculated flowers. Ganders (1974) found that the proportion of own-form pollen on stigmas of pin and thrum plants of *Jepsonia heterandra* was greatly reduced in plants on which all flowers had been emasculated. The exact amount of the reduction was obscured because emasculated flowers also received less opposite-form pollen. In another study, Ganders (1976) found that the emasculation of thrum flowers of *Amsinckia douglasiana* decreased the percentage of thrum pollen deposited on the stigmas from 24 to 16%. Assuming the total pollen loads are unaffected by emasculation (which was so in this study), it can be calculated that 41% of the thrum pollen on thrum stigmas comes from autogamy. Similarly, Nicholls (1985) estimated that 42% of the total pollen loads on stigmas of *Linum tenuifolium* comes from intraflower self-pollinations. Barrett and

Glover (1985) calculated that the fractions of pollen derived from autogamy on long-, mid-, and short-styled flowers of *Pontederia cordata* are 14, 64, and 13% respectively. The pollen loads on the stigmas of emasculated and intact flowers of *Primula vulgaris* (Piper and Charlesworth 1986) showed that autogamy contributed 52% of the pollen on thrum stigmas and 83% of that on pin stigmas. In contrast, Schou (1983) found that emasculated thrum flowers of *Primula elatior* had pollen loads that did not differ significantly from those of intact thrum flowers of the same age. Thus five of the six studies to date have shown that autogamy provides a sizeable proportion of the pollen load on intact stigmas, verifying Gander's criticism of the comparison of observed and "expected" loads.

Nevertheless, the view that the examination of pollen loads on stigmas offers only moderate support for Darwin's hypothesis of the promotion of legitimate crosses still prevails (e.g., Lewis 1982; Schou 1983; Ornduff 1983; Glover and Barrett 1986a). We suggest that the pollen load data have not been assessed adequately, and we reanalyze two data sets using a different method of calculation that allows the pollen loads to be examined in terms of male functions as well as the female function considered exclusively in the calculations described above. We include male functions because our models (below) show that selection for more proficient cross-pollination operates more often on male than on female fitness.

To compare legitimate and illegitimate pollen transfers from the viewpoints of both pollen donation and pollen receipt, data on stigma loads in a species are converted into the probabilities of a single pollen grain of each type being deposited on stigmas of each type. The probability of transfer of a single pollen grain of type i to any of the stigmas of type j, T_{ij}, is the total load of pollen of type i on all stigmas of type j divided by the total pollen production of type i.

$$\text{i.e., } T_{ij} = \frac{(\text{av. stigmaload})_{ij} \times (\text{no. flowers})_j}{(\text{pollen/flower})_i \times (\text{no. flowers})_i}, \tag{1}$$

where (av. stigma load)$_{ij}$ is the average number of i grains on each j stigma. Here i, j are any of the two morphs of distylous populations or three anther and stigma levels in tristylous populations. The probabilities take into account the frequencies of the morphs and their clonal growth and flower and pollen productions per ramet.

We reanalyse two of the three data sets for which the pollen loads of all morphs have been observed on stigmas of both intact and emasculated flowers – one distylous species and one tristylous species. Considering first the distylous species *Jepsonia heterandra* examined by Ganders (1974), Ganders (and also Ornduff 1971) observed that the morphs are equally frequent and produce the same number of flowers, but with different numbers of pollen grains per flower (Table 1). The pollen loads show that in both intact and emasculated flowers, pin stigmas receive more pollen than thrum stigmas in total. Table 1 also shows that more pin pollen than thrum pollen is deposited on stigmas generally, roughly in proportion to the relative numbers produced.

To remove the effect of unequal total flower and pollen numbers, the data are converted into probabilities of single pollen grains of each type being transferred to each type of stigma, using Eq. (1). Consider first the plants on which all flowers were emasculated. As maternal parents (comparisons in the same column), pin flowers are

Table 1. Pollen loads on stigmas and proficiencies of legitimate and illegitimate transfers in *Jepsonia heterandra*. (Data in Ganders 1974)

Pollen source		Average stigma loads (and probabilities of a pollen grain being deposited) on pin or thrum stigmas			
Morph[a]	Pollen no. per flower	Emasculated flowers		Intact flowers	
		Pin	Thrum	Pin	Thrum
Thrum	91×10^3	**406** ($\mathbf{4.46 \times 10^{-3}}$)	117 (1.29×10^{-3})	**700** ($\mathbf{7.69 \times 10^{-3}}$)	189 (2.08×10^{-3})
Pin	211×10^3	628 (2.98×10^{-3})	**572** ($\mathbf{2.71 \times 10^{-3}}$)	2822 (13.4×10^{-3})	**567** ($\mathbf{2.69 \times 10^{-3}}$)

[a] The two morphs are equally frequent and produce equal numbers of flowers. The numbers for legitimate pollinations are shown in bold type.

1.56 (4.46÷2.98) times as likely to receive any one thrum pollen grain as any one pin grain. Similarly, thrum stigmas are 2.1 (2.71÷1.29) times more likely to receive each pin grain than each thrum grain. Hence both morphs receive legitimate pollen at a considerably higher rate than illegitimate pollen. Examining the transfers from the viewpoint of paternal parents (comparing probabilities in the same row), a thrum pollen grain is 3.5 times more likely to be deposited on a pin stigma than on a thrum stigma, but a pin grain is slightly less likely (0.91 times) to be deposited on a legitimate stigma. Altogether, in three of the four comparisons for emasculated flowers, the proficiencies of the legitimate transfers exceed those of the illegitimate transfers. In the intact flowers, however, autogamous transfer of self-pollen is common and only two of the four comparisons show legitimate transfer probabilities exceeding illegitimate probabilities, and then by smaller amounts than in the corresponding emasculated flowers.

The average extent of the greater proficiency of legitimate transfers in *Jepsonia heterandra* can be measured as the geometric mean of the legitimate proficiencies divided by that of the illegitimate proficiencies. That is, the average extent of promotion of legitimate cross-pollination,

$$P = \left(\frac{d_{tp}d_{pt}}{d_{tt}d_{pp}}\right)^{\frac{1}{2}}, \tag{2}$$

where d_{tp} = probability of a thrum grain being deposited on a pin stigma (4.46×10^{-3} in this example), etc. From Table 1, the average extent to which the reciprocal herkogamy promotes legitimate crosses in emasculated flowers of *J. heterandra*, $P_{emasc.} = 1.78$. For the intact flowers, $P_{intact} = 0.86$. Thus when the distorting effects of self-pollinations are removed, the selective force postulated by Darwin provides a powerful, almost twofold advantage for legitimate pollinations. This advantage is not evident in the intact flowers because it is obscured by autogamous and geitonogamous self-pollinations.

The second species in which the pollen loads have been examined on emasculated and intact flowers of all morphs is the tristylous species, *Pontederia cordata*

(Barrett and Glover 1985). Table 2 shows the average pollen loads at Paugh Lake, Ontario, on stigmas of intact flowers and those of flowers of inflorescences on which all flowers had been emasculated. In both intact and emasculated flowers, the short stigmas receive more legitimate pollen grains (from short-level anthers) than they receive of either illegitimate type. For the mid- and long-styled stigmas, however, the legitimate pollen loads are in between the numbers of the two illegitimate types.

The unadjusted loads are influenced by the morph frequencies (Long = 0.324, Mid = 0.276, Short = 0.400, Barrett and Glover 1985), and by the number of flowers and pollen grains per flower produced by the morphs. The morphs do not differ in inflorescence production, flower number per inflorescence, or flowering phenology (Price and Barrett 1982). The number of pollen grains produced at each level in the morphs differs markedly, however. The numbers of pollen grains of each type were calculated as the averages of two samples of Paugh Lake plants (Price and Barrett 1982). These are given in Table 2 as the pollen number per flower (averaged over both relevant morphs). The number of pollen grains borne in short-level anthers is approximately four times that of the numbers in long- and mid-level anthers. Barrett and Glover (1985) compared the stigma loads with the numbers of pollen grains of the three types produced in the population in an ANOVA analysis. They concluded that in the emasculated flowers, the stigmas of long- and short-styled flowers received significantly more legitimate pollen, while in the intact flowers only the long-styled morph received significantly more legitimate pollen. A further chi-square analysis showed significant levels of legitimate pollination in emasculated flowers of all three morphs.

Table 2. Pollen loads on stigmas and proficiencies of legitimate and illegitimate pollen transfers in *Pontederia cordata*. (Data in Price and Barrett 1982, and Barrett and Glover 1985)

Pollen source (level borne at)	Pollen no. per flower	Average pollen loads (and probabilities of a pollen grain being deposited) on stigmas of the three morphs		
		Long-	Mid-	Short-
			Emasculated flowers	
Long	4573	**62.5 (6.6×10^{-3})**	26.0 (2.3×10^{-3})	10.3 (1.3×10^{-3})
Mid	6463	46.5 (3.2×10^{-3})	**74.9 (4.4×10^{-3})**	**31.0 (2.7×10^{-3})**
Short	23707	123.6 (2.8×10^{-3})	164.2 (3.2×10^{-3})	**138.7 (3.9×10^{-3})**
			Intact Flowers	
Long	4573	**70.6 (7.4×10^{-3})**	37.0 (3.3×10^{-3})	21.1 (2.7×10^{-3})
Mid	6463	65.6 (4.5×10^{-3})	**62.5 (3.7×10^{-3})**	42.1 **(3.6×10^{-3})**
Short	23707	145.1 (3.3×10^{-3})	607.7 (11.8×10^{-3})	**104.8 (2.9×10^{-3})**

The numbers for legitimate pollinations are shown in bold type.

The receipt and donation of pollen by the three morphs can be examined further by calculating the probabilities of a pollen grain of each type being deposited on a stigma of each type [Eq. (1)], using the morph ratios and pollen numbers given above. The results for emasculated flowers (Table 2) show that the stigmas of all three morphs have a higher probability of receiving single legitimate pollen grains than they have of receiving the illegitimate types (comparisons in the same column). On the male side, all three types of pollen are more likely to be deposited on a legitimate stigma than on either type of illegitimate stigma (comparisons in the same row). Hence legitimate pollen transfer occurs at a higher rate than illegitimate transfer in all 12 possible pairwise comparisons. The average extent of promotion of legitimate cross-pollinations can be calculated as the geometric mean of the probabilities of all (three) legitimate transfers divided by the geometric means of the (six) illegitimate combinations. Here $P_{emasc} = 1.94$. In intact flowers, which experience autogamous and geitonogamous self-pollinations, only the long-style morph is more likely to receive any legitimate pollen grain than either type of illegitimate pollen, and only the pollen produced at the long-level is more likely to land on a legitimate stigma than on either type of illegitimate stigma (Table 2). For intact flowers, $P_{intact} = 1.02$, roughly half the value for emasculated flowers.

The analysis of the probabilities of each type of transfer in *P. cordata* shows that when selfing is eliminated, the tristyly is highly effective in promoting legitimate cross-pollinations, as Darwin postulated. The approximate doubling of the average proficiency of legitimate transfers compared with the illegitimate transfers constitutes a powerful selective force maintaining reciprocal herkogamy in this species, as in *Jepsonia heterandra*.

The third study in which the stigma loads were compared on intact and emasculated flowers is that of Piper and Charlesworth (1986) on a population of *Primula vulgaris* containing pin, thrum, and long-homostyle plants. This study also provided strong support for Darwin's hypothesis, since emasculated flowers of both pin and thrum plants received significantly more legitimate than illegitimate pollen. Moreover, the presence of long-homostyles allows a comparison of the frequencies of autogamy in herkogamous pins and thrums and nonherkogamous homostyles, which have either anthers or stigmas at the same height as the two heterostyly morphs. The thrums and pins receive only a fifth as much autogamous pollen as the homostyles, showing that herkogamy does indeed reduce self-pollination dramatically.

Among the 16 distylous species other than *Jepsonia heterandra* in which pollen loads on intact flowers have been compared with the frequencies of the pollen types in the population (reviewed in Ganders 1979, plus Ornduff 1980a,b; Lewis 1982; Schou 1983), there is "random" or "excess" legitimate pollen receipt on at least one morph in all except four of the species – despite the bias from self-pollination. This strongly indicates that reciprocal herkogamy generally promotes legitimate crosses, but without emasculation studies the analyses of these species can go no further. Either the extent of cross-promotion or the degree to which it is obscured by self-pollination varies among species and between the morphs – thrum stigmas in general receive proportionately more legitimate pollen.

The proficiencies of pollen transfer in *Jepsonia* and *Pontederia* can also be used to determine the proportions of pollen of different types that are deposited on illegitimate stigmas (are "wasted"), and the sources of this wastage. Wastage of a pollen type, W, can be measured as the proportion of illegitimate depositions. For pin pollen of intact flowers of *Jepsonia heterandra*, from Table 1, $W_{intact} = 13.4/(13.4 + 2.69) = 0.83$. Most redeposited pollen is wasted. In the plants with emasculated flowers, wastage from geitonogamy and autogamy is prevented, and the reduced wastage, $W_{emasc} = 2.98/(2.98 + 2.71) = 0.52$, is entirely from illegitimate crosses. For thrum pollen, the corresponding figures are $W_{intact} = 0.21$ and $W_{emasc} = 0.24$. (The latter figure may be artificially low because the amount of legitimate pollen is also reduced on emasculated pin flowers – Table 1.)

Corresponding calculations of the proportions of wastefully deposited pollen in *Pontederia cordata* show that in intact flowers 0.45, 0.69 and 0.84 of the deposited grains from long-, mid-, and short levels respectively are illegitimately deposited. In plants with emasculated flowers, the figures are 0.36, 0.57, and 0.61. Thus the fraction of illegitimate depositions increases as the height of presented pollen decreases, as in *Jepsonia*. The discrepancies between the pollen from different levels persists in the plants with emasculated flowers (again as in *Jepsonia*), where it comes entirely from illegitimate crosses rather than from autogamy or geitonogamy.

When measured as the proportion of redepositions that are illegitimate, pollen wastage is considerable in both species. But we must recall that about 99% of pollen grains do not reach any type of stigma (Tables 1 and 2). We cannot determine from these experiments how much of the illegitimately deposited pollen would otherwise have been deposited on legitimate stigmas, and therefore how much the "wastage" really reduces paternal fitness. Moreover, these experiments provide no information as to whether wastage is reduced by the herkogamous separation of pollen and stigmas in each morph or by the reciprocity of positions in different morphs. To answer these questions, we would need to examine wastage and legitimate redeposition in homostylous mutants (cf. Piper and Charlesworth 1986) and monomorphic herkogamous populations.

The amounts of legitimate pollen deposited on *Jepsonia* and *Pontederia* stigmas are not consistently greater on emasculated flowers than on intact flowers (Tables 1 and 2). This suggests that the anthers of the two species, which are separated from the stigmas, do not obstruct the deposition of legitimate pollen to an appreciable extent, as Barrett and Glover (1985) recognized.

2.3 Seed Set on Self-Incompatible Homostyles

A unique piece of evidence in favor of Darwin's cross-promotion hypothesis was provided by Ganders' (1975) examination of a population of *Mitchella repens* which contained self-incompatible long-homostyles as well as pin and thrum plants. Like the thrum plants, the homostyles require pin pollen to be fertilized but they differ from thrum plants in having stigmas at the higher level rather than at the level of pin anthers. The homostyles had a lower seed set than thrum plants (78 versus 89%), as Darwin's hypothesis predicts.

2.4 Future Work

Although the studies of pollen loads on insects and stigmas reviewed above have provided impressive support for Darwin's hypothesis, further studies of pollen transfer are desirable. The first requirement of such studies is to remove the effects of autogamy by emasculation. Ideally, this would be done for all two or three morphs, as was done for *Jepsonia heterandra* (Ganders 1974) and *Pontederia cordata* (Barrett and Glover 1985), and would be accompanied by experiments in which all flowers are emasculated so that no geitonogamy is possible (Ganders 1974, 1976). Emasculation techniques that leave the anthers intact, so the floral structure is not disturbed, are desired. Emasculation should also reduce the advertising and reward functions of pollen as little as possible, so pollinators are not deterred.

The examination of total stigma loads in polymorphic populations does not allow the legitimate and illegitimate transfers to be studied separately, however, and does not allow the numbers of pollen grains transferred in single visits to be examined. These difficulties could be overcome by experimental manipulations in monomorphic field populations. Isolated populations or subpopulations containing only one morph occur in some species. In other species these could be created by removing on any one day all flowers of one morph or by planting monomorphic populations that have access to the natural pollinators. Suppose a population with only thrum flowers is available and some flowers are emasculated and screened from visitors. When the screens are removed (or a previously isolated plant is brought in), the amounts of pollen transferred from one thrum flower to another could be obtained after a single visit (q_{tt} in the models below), a certain duration of time, or over the entire period of anthesis of the flower (the latter is d_{tt} in the calculations above and the models below.) The amounts of pollen deposited could be compared with those of unemasculated single flowers and intact whole plants, to detect the separate effects of autogamy and geitonogamy. The introduction of a single emasculated pin plant would allow the proficiency of thrum to pin transfers, d_{tp}, to be evaluated. Converse experiments on a monomorphic pin population would provide information on the transfer of pollen from pin anthers in the presence and absence of autogamy and geitonogamy.

For species with pollinators that can be experimentally manipulated, comparable experiments can be done by presenting captive pollinators with flowers of the two morphs in a given order. In a pioneering set of experiments of this nature, Feinsinger and Busby (1987) studied pollen transfer by hummingbirds in legitimate pollinations of *Palicourea lasiorrachis* (Rubiaceae). They showed pollen carryover for up to 20 visits (the maximum studied) after pickup and demonstrated that the distribution of grains among successive stigmas is as important as the average number received by each stigma. Hopefully, future experiments of this kind will include illegitimate transfers, so the proficiencies of legitimate and illegitimate transfers can be compared.

The relative proficiencies of legitimate and illegitimate pollinations can also be examined by a genetical approach. The frequency of self-fertilization (autogamy and geitonogamy) can be determined by examining the morph distribution of allozymes or restriction enzyme polymorphisms (RFLP) among seed progeny of plants in natural populations. Furthermore, the frequency of geitonogamy can be separated from that of autogamy by examining morph frequencies in progeny from plants on which (a) only the target flower has been emasculated, preventing autogamy, and (b) all but the target flower have been emasculated, preventing geitonogamy. The families used for selfing estimates could also be examined for frequencies of thrum and pin plants. Subtracting the self-pollinations allows the relative frequencies of the four types of legitimate and illegitimate cross-fertilizations to be determined. If the numbers of plants, flowers per plant, and pollen grains per flower of the two morphs are known, the relative frequencies of legitimate and illegitimate fertilizations could be compared on a per plant, per flower, or per pollen grain basis. These frequencies would provide proficiency estimates in terms of actual fertilizations rather than the pollen depositions detected in observations of stigma loads.

This combination of genetical data has not been obtained for any distylous species, but there is some highly suggestive data for the more complex situation in a tristylous species, *Eichhornia paniculata*, obtained by Spencer Barrett and his colleagues. Glover and Barrett (1986b) used allozyme loci to estimate outcrossing rates; these ranged from 0.29 to 0.96 in various populations. In one population, B5, the outcrossing rates on long-, mid-, and short-styled plants were 0.98, 0.97, and 0.93 respectively. The progeny from natural pollinations in the same population were also scored for frequencies of the three heterostylous morphs (Barrett et al. 1987), providing estimates of the proportions of crosses between morphs, 0.90, 0.93 and 0.83 in the three morphs. The differences between the two sets of figures are the proportions of own-form crosses, 0.08, 0.04, and 0.10. The proportions among the cross-pollinations on a morph that occur with mates of their own form are then 0.08/0.98 = 8.2%, 4.1% and 10.8% for the three morphs. The calculations suggest that the success of legitimate crosses is even higher in terms of fertilizations achieved than is indicated in the stigma load data for pollen redeposition in *Pontederia cordata* discussed above. Differential germination or pollen-tube growth may reinforce reciprocal herkogamy in promoting legitimate crosses.

3 Models of Selection of Distyly

In this section, we develop quantitative phenotypic models of selection of the three principal aspects of distyly – reciprocal herkogamy, self-incompatibility and the ancillary morphological features. In particular, we examine the relative abilities of two selective forces, avoidance of self-fertilization and an increase in the proficiency of cross-fertilization, to support the evolution of the features of distyly. Selection against self-interference is unlikely to be sufficiently powerful to select reciprocal herkogamy or self-incompatibility by itself, and it is not explicitly treated in the models.

As Charlesworth and Charlesworth (1979) pointed out, the principal hurdle for models of the evolution of heterostyly is to find conditions under which a mutant is able to invade a previously monomorphic population without increasing to fixation, establishing a stable protected polymorphism. This requires that the relative fitnesses of the morphs are negatively frequency-dependent so that each morph is fitter than the other when rare but not when common.

The only previous quantitative model of the evolution of both self-incompatibility and reciprocal herkogamy, that of Charlesworth and Charlesworth (1979), assumed that the ancestor was homostylous and self-compatible. The Charlesworths postulated that the initial event in the evolution of heterostyly is the introduction of a pollen mutant that cannot fertilize itself or the original type. This mutant can spread because of its increased female fitness if it has a sufficient outbreeding advantage, but its increase is limited because it is functionally female until a second (stigma) mutant that is compatible with the mutant pollen type arises and establishes diallelic self-incompatibility. In their models, the reciprocal herkogamy arises after self-incompatibility. In contrast, our model starts with a herkogamous ancestor and examines the introduction of either reciprocal herkogamy or self-incompatibility first. It indicates that the reciprocal herkogamy is more likely to evolve first and is selected because it actively promotes cross-fertilization rather than because it prevents self-fertilization.

3.1 Selection of a Stigma-Height Polymorphism

We assume that the ancestor possessed approach herkogamy, in which the stigma protrudes beyond the anthers (Fig. 1). The plausibility of this assumption has been discussed in Chapter 6. To consider the establishment of a polymorphism in stigma height, we compare the fitness of an individual with approach herkogamy, w_a, with that, w_r, of an individual with reverse herkogamy in which the stigma is below the anthers. The anthers are in the same position in the two forms. The frequencies of the "approach" and "reverse" forms are a and r respectively, where $a + r = 1$. Both forms produce f flowers per plant, each with o ovules. Fractions u_a and u_r of the ovules in the two forms are self-fertilized autogamously during each insect visit. After visiting a flower, a pollinator has a probability, g, of visiting another flower of the same individual and causing geitonogamous self-pollination. When a pollinator visits an approach flower and picks up pollen, and then visits a reverse flower and redeposits the pollen, the quantity of pollen deposited is q_{ar}. Similarly, the proficiencies of the other types of pollen transfer are q_{aa}, q_{ra}, and q_{rr}. For simplicity, pollen is considered to be redeposited in the first flower after it is picked up (there is no pollen carryover). The number of ovules fertilized is kq_{ar}, where k represents the proportion of allogamously deposited pollen grains that fertilize ovules, assumed to be the same in all combinations. The fitness of progeny from outcrossing is one, and that of the inbred progeny from selfing is i, which is less than one if there is inbreeding depression. Pollinators visit approach and reverse plants at random and thus in the proportions in which they occur in the population (a, r).

The fitness of an individual is the sum of fitness gained from autogamous and geitonogamous self-pollinations (two gamete contributions to each offspring) and single contributions as a male and female parent of progeny from outcrossing. If there is a probability, p, that each flower is visited, i.e., there is insufficient pollination to fertilize all ovules (maternal fitness is pollinator-limited), the number of approach ovules cross-fertilized is $fopk(1 - g)\,(aq_{aa} + rq_{ra})$ and the success of approach pollen in cross-fertilization is $fopk(1 - g)\,(aq_{aa} + rq_{ar})$. Then the fitness of an approach individual,

$$w_a = fop\{2u_a i + 2gq_{aa}ki + k(1-g)[(aq_{aa} + rq_{ra}) + (aq_{aa} + rq_{ar})]\}$$

The fitness of a reverse individual,

$$w_r = fop\{2u_r i + 2qp_{rr}ki + k(1-g)[(aq_{ar} + rq_{rr}) + (aq_{ra} + rq_{rr})]\}$$

The fitness advantage of the reverse phenotype,

$$w_r - w_a = fop\{2i(u_r - u_a) + 2kgi(q_{rr} - q_{aa}) + k(1-g)\{a(q_{ar} + q_{ra} - 2q_{aa}) - r(q_{ar} + q_{ra} - 2q_{rr})\}. \quad (3)$$

Consider the effects of autogamy and geitonogamy first. It is evident from Eq. (3) that the ability of a reverse morph to invade an approach population is increased if the reverse morph has more frequent autogamy ($u_r > u_a$) or more efficient geitonogamous transfer of pollen ($q_{rr} > q_{aa}$). These advantages are not frequency-dependent, however, and so they could not cause a limited increase in the reverse morph, giving a polymorphism. Conversely, if the approach morph has higher self-fertilization, selection would cause it to be fixed in the population in the absence of differences between morphs in their outcrossing proficiencies.

Now consider the outcrossing component of Eq. (3) (inside the square brackets) when the two forms do not differ in the proficiencies of autogamous or geitonogamous pollinations. Then an individual with reverse herkogamy is fitter than one with approach herkogamy when

$$a(q_{ar} + q_{ra} - 2q_{aa}) - r(q_{ar} + q_{ra} - 2q_{rr}) > 0. \quad (4)$$

A mutant with reverse herkogamy can invade a population of plants with approach herkogamy (i.e., $w_r - w_a > 0$ when $a = 1, r = 0$) if

$$q_{ar} + q_{ra} > 2q_{aa}. \quad (5a)$$

Conversely, an approach mutant could invade a reverse population ($w_r > w_a$ when $r = 1, a = 0$) if

$$q_{ar} + q_{ra} > 2q_{rr}. \quad (5b)$$

Inequalities (5a) and (5b) together provide sufficient conditions for a protected polymorphism. A polymorphism is possible if each morph is more proficient in cross-pollinations with the other morph (on average as a maternal and a paternal parent) than with other plants of its own kind. When a morph is rare, it usually crosses with plants of the other morph rather than with its own kind and therefore has an advantage. The frequency-dependence of the morph fitnesses that is brought about by the superiority of legitimate pollen transfers leads to equilibrium when the two stigma-height morphs have equal fitnesses (Lloyd 1977); that is, $w_r - w_a = 0$. From Eq. (3), the equilibrium morph ratio is then

$$\frac{a}{r} = \frac{q_{ar} + q_{ra} - 2q_{rr}}{q_{ar} + q_{ra} - 2q_{aa}}. \quad (6)$$

If the illegitimate transfers are equally proficient, $q_{rr} = q_{aa}$, and the polymorphism is stable when the morphs are equally frequent.

The model shows that a stigma-height polymorphism can evolve through the selective advantage of legitimate cross-pollinations over illegitimate pollinations, as

Darwin (1862b, 1877) postulated, but not from an outbreeding advantage. Interestingly, the conditions for a polymorphism, [Eq. (5)], do not contain terms for the frequencies of autogamous or geitonogamous pollination (assuming the morphs do not differ in these respects). Hence the promotion of legitimate pollination could cause the evolution of a stigma-height polymorphism in a strictly outcrossing population or one that is partially selfing. It is evident from Eq. (3), however, that the size of the fitness advantage caused by Darwin's cross-promotion hypothesis depends on (l-g). Hence the polymorphism becomes more stable as the amount of outcrossing among individuals increases, although the equilibrium morph ratio is unaffected. The model can therefore explain the otherwise puzzling occurrence of (putative) multi-allelic self-incompatibility in heterostylous species of *Narcissus* and *Anchusa* (Bateman 1952; Fernandes 1964; Schou and Philipp 1984; Lloyd et al. 1990) and populations of *Narcissus* and *Epacris* with stigma-height polymorphisms (Dulberger 1964; Brown and Crowden 1984; O'Brien and Calder 1989).

Next, we consider the evolution of a stigma-height polymorphism from monomorphic herkogamy when pollinators are abundant, so maternal fitness is not pollinator-limited ($p = 1$) and paternal fitness is subject to pollen competition. Since we have shown above that neither geitonogamy nor autogamy can maintain a stigma-height polymorphism, we consider them here jointly as self-fertilization. Approach and reverse plants outcross in frequencies t_a, t_r and self-fertilize in frequencies $1-t_a$, $1-t_r$ respectively. In outcrossing, each individual mates with K plants, and as a pollen parent competes on the stigmas of each mate in a pollen pool from K plants, with approach and reverse pollen competitors in the proportions in which they occur in the population, a and r. (K is assumed to be large, so pollen competition is not local.) In each of v visits to a flower, the number of pollen grains deposited on the stigma depends on the type of pollen being redeposited and the level of the stigma. Thus on a single visit to an approach flower, for example, a quantity q_{ra} picked up from an immediately previous visit to a reverse flower is redeposited with no pollen carryover. Each approach flower therefore receives $v(aq_{aa} + rq_{ra})$ pollen grains, and the proportion of these deposited from a single approach pollen parent is $q_{aa}/K(aq_{aa} + rq_{ra})$. Each plant produces n seeds.

The fitness of a plant with approach herkogamy is the sum of the two gamete contributions to selfed ovules, the outcrossed ovule contribution and the outcrossed pollen contributions on the two kinds of mates. That is,

$$w_a = n\left\{2\,(1-t_a)\,i + t_a + \frac{Kv}{Kv}\left[\frac{at_a\,q_{aa}}{aq_{aa} + rq_{ra}} + \frac{rt_r q_{ar}}{aq_{ar} + rq_{rr}}\right]\right\}.$$

The fitness of a plant with reverse herkogamy,

$$w_r = n\left\{2(1-t_r)\,i + t_r + \frac{at_a\,q_{ra}}{aq_{aa} + rq_{ra}} + \frac{rt\,rq_{rr}}{aq_{ar} + rq_{rr}}\right\}.$$

$$\therefore w_r - w_a = n\left\{(t_r-t_a)(1-2i) + \frac{at_a\,(q_{ra}-q_{aa})}{aq_{aa} + rq_{ra}} + \frac{rt_r\,(q_{rr}-q_{ar})}{aq_{ar} + rq_{rr}}\right\}. \qquad (7)$$

An increase in cross-fertilization, t, by either morph gives that morph a selective advantage at all frequencies when inbreeding depression is sufficiently strong ($i < \frac{1}{2}$), assuming there is no difference in the cross-pollination proficiencies, i.e., $q_{aa} = q_{ra}$ and $q_{rr} = q_{ar}$. As in the case of pollinator-limited maternal fitness, selection based on an outbreeding advantage cannot maintain a stigma-height polymorphism.

Consider now the situation when the two forms have the same frequency of outcrossing ($t_r = t_a$), but the proficiencies of the various directions of cross-pollination differ. Plants with reverse herkogamy can invade a population of plants with approach herkogamy (i.e., $w_r > w_a$ when $a \approx 1$, $r \approx 0$) when, from Eq. (7),

$$q_{ra} > q_{aa}. \tag{8a}$$

Conversely, approach plants can invade a monomorphic population of reverse plants ($w_a > w_r$ when $r \approx 1$, $a \approx 0$) provided

$$q_{ar} > q_{rr}. \tag{8b}$$

Inequalities (8a) and (8b) provide conditions for a protected polymorphism when the fertilization of ovules is not pollinator-limited. In contrast to the situation with pollinator limitation [(5a) and (5b)], the selection of a stigma-height polymorphism here depends solely on each form being more proficient in donating pollen to the other form than the other form is in donating pollen to other flowers of its own type. The relative abilities of each form to receive pollen from the same form and the other form do not affect the outcome.

The measures of the proficiency of pollen transfer used in the models, q_{ij}, can readily be compared with the observed pollen loads on stigmas, d_{ij}, analyzed in the earlier section on the evidence for Darwin's cross-promotion hypothesis, since $d_{ij} = vq_{ij}$, where i and j represent the morphs or levels of a heterostylous species. In a species in which the amounts of pollen transferred in the various combinations have been observed on emasculated stigmas (eliminating the loads from selfing), one can therefore test whether there is sufficient promotion of legitimate crosses to account for the evolution of a herkogamy polymorphism. (Wolfe and Barrett (1989) have shown that pollen loads from single visits to flowers of the tristylous species, *Pontcdcria cordata*, arc in general agreement with loads from multiply visited flowers.) We will examine the pollen loads on emasculated stigmas of *Jepsonia heterandra*, described above. Admittedly we are dealing with a species that has already evolved reciprocal herkogamy, so we are really testing whether the promotion of cross-fertilization can maintain heterostyly rather than cause the evolution of a stigma-height polymorphism. From the data for emasculated flowers, the probabilities of pollen grains being transferred (Table 1, $\times 10^3$) are $d_{tp} = 4.46$, $d_{pt} = 2.71$, $d_{pp} = 2.98$, and $d_{tt} = 1.29$. Assuming that all flowers receive the same number of visitors, v, and all visits of one kind have the same effects, the total transfer probabilities (d_{ij}) are proportional to the corresponding quantities of pollen deposited per visit (since $d_{ij} = vq_{ij}$). Hence, $q_{tp} = 3.47\, q_{tt}$, $q_{pt} = 0.91\, q_{pp}$, and $\frac{1}{2}(q_{tp} + q_{pt}) = 2.78\, q_{tt} = 1.33\, q_{pp}$. Comparison with Eqs. (5) and (8) shows that the promotion of cross-pollination by itself can easily maintain the reciprocal herkogamy of *J. heterandra* if pollinators are not limiting, and it can explain the persistence of thrums but not quite that of pins if pollinators are limiting. The figures indicate that Darwin's cross-promotion hypothesis provides a sufficiently powerful selective force to account for the maintenance of the stylar polymorphism in this species.

3.2 Subsequent Selection of an Anther-Height Polymorphism and Ancillary Features

Suppose a stigma-height polymorphism has evolved through the promotion of cross-pollination as described above, and consider selection of an accompanying anther-height polymorphism to give the system of reciprocal herkogamy that defines heterostyly. For this purpose, we compare the fitness of the reverse form as given in the derivation of Eq. (7) when there is abundant pollination with the fitness of a mutant, h, which has the stigma height of the reverse form but an increased anther height that approximates the stigma height of the approach form (see Fig. 1c, Chap. 6). The mutant h therefore has the properties of a thrum morph of a heterostylous population. When h is rare, its fitness

$$w_h = n\left\{\cdot 2(1-t_h)\, i + t_h + \frac{at_a\, q_{ha}}{aq_{aa} + rq_{ra}} + \frac{rt_r\, q_{hr}}{aq_{ar} + rq_{rr}}\right\}.$$

Hence the fitness advantage of h over r,

$$w_h - w_r = n\left\{\cdot(t_h - t_r)(1-2i) + \frac{at_a(q_{ha} - q_{ra})}{aq_{aa} + rq_{ra}} + \frac{rt_r(q_{hr} - q_{rr})}{aq_{ar} + rq_{rr}}\right\}. \quad (9)$$

When $i < \frac{1}{2}$, the thrum form will have an advantage if it increases the rate of cross-fertilization ($t_h > t_r$) through increasing the separation between the pollen and stigmas in each flower. In addition, the increase in anther height in the thrum flowers brings the anthers to the height of the stigmas in approach flowers and may well increase the donation of pollen to these flowers ($q_{ha} > q_{ra}$) and decrease pollen donation to reverse flowers ($q_{hr} < q_{rr}$). These effects would increase the amounts of legitimate pollen donation [the second major term in Eq. (9)] and decrease the amount of illegitimate pollen donation (the third term). Recall the result of the previous model, that the stigma-height polymorphism can be established when pollinators are abundant only if legitimate pollen donation exceeds illegitimate donation [(8a), (8b)]. Hence it is likely that the increase in anther height will cause the positive effect on heterostyly donation to exceed the negative effect on donation in illegitimate cross-pollinations. Thus the thrum form could replace the reverse form either because it increases the outcrossing rate or because it causes a net increase in the proficiency of pollen donation in cross-pollinations. The addition of a pollen-height polymorphism to an existing stigma-height polymorphism appears to be easy to achieve.

Similar models can be applied to the selection of ancillary features of heterostyly, such as morph differences in pollen size or sculpturing or characters of the stigmatic papillae. The ancillary feature can readily be selected if a change in an ancillary feature of one morph decreases self-fertilization or increases its proficiency in legitimate pollination as a pollen donor or receiver (the latter only when pollinator activity limits fertilization).

3.3 Selection of Self-Incompatibility in the Absence of Reciprocal Herkogamy

We now consider the selection of the diallelic self-incompatibility of distylous species, first in the absence of reciprocal herkogamy and then in its presence. There are a considerable number of potential scenarios for the initial evolution of self-incompatibility. These may involve either a reduction of the number of alleles from a multiallelic system (Crowe 1964; Muenchow 1982), the evolution of a new polymorphism with separate pollen and style loci (Charlesworth and Charlesworth 1979), modification of pollen growth patterns in styles of different length (Darwin 1877 and below), or even the modification of a preexisting polymorphism that previously controlled another aspect of pollen growth and is converted to a role in self-incompatibility. Moreover, if self-incompatibility arises anew, the strength of the sterility reaction may increase gradually or the sterility may be complete from the start. We examine the evolution and maintenance of self-incompatibility by considering the selection of phenotypes that affect the strength of self-incompatibility. In this way, we can quantify conditions for the selection of self-incompatibility in the presence or absence of reciprocal herkogamy and compare these conditions with those already known for the maintenance of multiallelic self-incompatibility. The models apply to gradual or discrete changes in the strength of self-incompatibility and to systems that are controlled by specific self-incompatibility loci or by the polygenic modification of the ability of pollen types to grow in styles of different lengths. They are not relevant, however, to an origin of diallelic self-incompatibility from a multiallelic system.

We will examine selection of the strength of self-incompatibility in one of the self-incompatibility morphs of a dimorphic system, independently of the strength of self-incompatibility in the other morph. The assumption of separate modification of self-incompatibility in different morphs is plausible for a two- or three-morph system, as discussed in the previous article, and is supported by observations that the morphs of heterostylous populations often differ in the strength of self-incompatibility (Charlesworth and Charlesworth 1979; Barrett and Anderson 1985; Gibbs 1986).

Suppose a population that lacks reciprocal herkogamy possesses two phenotypes, 1 and 2, in frequencies f_1 and f_2, where $f_1 + f_2 = 1$. The pollen of each phenotype grows less well in styles of the same phenotype than in styles of the other phenotype. The phenotypes undergo frequencies s_1 or s_2 of autogamous self-fertilization and t_1 or t_2 of cross-fertilization ($s_i + t_i = 1$). Again for simplicity we exclude geitonogamy, which does not affect the major features of the model. Pollinator visits are frequent, so the number of seeds per plant, n, is solely resource-limited. The effective amount of pollen transferred during a visit from one flower of type i to a single flower of type j is e_{ij}, determined by the amount of pollen deposited and its relative growth rate. Each plant mates with K other plants as both a pollen and seed parent and K is large.

The fitness of a plant of type 1 is the sum of contributions from selfing, its outcrossed seeds, and pollen success in crosses with mates of types 1 and 2. Thus

$$w_1 = n\left\{2s_1 i + (1-s_1) + \frac{K}{K}\left[\frac{f_1 t_1 e_{11}}{f_1 e_{11} + f_2 e_{21}} + \frac{f_2 t_2 e_{12}}{f_1 e_{12} + f_2 e_{22}}\right]\right\}.$$

Suppose there is a rare mutant, y, which alters the deposition or growth of pollen of plants of phenotype 1 on stigmas of types 1 and 2 so that s_1, e_{11}, and e_{12} are changed to s_y, e_{y1}, and e_{y2}. The fitness of plants expressing y is

$$w_y = n\left\{2s_y i + (1-s_y) + \frac{f_1 t_1 e_{y1}}{f_1 e_{11} + f_2 e_{21}} + \frac{f_2 t_2 e_{y2}}{f_1 e_{12} + f_2 e_{22}}\right\}.$$

The fitness advantage of y compared with 1,

$$w_y - w_1 = n\left\{\underset{\text{I}}{(s_y-s_1)(2i-1)} + \underset{\text{II}}{\frac{f_1 t_1 (e_{y1}-e_{11})}{f_1 e_{11} + f_2 e_{21}}} + \underset{\text{III}}{\frac{f_2 t_2 (e_{y2}-e_{12})}{f_1 e_{12} + f_2 e_{22}}}\right\}. \tag{10}$$

The third major term (III) in Eq. (10) is the effect of a change in the rate of pollen growth in legitimate cross pollinations. Assuming that y does not affect legitimate pollinations ($e_{y2} = e_{12}$), (10) becomes

$$w_y - w_1 = n\left\{\underset{\text{I}}{(s_y-s_1)(2i-1)} + \underset{\text{II}}{\frac{f_1 t_1 (e_{y1}-e_{11})}{f_1 e_{11} + f_2 e_{21}}}\right\}. \tag{11}$$

The first term (I) is the effect of an alteration in the strength of self-incompatibility on the fitness components from selfing and outcrossed seeds. The quantity $n(s_y-s_1)(2i-1)$ expresses the conditions for the selection of self- and cross-fertilization when all cross-fertilizations are fully compatible (approximated in multiallelic systems). If inbreeding depression is strong ($i < \frac{1}{2}$), selection favours the total failure of self-fertilization, as already derived for homomorphic populations [cf. Lloyd 1979, Equation (1)].

Equation (11) also shows that in a diallelic system, self-incompatibility has an extra disadvantage compared with multiallelic self-incompatibility because it reduces the success of illegitimate cross-pollinations (term II). These are half of all cross-pollinations on average. The magnitude of this extra disadvantage can be seen by making further simplifying assumptions. Suppose that plants of phenotypes 1 and 2 are completely self-incompatible [i.e., in (11), $e_{11} = 0$, $s_1 = 0$, $t_1 = 1$] whereas the mutant y is completely self-compatible ($e_{y1} = e_{21}$) although not necessarily always self-fertilized. Then if the self-incompatibility morphs 1 and 2 are equally frequent, Eq. (11) reduces to

$$w_y - w_1 = n[s_y(2i-1) + 1], \tag{12a}$$

and if i = 1 (no inbreeding depression),

$$w_y - w_1 = n(s_y + 1) \tag{12b}$$

A self-compatible mutant, y, in a diallelic self-incompatible population has a twofold advantage. It gains one gamete contribution for every selfed progeny [the well-known "automatic advantage of self-pollination" or cost of outcrossing that is also present in multiallelic self-incompatibility systems and is represented by s_y in Eq. (12b)], and its outcrossing pollen contribution is doubled because it is cross-compatible with all plants [represented by the 1 in Eq. (12b)]. The second factor is present in diallelic systems but not in multiallelic systems, where cross-incompatibilities are rare. It may often confer a greater advantage than increased selfing because morphological features may limit the frequency of self-fertilization and the automatic advantage is diminished by the lower success of selfed progeny [Eq. (12a)]. These calculations demonstrate the force of Darwin's argument that self-incompatibility mechanisms in a distylous species, unlike reciprocal herkogamy, could not be selected as a mechanism encouraging cross-fertilization because it results in the cross-sterility of each plant with "half its brethren" (Darwin 1877).

The extra disadvantage of cross-incompatibility in a diallelic system, especially one unaccompanied by reciprocal herkogamy, allows self-fertile mutants to invade much more readily. This factor may explain why self-fertile mutants invade heterostylous species so frequently. It also helps to explain the absence of reports of diallelic self-incompatibility systems in groups without heterostyly, and it provides a strong argument against the evolution of self-incompatibility before that of reciprocal herkogamy in the evolution of heterostyly.

3.4 Selection of Self-Incompatibility in a Population with Reciprocal Herkogamy

We now examine selection of diallelic self-incompatibility in a distylous population. The herkogamous morphs are assumed to be intrasterile and interfertile, as in many distylous populations. We initially examine mutants that affect the strength of self-incompatibility in only one of the two morphs. That is, the incompatibility effects of mutants are "morph-limited" and parallel to those of sex-limited genes. The previous model for self-incompatibility in a nonherkogamous population can be transferred directly to this situation by considering that pin and thrum morphs correspond to forms 1 and 2 and a mutant z affects the degree of self-incompatibility of pin plants only. Substituting the subscripts p (for pin), t (for thrum) and z for 1, 2 and y respectively in Eq. (11) for the situation where the mutant has no effect on the pollen growth of legitimate pollinations gives

$$w_z - w_p = n\left\{\underbrace{(s_z - s_p)(2i-1)}_{\text{I}} + \underbrace{\frac{f_p t_p (e_{zp} - e_{pp})}{f_p e_{pp} + f_t e_{tp}}}_{\text{II}}\right\}. \tag{13}$$

Although Eqs. (11) and (13) are mathematically equivalent, they apply to very different biological situations. In (11), term II for the effect of a mutant on illegitimate crosses is a large factor constituting a major disadvantage for a self-incompatibility mechanism because incompatible cross-pollinations are abundant. In (13), however, term II will be much less important whenever reciprocal herkogamy reduces the transfer of pollen from low anthers to high stigmas. Hence the presence of reciprocal herkogamy makes the origin and maintenance of self-incompatibility much more likely to be selectively favored because it reduces the special liability of a diallelic system, frequent cross-incompatibility. This conclusion is additional to the argument in the preceding article that diallelic self-incompatibility is much easier to evolve in physiological terms in the presence of reciprocal herkogamy since the pollen growth patterns can be based on existing morphological differences. Both factors lead to the conclusion that self-incompatibility is more likely to evolve in heterostylous species after the evolution of reciprocal herkogamy than before.

This conclusion differs from Darwin's views (1877) on the origin of heterostyly in an instructive way. We concur with Darwin's proposal that intramorph sterility cannot be selected as a mechanism that promotes cross-pollination and his suggestion that the evolution of the sterility mechanisms follows that of reciprocal herkogamy. Darwin (1877, p. 265), however, further suggested that the intramorph sterility is "an incidental and purposeless result", what we would now describe as a pleiotropic byproduct of the evolution of reciprocal herkogamy. Darwin was forced into this conclusion because he did not make a firm distinction between the avoidance of self-fertilization and the active promotion of cross-fertilization. But if we make that distinction, we may postulate that the sterility mechanism has been selected because it restricts self-fertilization even when it does not directly promote cross-fertilization. Our model supports this suggestion.

Consider now the selection of a mutant pollen type that restricts the growth of pollen in the styles of any plant in which it occurs, rather than in only one morph as considered above. Such a mutant might be unlinked to the locus controlling reciprocal herkogamy, or it might be linked with varying amounts of recombination and linkage disequilibrium. The selective advantage of this mutant over phenotype 1 would be similar to that in Eq. (10) if the mutant is expressed equally in the two morphs, and identical to that in Eq. (11) if the mutant is completely linked to the herkogamy locus. Intermediate degrees of linkage disequilibrium would confer intermediate selective pressures on the mutant. If linkage is sufficiently tight, a morph-linked incompatibility mutant could overcome the disadvantage of cross-incompatibility and be selected.

There are therefore two kinds of genes that could be responsible for the evolution of self-incompatibility in a heterostylous population, morph-limited and morph-linked genes. There is at present no critical information that would allow us to discriminate between the two possibilities. Furthermore, we suggest that a choice between the two alternatives is simplistic. They may both be involved in different species, or they may well have participated jointly in the same species, as discussed in the previous article. The appearance de novo of either a strictly morph-limited gene or a tightly linked gene may be unlikely. Suppose instead that a partly morph-limited mutant arose that restricted, say, the growth of pin pollen on pin styles but had a smaller effect on the growth of pin pollen on thrum styles (or thrum pollen on pin styles). If the former effect sufficiently outweighed the latter, the mutant could be selected, from Eq. (10). Subsequently, selection could act to increasingly limit the action of the gene to illegitimate pollinations or to increase its association with one morph by introducing and tightening linkage with the herkogamy gene. This sequence of events is less stringent than a requirement for strict morph limitation or tight linkage from the beginning. It could also explain two of the principal features of the inheritance of the heterostyly super-gene in *Primula*, the failure to detect a separate gene for a female self-incompatibility reaction and the tight linkage between the locus for the pollen reaction and the loci controlling stigma and anther heights (Ernst 1955; Dowrick 1956; Charlesworth and Charlesworth 1979).

3.5 The Roles of Various Selective Forces

The preceding models lead to several general conclusions about the selection of the features of heterostylous species:

1. If a herkogamous ancestor is assumed rather than a homostylous one, it is not difficult to obtain plausible conditions for the evolution of stigma- and anther-height polymorphisms without support from a self-incompatibility mechanism.
2. The principal selection agent in this case is that proposed by Darwin (1862b, 1877), the promotion of cross-pollination, rather than avoidance of self-fertilization.
3. The ancestral monomorphic population may therefore have been strictly outcrossing or partly self-fertilizing.
4. The self-incompatibility mechanism hinders (illegitimate) *cross-fertilization*, and is therefore more likely to evolve after the evolution of reciprocal herkogamy than before it.
5. The self-incompatibility mechanism of heterostylous populations is selected when it is advantageous to avoid selfing.
6. Self-incompatibility can evolve in heterostylous populations solely by the action of pollen mutants that cause either type of pollen to grow less well in one or other stylar environment. There need not be a locus for style reactions other than that controlling the stigma-height polymorphism.
7. The ancillary features of heterostylous populations can evolve readily after the evolution of reciprocal herkogamy if they confer an advantage from either reducing self-fertilization or promoting cross-fertilization, or both.
8. The mutations for self-incompatibility or ancillary features may be morph-limited or morph-linked, or the evolution of these features may involve a combination of both types of association.

The models presented above involved choices of some additional features. In particular, the models assumed that the ancestors of heterostylous populations had approach herkogamy rather than reverse herkogamy, and that the first step in the evolution of heterostyly is the evolution of a stigma-height polymorphism rather than an anther-height polymorphism. We have given reasons for these assumptions in Chapter 6, but the same selective forces would work equally well with the opposite assumptions.

4 A Possible Mechanism Promoting Legitimate Cross-Pollination

We have seen that there is considerable evidence that reciprocal herkogamy does promote legitimate cross-fertilization. Moreover, the models presented above have confirmed that a stigma-height polymorphism, and subsequently reciprocal herkogamy, can be selected from an ancestor with approach herkogamy if the transfer of pollen between flowers of different morphs is more proficient than transfer between flowers of the same type. Here we examine exactly how intermorph transfer may be enhanced (Fig. 1).

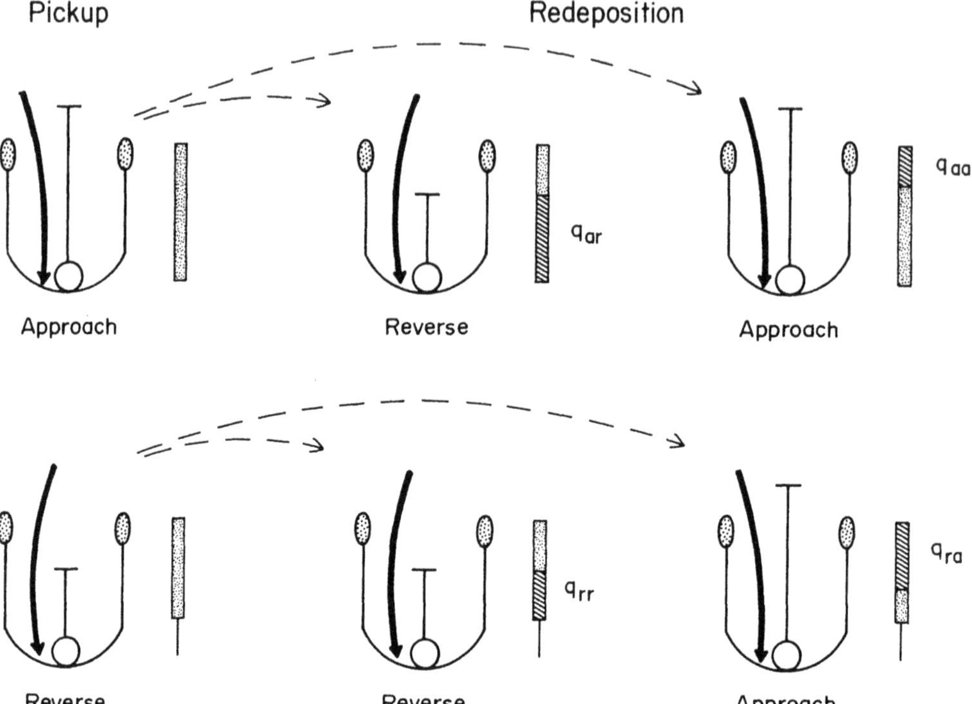

Fig. 1. A possible mechanism by which the pollen of two stigma-height morphs is segregated on a pollinator's probe and transferred more proficiently onto legitimate stigmas. The point of the probe traces a curved "free path" mid-way between the stamens and carpels, which differs in the approach (*above left*) and reverse (*below left*) morphs. The segment of the probe over which pollen is picked up is shown in the *stippled columns on the left*. The segment of the probe from which pollen is redeposited on reverse and approach stigmas is shown in the *cross-hatched columns in the centre and right-hand diagrams*. The parameter q_{ij} is the quantity of pollen transferred from anthers of morph i to a stigma of morph j during a single visit. Here $q_{ar} > q_{aa}$, $q_{ra} > q_{rr}$

Darwin (1862b, 1865, 1877) and subsequently others have assumed that the promotion of legitimate crosses comes about because pollen and stigmas presented at the same level come into contact with the same part of a pollinator's body, facilitating the transfer of pollen. The organs at different levels in a flower experience "segregated contact" on pollinators' bodies. In contrast, in a monomorphic population pollen and stigmas of each flower must touch the same part of a pollinator's body. When the pollen and stigmas are separated in flowers, as they are in the postulated ancestor with approach herkogamy, pollen and stigmas undergo "sequential contact". Pollinators entering approach-herkogamous flowers are likely to contact the stigmas and deposit pollen picked up from other flowers before they touch the anthers further in (Webb and Lloyd 1986). The evolution of a stigma-height polymorphism, and subsequently reciprocal herkogamy, therefore involves *a change from sequential contacts of stigmas and anthers to segregated contacts*.

We postulate that the likely circumstances for such a change to be beneficial occur when the ancestral monomorphic, approach-herkogamous system is *operating inefficiently because the parts of an insect's body that contact the pollen and stigmas do not entirely coincide.* This situation makes a monomorphic herkogamous population vulnerable to invasion. If a mutant (reverse herkogamy) morph has flowers structured so that the pollen is loaded onto the parts of its body that coincide more closely with the parts that contact the stigmas of flowers with approach herkogamy, transfer from reverse to approach flowers will be more proficient than that between approach flowers (i.e., $q_{ra} > q_{aa}$). This satisfies the requirement for a reverse mutant to invade an approach population when maternal fitness is solely resource-limited [Eq. (8a)]. Furthermore, if its stigmas coincide better with the part of the pollinator's body that approach pollen is loaded onto, pollen transfer to reverse stigmas will also be enhanced (i.e., $q_{ar} > q_{aa}$). When both directions of transfer are improved, $q_{ar} + q_{ra} > 2q_{aa}$, and the less stringent condition in Eq. (5a) for invasion of a reverse mutant with pollinator-limited maternal fitness is satisfied.

There are several ways in which the required segregation of contacts between approach and reverse flowers might promote legitimate crosses:

1. The entry/exit path of a probe (proboscis, proboscis plus head, beak, etc.) may be curved and take a slightly different course in the two morphs.
2. The probe may vary in diameter throughout its length (e.g., taper towards the tip).
3. The entry and exit paths of the probe may differ slightly.

We will examine how the first factor might operate. The available space for a probe to enter a flower is different in the flowers of approach and reverse morphs (and in pin and thrum flowers) because of the position of the anthers and styles/stigmas. A straight probe entering an approach (or pin) flower with a protruding stigma has an initial available space in the form of a ring bounded on the inside by the stigma and on the outside by the corolla. If the probe is inserted so that its tip is always in the centre of the available space, the tip traces out a "free path" that is slightly curved as it enters the flower (Fig. 1). The probe would be in contact with an anther over the entire inserted length of the probe up to the position of the anthers. The stigma, however, would come into contact only with the upper part of the probe in the later stages of insertion as the probe becomes oriented more vertically.

A probe entering a reverse (or thrum) flower has a different available space at the flower entrance, in the form of a solid circle bounded only by the anthers and corolla on the outside. Its free path during insertion into the flower is curved in the opposite direction (Fig. 1), causing it to contact the stigma during the whole process of insertion, but contact with the anthers is delayed until the probe is partly inserted.

The differently curved paths into approach and reverse flowers can therefore result in segregated contacts, as required for Darwin's hypothesis. Considering the contact positions on probes as postulated, it is possible that the degree of overlap in anther and stigma contacts may be greater for legitimate transfers than for illegitimate transfers (Fig. 1). Each morph could benefit both in pollen donation ($q_{ar} > q_{rr}$, $q_{ra} > q_{aa}$) and pollen receipt ($q_{ar} > q_{aa}$, $q_{ra} > q_{rr}$), giving a protected polymorphism.

We do not know whether this mechanism is how the segregated contacts necessary for Darwin's hypothesis are actually brought about, or whether the segregation depends on the probe shape or different entry and exit paths. There is no empirical data that would allow us to discriminate among the alternative mechanisms, which are not mutually exclusive. Experimental populations, such as those suggested above to examine the proficiencies of different types of transfer, could be used to record the quantities and positions of pollen picked up and redeposited from insect bodies in the different combinations. A careful comparison of the dimensions of probes and flowers at different levels, and the use of videotapes to determine entry and exit paths, would also help to elucidate the mechanism by which the pollinator contacts are segregated in thrum and pin flowers.

5 Discussion

Darwin (1877) proposed that the promotion of legitimate cross-pollination was the principal selective force responsible for the evolution of reciprocal herkogamy, which he believed to be the initial step in the evolution of heterostyly. The studies of *Jepsonia heterandra* and *Pontederia cordata* show that the average proficiencies of legitimate donation and receipt are approximately twice those of the corresponding illegitimate combinations. The twofold advantages detected in these two species are strong selective forces, even for polymorphic traits (Endler 1986). Far from there being little evidence in favor of Darwin's hypothesis, as most recent authors have concluded, few selection hypotheses can claim this degree of support.

Darwin's hypothesis that heterostyly promotes cross-pollination is one of the most testable hypotheses of natural selection. The selective forces involved are powerful and readily estimated by counts of pollen loads on stigmas, parameters that relate directly to reproductive fitness. The cause of the selection differentials, the segregation of pollen types on the bodies of pollinators, is equally amenable to observation. Furthermore, the overall proficiencies of pollen donation and receipt over the full period that a flower is open is easily broken down into the elementary causative events, single pollinator visits. Manipulative experiments can be conducted with manageable pollinators (cf. Feinsinger and Busby 1987) or by controlling the numbers of flowers of different types in natural populations, as discussed above. The success of pollinations in these situations could be measured by counts of the seeds produced, a direct measure of reproductive fitness. Altogether, the comparison of legitimate and illegitimate pollen transfer offers unsurpassed opportunities for the analysis of natural selection.

The validity of the selection models presented above depends on the sequence in which the principal features of heterostylous species have appeared. If the ancestors of the heterostylous species in any family were not herkogamous or if self-incompatibility evolved before reciprocal herkogamy, as suggested for the Plumbaginaceae by Baker (1966), selection against self-fertilization is likely to have played a much more important role than we have assigned to it (cf. Charlesworth and Charlesworth 1979).

It is well known that Darwin favored the evolution of characters by gradual change rather than by large sudden steps. In the case of heterostyly, Darwin (1877, p. 260 on) postulated a gradual divergence, from a monomorphic ancestor, of two types differing in style and stamen lengths. According to our adaptive hypothesis, however, the first step in the origin of heterostyly involves the invasion of a reverse herkogamy mutant that differs discretely in style length from the ancestral condition. If our hypothesis is correct, a mutant with only a slightly shorter style would not have the same advantage and would be unable to invade the ancestral population (see Fig. 1 in Chap. 6). A small change would bring anthers and stigmas closer together and thus increase self-fertilization without making cross-pollination much more effective [without appreciably increasing q_{ar} or q_{ra} in Eqs. (5) and (8)]. On our hypothesis of the evolution of heterostyly, the first change therefore involves a discrete step across an adaptive valley to a new adaptive peak. The invasion of strictly female mutants into a hermaphrodite population, establishing gynodioecy, is another example where only a large sudden change can be selected (Lloyd and Venable 1991).

In the selection models presented above, we have attempted to cover the principal factors governing the selection of the various features of heterostyly polymorphisms. The models could undoubtedly be developed further, however. For example, the precise effects of linkage relationships could be examined, as in the models of Charlesworth and Charlesworth (1979).

One combination of factors we have not considered is the joint interaction of pollen carryover and successive visits of pollinators to a number of flowers of the same morph. Spencer Barrett and his colleagues have noted that the occurrence of many flowers on a plant at one time, particularly in species with extensive clonal growth, will result in pollinators often making a number of successive visits to the same genotype. They have proposed that reciprocal herkogamy may reduce pollen wastage caused by redeposition on flowers of the same plant and hence "increase the distance that pollen may be transported by pollinators" (Glover and Barrett 1986a; also Price and Barrett 1984; Wolfe and Barrett 1989). It may well be that the herkogamy of single morphs and the herkogamy polymorphism may reduce pollen wastage, but it seems likely that the benefits will derive primarily from leaving more pollen to be available for crosses rather than from the increased distance that pollen travels. There is, however, a limit to the extent to which increasing pollen carryover aids paternal fitness. Assuming that the amount of pollen on an insect's body remains approximately constant with continued visits, an increase in carryover entails a decrease in the rate of pollen pickup. If there is too much pollen carryover, paternal fitness may be limited by the rate of pollen pickup, particularly when there are not many successive visits to flowers of one plant.

In self-compatible plants, pollen carryover may also have an important effect on maternal fitness. Alastair Robertson (unpubl.) has shown that increased pollen carryover decreases geitonogamous self-fertilization by a considerable amount, even when there are many successive visits to flowers of the same plant. Hence pollen carryover has at least four effects – on pollen pickup, pollen wastage, pollen outcrossing distances, and self-fertilization. In monomorphic plants, the selected amount of carryover represents the optimal compromise among these factors. In heterostylous populations, the segregated transfer of pollen in the different directions of legitimate pollinations means that the processes of pollen pickup and legitimate redeposition in one flower are largely separated from each other. Selection for the optimal rate of pickup is therefore largely decoupled from selection for legitimate redeposition. To the extent that the segregation of pollen on insects is effective, optimal pickup and optimal redeposition can be selected independently. It is evident that we know little of the complex processes involved in the pollination of heterostylous species. The ecological aspects of pollination dynamics warrant much more attention than they have hitherto received.

In an earlier paper (Lloyd and Webb 1986), we proposed that outcrossed flowers in general experience a conflict between two opposed selection forces. Selection for economical pollination favors the placement of anthers and stigmas in similar positions within flowers, but selection to avoid self-fertilization and self-interference favors the separation of the pollinating surfaces in space or time. Heterostyly may be viewed as a unique way of resolving this conflict and combining more precise pollination with the minimization of self-fertilization and self-interference. If the evolution of heterostyly has followed this evolutionary scenario and resulted from the selection forces we have proposed, the resolution of the conflict has occurred in several steps. Self-fertilization, as well as pollen wastage, stigma clogging, and other aspects of the mutual obstruction of stamens and carpels were presumably already alleviated in the monomorphic herkogamous ancestors of heterostylous species, if they were herkogamous, as much as they are in their dimorphic descendants (cf. Ganders 1979). The introduction of reciprocal herkogamy provided a more economical transfer of outcrossed pollen. Subsequently, the evolution of self-incompatibility further reduced self-fertilization, while the divergence of pollen and stigma characteristics of the morphs may have involved all three selective forces.

Although we have not endorsed his selection hypotheses in all respects, we believe Darwin (1877) was more correct than many of his successors in postulating that the promotion of cross-pollination has been the selective force responsible for the evolution of reciprocal herkogamy. This parallels our conclusion in Chapter 6 that Darwin correctly identified the evolution of reciprocal herkogamy as the key initial step in the evolution of the heterostyly syndrome. Much more work still needs to be done, however, on both the evolutionary and adaptive aspects of heterostyly. The techniques for doing this work are available, and we expect that heterostyly will continue to be as favored in the next century as it has been in the last as a model system for studying the behavior of plant populations.

6 Historical Appendix

The pollen wastage hypothesis, that heterostyly reduces the loss of pollen on illegitimate stigmas, is generally attributed to Darwin (1877) but we can find no mention of it in that work. In at least two places in *The Different Forms of Flowers* (pp. 138, 263), Darwin refers to "wastage" or "loss" of pollen, but these passages refer to the transfer of pollen between anthers and stigmas at the same height – that is, to Darwin's hypothesis of the promotion of legitimate cross-pollination. On the other hand, in the *Fertilisation of Orchids*, Darwin (1862a, p. 114) mentioned that if a pollinator moved from one flower where it picks up pollen to another lower down on the same inflorescence that was presenting a stigma, "the pollen masses would be brushed off her proboscis and wasted. But nature suffers no such wastage." This passage can be interpreted as a recognition of the principle that selection may operate to prevent pollen being wastefully deposited where it cannot achieve fertilization. Hence the "pollen wastage" hypothesis may be attributed to Darwin, but to a different work from the one usually cited. If this is done, Herbert Baker (1954, 1964) should be given credit for rediscovering the pollen wastage hypothesis, first using it in the modern sense of a loss of pollen on self-incompatible stigmas, and applying it to illegitimate pollinations in heterostylous species. Moreover, Baker was the first modern author to propose any form of selection to avoid self-interference (as interpreted in Lloyd and Webb 1986).

Acknowledgments. We are grateful to Deborah Charlesworth and Spencer Barrett for their careful reviews of the manuscript and to Fred Ganders and Bob Ornduff for helpful comments.

References

Baker HG (1954) Dimorphism and incompatibility in the Plumbaginaceae. Rapp Comm VIIIe Congr Int Biol Paris Sect 10:133–134
Baker HG (1964) Variation in style length in relation to outbreeding in *Mirabilis* (Nyctaginaceae). Evolution 18:507–512
Baker HG (1966) The evolution, functioning and breakdown of heteromorphic incompatibility systems. I. The Plumbaginaceae. Evolution 20:349–368
Barrett SCH, Anderson JM (1985) Variation in expression of trimorphic incompatibility in *Pontederia cordata* L. (Pontederiaceae). Theor Appl Genet 70:355–362
Barrett SCH, Glover DE (1985) On the Darwinian hypothesis of the adaptive significance of tristyly. Evolution 39:766–774
Barrett SCH, Wolfe LM (1986) Pollen heteromorphism as a tool in studies of the pollination process in *Pontederia cordata* L. In: Mulcahy DL, Mulcahy GB, Ottaviano E (eds) Biotechnology and ecology of pollen. Springer, Berlin Heidelberg New York, pp 435–442
Barrett SCH, Brown AHD, Shore JS (1987) Disassortative mating in tristylous *Eichhornia paniculata* (Pontederiaceae). Heredity 58:49–55
Bateman AJ (1952) Trimorphism and self-incompatibility in *Narcissus*. Nature (Lond) 170:496–497
Brown N, Crowden RK (1984) Evidence of heterostyly in *Epacris impressa* Labill. (Epacridaceae). In: Williams EG, Knox RB (eds) Pollination 1984: Proceedings of a Symposium held at the Plant Cell Biology Research Centre, University of Melbourne. Melbourne Univ Press, Melbourne, pp 187–193
Charlesworth D, Charlesworth B (1979) A model for the evolution of distyly. Am Nat 114:467–498
Crowe LK (1964) The evolution of outbreeding in plants. I. The angiosperms. Heredity 19:435–457
Darwin C (1862a) On the various contrivances by which British and foreign orchids are fertilised by insects. Murray, Lond
Darwin C (1862b) On the two forms, or dimorphic condition in the species of *Primula*, and on their remarkable sexual relations. J Linn Soc Lond Bot 6:77–96
Darwin C (1865) On the sexual relations of the three forms of *Lythrum salicaria*. J Linn Soc Lond Bot 8:169–196
Darwin C (1877) On the different forms of flowers on plants of the same species. Murray, Lond
Dowrick VPJ (1956) Heterostyly and homostyly in *Primula obconica*. Heredity 10:219–236
Dulberger R (1964) Flower dimorphism and self-incompatibility in *Narcissus tazetta* L. Evolution 18:361–363
Endler JA (1986) Natural selection in the wild. Princeton Univ Press, Princeton
Ernst A (1955) Self-fertility in monomorphic primulas. Genetica 27:391–448

Feinsinger P, Busby WH (1987) Pollen carryover: experimental comparisons between morphs of *Palicourea lasiorrachis* (Rubiaceae), a distylous, bird-pollinated, tropical treelet. Oecologia 73:231–235
Fernandes A (1964) Contribution à la connaissance de la génétique de l'hétérostylie chez le genre *Narcissus* L. I. Résultats de quelques croisements. Bol Soc Broteriana 38:81–96
Ganders FR (1974) Disassortative pollination in the distylous plant *Jepsonia heterandra*. Can J Bot 52:2401–2406
Ganders FR (1975) Fecundity in distylous and self-incompatible homostylous plants of *Mitchella repens* (Rubiaceae). Evolution 29:186–188
Ganders FR (1976) Pollen flow in distylous populations of *Amsinckia* (Boraginaceae). Can J Bot 54:2530–2535
Ganders FR (1979) The biology of heterostyly. NZ J Bot 17:607–635
Gibbs PE (1986) Do homomorphic and heteromorphic self-incompatibility systems have the same sporophytic mechanism? Plant Syst Evol 154:285–323
Glover DE, Barrett SCH (1986a) Stigmatic pollen loads in populations of *Pontederia cordata* from the southern U.S. Am J Bot 73:1607–1612
Glover DE, Barrett SCH (1986b) Variation in the mating system of *Eichhornia paniculata* (Spreng.) Solms. (Pontederiaceae) Evolution 40:1122–1131
Lewis D (1982) Incompatibility, stamen movement and pollen economy in a heterostyled tropical forest tree, *Cratoxylum formosum* (Guttiferae). Proc R Soc Lond Ser B 214:273–283
Lloyd DG (1977) Genetic and phenotypic models of natural selection. J Theor Biol 69:543–560
Lloyd DG (1979) Some reproductive factors affecting the selection of self-fertilisation in plants. Am Nat 113:67–79
Lloyd DG, Venable DL (1991) Some properties of natural selection with single and multiple constraints. Theor Pop Biol (in press)
Lloyd DG, Webb CJ (1986) The avoidance of interference between the presentation of pollen and stigmas in angiospermis. I. Dichogamy. NZ J Bot 24:135–162
Lloyd DG, Yates JMA (1982) Intrasexual selection and the segregation of pollen and stigmas in hermaphroditic plants, exemplified by *Wahlenbergia albomarginata* (Campanulaceae). Evolution 36:903–913
Lloyd DG, Webb CJ, Dulberger R (1990) Heterostyly in species of *Narcissus* (Amaryllidaceae) and *Hugonia* (Linaceae) and other disputed cases. Plant Syst Evol 172:215–227
Muenchow G (1982) A loss-of-alleles model for the evolution of distyly. Heredity 49:81–93
Nicholls MS (1985) Pollen flow, population composition, and the adaptive significance of distyly in *Linum tenuifolium* L. (Linaceae). Biol J Linn Soc 25:235–242
Nicholls MS (1987) Pollen flow, self-pollination and gender specialisation: factors affecting seed-set in the tristylous species *Lythrum salicaria* (Lythraceae). Plant Syst Evol 156:151–157
O'Brien S, Calder DM (1989) The breeding biology of *Epacris impressa*. Is this species heterostylous? Aust J Bot 37:43–54
Olesen JM (1979) Floral morphology and pollen flow in the heterostylous species *Pulmonaria obscura* Dumort (Boraginaceae). New Phytol 82:757–767
Ornduff R (1971) The reproductive system of *Jepsonia heterandra*. Evolution 25:300–311
Ornduff R (1980a) Pollen flow in *Primula veris* (Primulaceae). Plant Syst Evol 135:89–94
Ornduff R (1980b) Heterostyly, population composition, and pollen flow in *Hedyotis caerulea*. Am J Bot 57:1036–1041
Ornduff R (1983) Interpretations of sex in higher plants. In: Meudt WJ (ed) Strategies of plant reproduction. Allanheld Osmum, Lond, pp 21–33
Piper J, Charlesworth B (1986) The evolution of distyly in *Primula vulgaris*. Biol J Linn Soc 29:123–137
Price SD, Barrett SCH (1982) Tristyly in *Pontederia cordata*. Can J Bot 60:897–905
Price SC, Barrett SCH (1984) The function and adaptive significance of tristyly in *Pontederia cordata* L. (Pontederiaceae). Biol J Linn Soc 21:315–329
Rosov SA, Screbtsova ND (1958) Honeybees and selective fertilisation of plants. XVII Int Beekeeping Congr 2:494–501
Schou O (1983) The distyly in *Primula elatior* (L.) Hill (Primulaceae), with a study of flowering phenology and pollen flow. Bot J Linn Soc 86:261–274

Schou O, Philipp M (1984) An unusual heteromorphic incompatibility system. III. On the genetic control of distyly and self-incompatibility in *Anchusa officinalis* L. (Boraginaceae). Theor Appl Genet 68:139–144
Shore JS, Barrett SCH (1984) The effect of pollination intensity and incompatible pollen on seed set in *Turnera ulmifolia* L. (Tuneraceae). Can J Bot 62:1298–1303
Webb CJ, Lloyd DG (1986) The avoidance of interference between the presentation of pollen and stigmas in angiosperms. II. Herkogamy. NZ J Bot 24:163–178
Wolfe LM, Barrett SCH (1989) Patterns of pollen removal and deposition in tristylous *Pontederia cordata* (Pontederiaceae). Biol J Linn Soc 36:317–329
Yeo PF (1975) Some aspects of heterostyly. New Phytol 75:147–153

Chapter 8

The Application of Sex Allocation Theory to Heterostylous Plants

B.B. CASPER[1]

1 Introduction

Sex allocation theory predicts the evolutionarily stable sex ratio in dioecious species, the allocation of resources to male vs. female function in simultaneous hermaphrodites, and the time and order of sex change in sequential hermaphrodites (Charnov 1982). In all of these problems, phenotypic fitness is frequency-dependent, and the equilibrium proportion of phenotypes, the evolutionarily stable strategy (Maynard Smith 1982), often satisfies a certain optimality principle (Charnov 1979, 1982, 1987; Charnov and Bull 1986). Casper and Charnov (1982) and Taylor (1984) applied sex allocation theory to heterostylous plants. Because heterostylous populations consist of two or three hermaphroditic (style length) morphs, there are two sex allocation problems in this system – the optimal allocation to pollen and seed production in each morph and the equilibrium morph ratio.

This chapter traces the development of the idea that in heterostylous species fitness through male and female functions might differ between morphs and briefly reviews genetic models of equilibrium morph ratios. It then summarizes the sex allocation models, discusses relevant empirical data from natural populations, and considers possible mechanisms through which morph ratios may be controlled.

2 Considerations of Gender in Heterostylous Plants

Well before the development of sex allocation theory, the observation that style length morphs sometimes produce different quantities of seeds stimulated thinking about their functional gender. Darwin (1877) originally thought heterostylous plants were developing dioecy. He explained that in *Primula* the greater seed production of thrums contradicted his first impression that the long-styled pin form, with its larger stigma, shorter stamens, and smaller pollen, was the more feminine. He considered the mid-style length morph of *Lythrum salicaria* less masculine than the other two and reasoned that through continued deterioration of male organs in that morph the species could eventually consist of two heterostylous hermaphrodites and a female.

The repeated evolution of dioecy from distyly is now well documented (Wyatt 1983). Such a transition, of course, requires a divergence in sexual function such that

[1]Department of Biology, University of Pennsylvania, Philadelphia, PA 19104–6018, USA

Monographs on Theoretical and Applied Genetics 15
Evolution and Function of Heterostyly (ed. by S.C.H. Barrett)

one morph reproduces primarily through seeds and the other through pollen. Some workers believe that, contrary to the pattern in *Primula*, the pin morph usually becomes the pistillate parent (Ganders 1979; Beach and Bawa 1980). Baker (1958) and Beach and Bawa (1980) explained that a change in pollinators could disrupt reciprocal pollen flow between morphs and more likely result in pollen transferring primarily from the anthers of thrums, which are positioned high in the corolla tube, to the exerted stigmas of pins. Muenchow and Grebus (1989) feel, however, that the morphological evidence for pin plants normally evolving into females is inconclusive.

During the 1970s, workers began to consider selection pressures on male and female functions in plants as potentially very different (Charnov et al. 1976; Horovitz 1978; and Willson 1979). Horovitz and Willson applied this idea to heterostylous species. Willson viewed heterostyly as a possible means of achieving different probabilities of pollen receipt and donation and thus expected to find gender specialization between morphs. Horovitz, on the other hand, speculated that equal division of fitness between male and female function might be adaptive and suggested that heterostyly could be a means of equalizing pollen receipt and donation.

Lloyd (1979) first quantified gender specialization in heterostylous plants. He used a formula based on morph frequencies, numbers of seeds produced in the two morphs, the amounts of self- and illegitimate fertilizations, and the relative fitness of selfed offspring to calculate the proportion of a plant's genes that are transmitted through pollen vs. ovules. When self- or illegitimate matings do not occur, functional gender can be calculated from seed production data alone. In such case the average femaleness of a long-styled morph ($\bar{G}_l$) equals

$$\frac{n_l\bar{d}_l}{n_l\bar{d}_l + n_s\bar{d}_s}\,, \tag{1}$$

where n = numbers of individuals, $\bar{d}$ = mean number of seeds per individual and the subscripts s and 1 represent short- and long-styled morphs respectively. Similarly,

$$\bar{G}_s = \frac{n_s\bar{d}_s}{n_l\bar{d}_l + \bar{n}d_s}\,. \tag{2}$$

If only legitimate fertilizations occur, the average pollen fitness among individuals of one morph ($\bar{A}_s$ or $\bar{A}_l$) equals the average seed fitness of the other. For example, $\bar{G}_s = \bar{A}_l = 1 - \bar{G}_l$. Sex allocation theory extends this concept of gender to models of equilibrium morph ratios and optimal distribution of resources to pollen and seeds.

3 Genetic Models of Style Morph Frequency

The literature contains two approaches to understanding equilibrium morph ratios. The first derives the equilibrium ratio under the assumption of a specific sex-determining mechanism (Finney 1952; Spieth 1971; Heuch 1979ab; Heuch and Lie 1985). The second approach, used by the sex allocation models described here, predicts that autosomal control of morph ratios will result in an equilibrium where all individuals have equal fitness. The second is more general in that it applies where fitness is a phenotypic trait and is uninfluenced by the particular genes at the sex determining locus (see Lloyd 1977; Charnov 1982 pp. 10–12 for applications to morph ratios under gynodioecy.) The two models overlap in that many specific genetic models satisfy the "equalization of fitness" equilibrium (Lloyd 1977).

The equilibrium proportions of style length morphs in tristylous species has long been of theoretical interest (Fisher 1941, 1944). Finney (1952) defined equal representation of style length morphs as isoplethy, and then, using hypothetical genetic systems, showed that stable anisoplethy can exist. His and subsequent models are constructed using two different mating patterns, either: (1) pollen elimination in which the egg genotype is drawn at random, and the pollen genotype is drawn at random from all compatible pollen or (2) zygote elimination which assumes that legitimate matings occur with frequencies proportional to the product of the frequencies of the particular morphs involved. In hypothetical tristylous systems, these two approaches can yield different anisoplethic equilibria (Spieth 1971).

Heuch (1979a) demonstrated that in tristylous species with complete legitimate matings and no initial fitness differences among morphs, isoplethy is stable. He showed that differences in total fitness among morphs or asymmetric mating patterns, such as differing degrees of self-pollination, can result in anisoplethy (Heuch 1979b). Heuch used a known genetic basis for tristyly and his models satisfied the equalization of fitness equilibrium.

4 Sex Allocation Models

Sex allocation theory for autosomal genes is based on the simple axiom that every diploid organism has one father and one mother (Fisher 1930). Half the autosomal genes in any zygote come from each parent. Because of this, in large, randomly mated populations of males and females, the less common sex will always be selected for, and frequency-dependent natural selection will maintain equal numbers of the two sexes (Charnov 1982). This evolutionarily stable strategy (ESS) (Maynard Smith 1982) equalizes fitness through male and female offspring. Unequal numbers of sexual morphs are predicted if males and females cost different amounts of resource to produce (Fisher 1930) or if the environment or spatial structuring of the population affects fitness in the two sexes differentially (Hamilton 1967; Charnov 1982).

The allocation of resources between the production of male and female gametes in hermaphrodites can be treated similarly. Sex allocation theory considers the set of possible fitness values obtainable by allocating different proportions of resource to male vs. female function as defined by the male/female fitness set (Fig. 1). The curve shows, for example, how much fitness in male function (m) is gained through the ability to produce more pollen if some increment of fitness through female function (f) is given up because a plant produces fewer seeds. Fitness is obviously a complex function of gamete numbers, gamete quality (such as seed quality), and how gamete success is affected by the particular environment (see Charnov and Bull 1986). ESS allocation to male and female fitness is the point on the curve that maximizes total reproduction through both male and female function (Charnov 1982). That point is stable to the invasion of any genotype with any other combination of m and f. Hermaphroditism is selected for if the fitness set is convex because the production

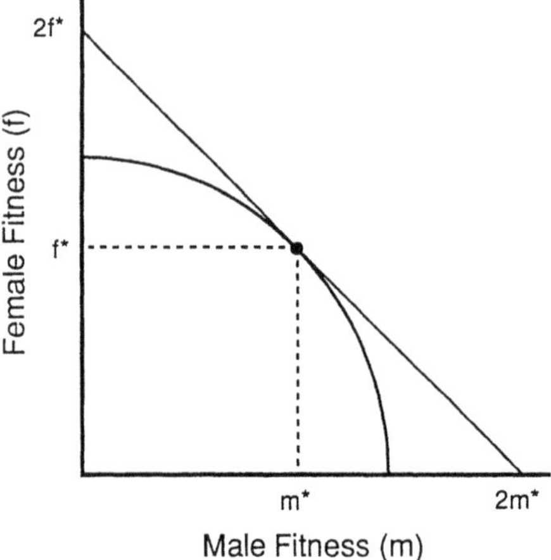

Fig. 1. Hypothetical male (m)/ female (f) fitness tradeoff for an hermaphrodite. The tangent line (defined as $\frac{\partial f}{\partial m} = \frac{-f}{m}$) gives the ESS allocation (m^*, f^*), the point on the fitness set at which reproduction through male and female function is maximized. This tangent line intercepts the m and f axes at $2m^*$, $2f^*$ respectively. (See further discussion in Charnov 1982)

of increasingly more pollen (or increasingly more seeds) yields diminishing returns in total fitness. If the fitness set is concave, the ESS consists of separate male and female individuals (Charnov et al. 1976).

The sex allocation model derived by Casper and Charnov (1982) applies to self-incompatible distylous species in which a pin morph produces F_p seeds and M_p pollen grains, while the thrum morph produces F_t and M_t. In calculating the stable equilibrium morph ratio, a rare mutant at a locus unlinked to the heterostyly locus (supergene) is assumed to produce $\hat{q}$ proportion pins among its progeny while the wild type produces q pins. The fitness (W) for the mutant individual in a large population of size N is the sum of the following four terms [Eq. (3)]:

$\hat{q} \cdot F_p$ (fitness through seeds of its pin progeny)

$(1 - \hat{q})F_t$ (fitness through seeds of its thrum progeny)

$\left(\frac{(1-\hat{q})\, M_t}{(1-q)\, N \cdot M_t}\right)(q \cdot N \cdot F_p)$ (fitness through pollen of its thrum progeny)

$\left(\frac{\hat{q} \cdot M_p}{q \cdot N \cdot M_p}\right)[(1-q)N \cdot F_t]$ (fitness through pollen of its pin progeny).

The ESS morph ratio among zygotes is found by setting $\partial W/\partial q = 0$ and solving for q, which equals 1/2 (Casper and Charnov 1982). The ESS is independent of the F and M values, which means that q = 1/2 even if pins and thrums suffer differential mortality to adulthood or produce different quantities of seeds or pollen. This model assigns control of morph ratios to the maternal plant, but the same equilibrium of 1/2 would result even if control were assigned to the zygote (Taylor 1984). This equilibrium is where the fitness of a pin equals that of a thrum; each individual passes genes to $F_p + F_t$ seeds.

The model assumes that the equilibrium is maintained by selection acting on autosomal genes even though equal proportions of style length morphs in the ratio expected from the genetics at the style length locus (Ganders 1979). Casper and Charnov (1982) suggested that morph ratios could be controlled through either pre-zygotic selection among pollen tubes or through selective embryo abortion.

Because the model is comparable to models of primary sex ratio in dioecious species (Shaw and Mohler 1953; Leigh 1970; Charnov 1975), there are several conditions under which deviations from equal numbers of the two morphs would be predicted: (1) if seeds carrying the pin and thrum genotypes disperse different distances (Clark 1978; Bulmer and Taylor 1980); (2) under conditions of local mate competition, if one morph specializes in the production of seed and the other pollen (Hamilton 1967; Werren 1980); (3) if seeds carrying the pin and thrum genotypes received different quantities of resources from the parent (Fisher 1930; Maynard Smith 1978), or (4) if some intramorphic matings occurred with different frequencies in the two morphs (Casper and Charnov 1982).

The other sex allocation problem in the system is the evolutionarily stable allocation to pollen and seeds in each morph. The model assumes that a rare mutant autosomal gene in a distylous population which when present in a style-length morph (here a pin) alters its resource allocation between pollen and seeds from F_p, M_p to $\hat{F}_p$, $\hat{M}_p$. In a large, randomly mated population of size N, the mutant individual will have fitness (W_p):

$$W_p = \hat{F}_p + \left(\frac{\hat{M}_p}{1/2 \cdot N \cdot M_p}\right)(1/2 \cdot N \cdot F_t)\,. \tag{3}$$

The equation reduces to:

$$W_p = \hat{F}_p + F_t \cdot \left(\frac{\hat{M}_p}{M_p}\right), \tag{4}$$

with the analagous fitness for a rare thrum mutant being:

$$W_t = \hat{F}_t + F_p \cdot \left(\frac{\hat{M}_t}{M_t}\right). \tag{5}$$

Consider W_p; if the mutant is the same as the wild-type ($\hat{M}_p = M_p$, $\hat{F}_p = F_p$), $W_p = F_p + F_t$. Selection will favor an altered allocation ($\hat{M}_p$ or $\hat{F}_p$ that makes $W_p > F_p + F_t$. The same holds for W_t. Thus at ESS, the values for F_t, F_p, M_t, M_p are such that fitness W_t and W_p cannot be increased by any other values. The result is graphically illustrated in Fig. 2. The ESS allocation to F and M in each morph is the point on the tradeoff curve where a line intercepting the F axis at $F_p + F_t$ is tangential. The tangent finds the ESS because all other possible combinations of m and f, as defined by the tradeoff curve, fall to the inside of the line. For any mutant to make W_p or $W_t > F_p + F_t$, the mutant would have to fall to the outside of the respective tangent and outside the boundary of the tradeoff curve. Casper and Charnov (1982) suggested that the male/female tradeoff relations (as in Fig. 1) may well be shaped differently for the two morphs. In the illustration presented here, the thrum morph specializes in the

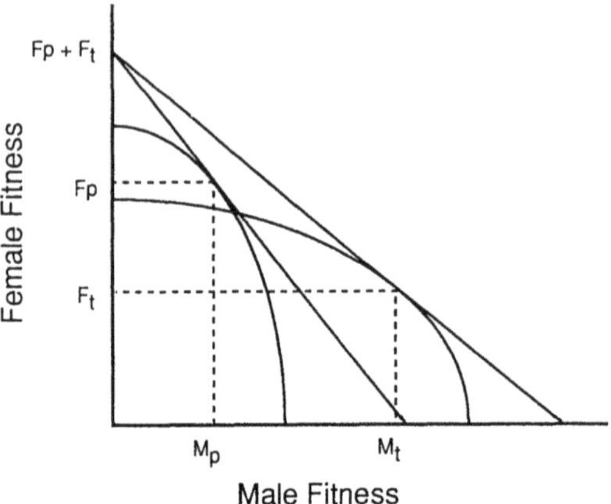

Fig. 2. Hypothetical male, female tradeoffs for two style length morphs, pins (*p*) and thrums (*t*). The tangent lines give a different ESS allocation for each morph. (Corrected by P. Taylor from Casper and Charnov 1982)

production of pollen and thus reproduces mostly through male function while the pin reproduces mostly through seeds.

Sex allocation models for species with more than two style length morphs are necessarily more complicated (Casper and Charnov 1982). The morphs may have different male/female tradeoffs and exhibit various mating relations with each other. Taylor (1984) simplified his model by assuming that all morphs produce the same number of seeds so that increasing investment in female function only increases seed size, but he used a constant pollen grain size so that pollen numbers vary with expenditure. Taylor supposed that (1) each individual has a fixed amount of energy (e) to allocate to reproduction, (2) a plant spending x units of energy on ovules has a female fitness of F(x), and (3) a plant spending y units on type j pollen contributes a mass $V_j(y_{ij})$ to the type j pollen pool. Type j pollen only fertilizes the ovules of morph type j. This assumption effectively says that, in this case, a plant of type i can mate with plants of type j simply by producing type j pollen (except $j \neq i$).

Where a type i flower spends x_i on seeds and y_{ij} on type j pollen ($j \neq i$),

$$x_i + \sum_{j \neq i} y_{ij} = e. \tag{6}$$

If f_i is the frequency of type i in the population, the total expected fitness of a type i plant (W_i) is

$$W_i = F(x_i) + \sum_{j \neq i} f_j F(x_j) V_j (y_{ij}) / \bar{V}_j \ , \tag{7}$$

where $\bar{V}_j$ is the average contribution to the type j pollen pool and

$$\bar{V}_j = \sum_{i \neq j} f_i V_j (y_{ij}). \tag{8}$$

Since at equilibrium (the ESS) type frequencies all individuals will have the same fitness (Lloyd 1977; Heuch 1979a), Eq. (7) equals two times the average fitness through ovules ($2\bar{F}$) so that

$$2\bar{F} = F(x_i) + \sum_{j \neq i} f_j F(x_j) V_j (y_{ij}) / \bar{V}_j \,, \tag{9}$$

where $\bar{F}$ =

$$\sum f_i F(x_i). \tag{10}$$

A rare mutant allele which alters its X_i and y_{ij} values will have fitness

$$\hat{W}_i = F(\hat{x}_i) + \sum_{j \neq i} f_j F(x_j) V_j (\hat{y}_{ij}) / \bar{V}_j \,. \tag{11}$$

For evolutionary stability, this must not exceed W_i in (7), and resources allocated to x_i and all y_{ij} must not exceed e. The Lagrange multiplier condition (see Taha 1971) says that at an interior maximum of W_i subject to Eq. (6), there is some λ_i for which

$$\partial \hat{W}_i / \partial \hat{x}_i = \partial W_i / \partial \hat{y}_{ij} = \lambda_i \tag{12}$$

for all $j \neq i$ at the point (x_i, y_{ij}). Using Eq. (11), the result is that if $j \neq i$ (and if " ' " refers to a derivative):

$$F'(x_i) = f_j F(x_j) V_j'(y_{ij}) / \bar{V}_j = \lambda_i, \tag{13}$$

which says that marginal fitness of a type i plant through its seeds is equal to its marginal fitness through type j pollen for all $j \neq i$.

Taylor noted that given the female fitness function W(x) and pollen effectiveness function $V_j(y)$, the Eqs. (6), (9), and (13) can be solved for the variables f_i, x_i, y_{ij}, and λ_i. He pointed out that these conditions are necessary but not sufficient for an ESS, but showed how solutions of the equations can be analyzed to determine if an ESS is achieved.

Taylor applied his model to data of Price and Barrett (1982) for tristylous *Pontederia cordata*. He assumed that pollen effectiveness $V_j(y)$ is linear so that the contribution of a plant to a pollen pool (and fitness through male function) is proportional to the number of grains of that pollen type produced. Using estimates of pollen and seed production values from natural populations, he calculated theoretical equilibrium frequencies of the three style length morphs.

The morph frequencies Taylor obtained are not as close to the actual population values as those calculated by Barrett et al. (1983), using Heuch's (1979a) model, who assumed a somewhat simpler pattern of pollen type production. Taylor listed possible reasons why his results do not fit the data better. Among them is the possibility that the pollen effectiveness function is not linear. There is reason to suspect that in general male fitness is not a linear function of pollen produced for most insect-pollinated – and perhaps wind-pollinated – plants (Charnov 1979, 1982, 1987).

Charlesworth (1989) used the approach of Casper and Charnov to model the evolution of dioecy from distylous plants under particular conditions. She assumed complete incompatibility and made seed production a concave function of the number of compatible pollen grains received. She showed that if one morph receives inadequate pollen for full seed set, then that morph should allocate more resources to male function and the other morph more to female function. She concluded that the two morphs should exhibit large differences in their ESS allocation to pollen vs. seeds with only weakly assymmetrical pollen flow.

Plant ecologists are beginning to acquire information on male fitness characteristics in plants (Goldman and Willson 1986; Snow 1986; Stanton et al. 1986). Still, measuring fitness in a way that defines the shape of the male/female tradeoff curve is difficult, although there may be ways to determine it indirectly (Charnov 1987). Lacking better data on male function, sex allocation theory, therefore, is perhaps best used at the moment for comparisons among species and qualitative predictions of optimal allocation to sexual functions and equilibrium morph ratios. An important role of the theory is that it forces clear thinking about reproductive characteristics, and ESS calculations, like Charlesworth's, reveal what factors are important in determining the optimal allocation patterns.

5 Gender and Morph Ratio Data

Because we know so little about fitness through male function in heterostylous plants (but see Kohn and Barrett 1992), the evidence that morphs might specialize in functional gender mostly comes from differences in seed production. Different style length morphs sometimes produce different numbers of pollen grains but this usually accompanies a pollen size dimorphism (Ganders 1979). It is not clear that these pollen characters represent different amounts of invested resources or have any significance for sex allocation theory, and thus they will not be considered here. Additional components of fitness through female function, such as seed germinability (Nicholls 1987), might also differ, of course, but these are seldom measured.

One type of evidence for gender specialization comes from crossing experiments in which more seeds are obtained following an intermorph cross in one direction than in the reciprocal. Greater seed set occurs in the thrum following intermorph pollinations in all six distylous species of *Primula* (Darwin 1877; Dowrick 1956) and in *Nymphoides indica* (Barrett 1980), while pin × thrum crosses yield more seeds in *Oldenlandia umbellata, O. scopulorum*, and *Hedyotis nigricans* (Bahadur 1963, 1966, 1970). Darwin noted that the mid-styled form of tristylous *Lythrum salicaria* produces more seeds than the other morphs following all types of crosses.

Some heterostylous species exhibit such extreme gender specialization that they are functionally dioecious. Perhaps the most frequently cited examples are in the genus *Cordia*. In three nearly dioecious species, Opler et al. (1975) considered the morph that produces almost no seeds the thrum and that with aborted anthers or inviable pollen the pin. Muenchow and Grebus (1989) pointed out, however, that the shortened corolla tubes in these species make assignment of pin and thrum ancestry problematical. Lack and Kevan (1987) considered *Sarcotheca celebica* (Oxalidaceae) to be at an intermediate stage between distyly and dioecy because pins, which produce very little pollen, yield more seeds per fruit following natural pollinations and are more likely to set fruit following intermorph crosses than are thrums.

Less pronounced gender specialization has been reported for naturally pollinated individuals of *Villarsia capitata*, *V. lasiosperma*, and *V. parnassiifolia* (Menyanthaceae) (Ornduff 1986), *Linum perenne* (Linaceae) (Nicholls 1986), and *Mitchella repens* (Rubiaceae) (Hicks et al. 1985). Ornduff found variability in the seed set pattern for all three species of *Villarsia*; there was no differences between morphs in some populations, but pins produced more seeds in others. Intermorph crosses with *V. capitata* produced a mean of 10.2 seeds per pollination in pins and 3.3 in thrums (Ornduff 1982). Naturally pollinated pins also produce more seeds in *L. perenne*. Hicks et al. (1985) reported equal fruit set in the two morphs in one population of *M. repens* but higher fruit set in thrums in another, obviously a case where the pin morph does not specialize in female function.

The above examples are not intended to convey that gender specialization is a regular feature of heterostyly. It is also clear that repeated samplings of the same natural populations or a combination of field surveys and hand pollination experiments are needed to demonstrate consistent differences in seed production. Such data are rarely available. Similar cautions apply to population morph ratios because environmental stochasticity or colonization events could result in anisoplethy (Barrett et al. 1983; Morgan and Barrett 1988).

Stable anisoplethic populations are of special interest in testing sex allocation theory since we can look for factors that are predicted to alter ratios from 1/2. Table 1 contains a list of species (Table 1) exhibiting anisoplethy for which information is available on their incompatibility, gender, and growth form. For most it is not possible to determine if morph ratios are stable, although a few studies document general anisoplethy after sampling a large number of populations (e.g., Ornduff 1980 for *Hedyotis caerulea*; Haldane 1936; Schoch-Bodmer 1938, and Halkka and Halkka 1974 for *Lythrum salicaria*; and Price and Barrett 1982, and Barrett et al. 1983 for *Pontederia cordata*.

The characters listed in Table 1 were selected as potential sources of clues for why the species exhibit anisoplethy. The complete or partial breakdown of an incompatibility system, for example, together with asymmetric pollen flow could cause unequal fitnesses among morphs. Anisoplethy should exist under those conditions (Heuch 1979b; Casper and Charnov 1982). In *Palicourea petiolaris*, populations are increasingly biased towards pins the weaker the incompatibility (Sobrevila et al. 1983). Gender specialization could automatically result in anisoplethy in tristylous (but not distylous) species (Casper and Charnov 1982), and any heterostylous species with gender specialization could be subject to the same sorts of selection pressures as those causing skewed sex ratios in dioecious organisms (Charnov 1982).

Finally, the growth form of the plants is also relevant. The models are built for species with non-overlapping generations and predict the equilibrium morph ratio among zygotes – not flowering adults. A perennial habit might not change the expected morph ratio, but differences among morphs in their ability to clone would skew ratios and essentially function as a form of differential mortality (Charnov, pers. commun.).

No striking patterns emerge from the data summarized in Table 1. Anisoplethy occurs in species exhibiting incompatibility and those lacking it. Among distylous species, pins and thrums are equally likely to be present in excess. Among species that specialize in gender, there is no tendency for a particular sexual morph to be

Table 1. Characteristics of species with anisoplethic populations

Species	Morph in excess	Incompatibility	Gender differences ("♀" morph)	Growth form	Reference
Boraginaceae					
Amsinckia douglasiana	Pin	Cryptic	No	Annual	Ganders (1976); Casper et al. 1988; Weller and Ornduff 1977)
Amsinckia vernicosa var. *furcata*	Thrum	No	No	Annual	(Ganders 1976)
Cordia collococca	Thrum	Functionally dioecious	Pin[a]	Tree	(Opler et al. 1975)
Cordia curassavica	Varies	Yes	Pin	Clonal shrub	(Opler et al. 1975)
Cordia inermis	Pin	Functionally dioecious	Pin[a]	Clonal shrub	(Opler et al. 1975)
Cordia panamensis	Pin	Functionally dioecious	Pin[a]	Tree	(Opler et al. 1975)
Cordia pringlei	Thrum	Yes	No	Clonal shrub	(Opler et al. 1975)
Lithospermum caroliniense	Thrum	Incomplete in thrums	Thrum	Perennial herb	(Levin 1968, 1972)
Lythraceae					
Lythrum salicaria	Long[b]	Incomplete	Mid[c]	Clonal perennial herb	(Darwin 1877; Haldane 1936; Schoch-Bodmer 1938; Halkka and Halkka 1974; Nicholls 1987)
Menyanthaceae					
Villarsia parnassiifolia	Usually thrum	Yes	No or pin	Perennial herb	(Ornduff 1986)
Oxalidaceae					
Oxalis priceae ssp. *priceae*	Pin	No	?	Clonal herb	(Mulcahy 1964)
Oxalis alpina	Long, short[b]	Modified	Slight	Clonal herb	(Weller 1981a,b, 1986)
Pontederiaceae					
Pontederia cordata	Short[b]	Differs among morphs	No[d]	Clonal aquatic	(Ornduff 1966; Price and Barrett 1982; Barrett et al. 1983)
Rubiaceae					
Hedyotis caerulea	Usually pin	Yes	No	Perennial herb	(Ornduff 1980)
Hedyotis nigricans	Pin	Yes	No	Perennial herb	(Levin 1974)
Mitchella repens	Thrum	Yes	Thrum	Clonal perennial herb	(Hicks et al. 1985)
Oldenlandia umbellata	Usually pin	Yes	Pin	Annual	(Bahadur 1963)

Table 1. *(Continued)*

Species	Morph in excess	Incom-patibility	Gender differences ("♀" morph)	Growth form	Reference
Palicourea petiolaris	Often pin	Varies	No	Shrub	(Sobrevila et al. 1983)
Santalaceae					
Quinchamalium chilense	Thrum	No	Pin	Parasitic herb	(Riveros et al. 1987)

[a]But see Muenchow and Grebus (1989).
[b]Tristylous species.
[c]Germination lowest in short-styled morph.
[d]No gender specialization among morphs through seeds; may be differences in pollen production.

more common. Perhaps it is noteworthy that only two of the 19 species exhibiting anisoplethy, *Hedyotis caerulea* and *H. nigricans*, have a strong incompatibility system and apparently lack both gender specialization and clonal growth.

6 Control of Morph Ratios

Recall that a major assumption of sex allocation models is that plants have the ability to control offspring morph ratios through autosomal genes, separate from the heterostyly locus. Is there any evidence that such control is achieved through differential pollen tube growth or selective embryo abortion, as Casper and Charnov (1982) suggested?

Possible evidence comes from a study of tristylous *Oxalis alpina* by Weller (1986). The mid-style length morph is absent in some populations and varies in frequency from 1 to 46% in others. The modified incompatibility system results in the long- and short-styled morphs being completely interfertile. This, together with modification of the mid-length stamens in shorts and longs in a way that should facilitate pollen transfer between them suggested to Weller that selection would operate against the mid-style length morph. Computer simulations of the reproductive system (Charlesworth 1979) also showed that mids should become extinct. Weller tested the progeny of naturally pollinated morphs for evidence of selection against the mid form. He found the mid-styled morph to be more common than predicted based on the amount of pollen produced by each morph, the pattern of pollen flow, and the genetics of the system. He concluded that gametophytic selection might be operating to maintain mids in the population.

Gametophytic selection might also occur in *Amsinckia douglasiana*. Casper et al. (1988) found that pin × thrum (and sometimes thrum × pin) crosses yield an excess of pin progeny. Because the pin morph is thought to be homozygous recessive at the style length locus (*ss*) and the thrum heterozygous (*Ss*) (Ray and Chisaki 1957), the dominant thrum allele could be selected against – either in the style or following fertilization.

Another interpretation for why excess pins sometimes occur is that deleterious recessive genes become linked to the dominant style length (thrum) allele because it is never (or seldom) homozygous in the sporophyte (Mather and DeWinton 1941; Baker 1975). It is thought that such genes could cause lethality of *SS* zygotes (Mather and DeWinton 1941), and if they were expressed in the haploid gametophyte, could cause pollen carrying the *S* allele to be competitively inferior in the style (Baker 1975). Baker (1975) reviewed studies showing distorted ratios of style length morphs following hand pollinations of *Primula* but concluded that none of them distinguishes between differential pollen tube growth and differential zygote viability.

By the same reasoning, the action of linked deleterious recessive genes might also cause the excess of the mid-styled morph in *O. alpina*. In that system, there are two controlling loci. The *S* allele, which codes for the short-styled morph, is epistatic to the linked locus coding for the long or mid phenotype. The presence of a dominant (*M*) allele at the second locus in the absence of *S* results in a mid-styled morph. The dominant *S* allele is never homozygous, but the dominant *M* allele may be. Thus the genotypes of shorts, mids, and longs are *Ss*--, *ssM*-, and *ssmm*, respectively. The excess of mids in Weller's study largely (but not totally) results from a deficiency of short morphs in the progeny of mids. Deleterious recessive genes linked to the *S* allele could be expressed in the gametophyte and result in competitively inferior pollen and a deficiency of the short-styled phenotype.

Self-pollinations of the short-styled morph in *Eichhornia paniculata* also result in lower mean seed set than self-pollinations of the other two morphs, and some families are deficient in short-styled progeny (Barrett et al. 1989). However, no evidence suggests that gametophytes or zygotes carrying the *S* allele are at a disadvantage in intermorph crosses in that species.

Any post-pollination selection that results from genes linked to the heterostyly locus would violate the assumptions of the sex allocation models. It is also possible, of course, that the above examples represent either gametophytic or post-zygotic selection that is under control of other maternal (autosomal) genes.

Clearly, sex allocation models of heterostyly have not yet been subjected to rigorous empirical tests. The models suggest a wide range of studies addressing the selective pressures affecting seed and pollen production and the mechanisms controlling morph ratio. By focusing on aspects of resource allocation, the theory directs considerations of how heterostyly may function in ways that are often unrelated to its role as an incompatibility system.

Acknowledgments. I thank E. Charnov for his careful attention to an earlier version of this chapter and his helpful suggestions. The comments of J. Kohn, R. Morris, R. Niesenbaum, and D. Vann also improved the manuscript. I thank P. Taylor for pointing out the correct form of Fig. 2. Figure 2 and the models for distyly are used with permission from J. Theor. Biol. (Casper and Charnov 1982). This work was partially funded by NSF grant BSR-8507206.

References

Bahadur B (1963) Heterostylism in *Oldenlandia umbellata* L. J Genet 58:429–439
Bahadur B (1966) Heterostyly in *Oldenlandia scopulorum* Bull. J Genet 59:267–273
Bahadur B (1970) Heterostyly in *Hedyotis nigricans* (Lam.) Fosb. J Genet 60:175–177
Baker HG (1958) Studies in the reproductive biology of West African Rubiaceae. J West Afr Sci Assoc 4:9–24
Baker HG (1975) Sporophyte-gametophyte interactions in *Linum* and other genera with heteromorphic self-incompatibility. In: Mulcahy DL (ed) Gamete competition in plants and animals. North-Holland, Amsterdam, pp 191–200
Barrett SCH (1980) Dimorphic incompatibility and gender in *Nymphoides indica* (Menyanthaceae). Can J Bot 58:1938–1942
Barrett SCH, Price SD, Shore JS (1983) Male fertility and anisoplethic population structure in tristylous *Pontederia cordata* (Pontederiaceae). Evolution 37:745–759
Barrett SCH, Morgan MT, Husband BC (1989) The dissolution of a complex genetic polymorphism: the evolution of self-fertilization in tristylous *Eichhornia paniculata* (Pontederiaceae). Evolution 43:1398–1416
Beach JH, Bawa KS (1980) Role of pollinators in the evolution of dioecy from distyly. Evolution 34:1138–1143
Bulmer MG, Taylor PD (1980) Dispersal and the sex ratio. Nature (Lond) 284:448–449
Casper BB, Charnov EL (1982) Sex allocation in heterostylous plants. J Theor Biol 96:143–149
Casper BB, Sayigh LS, Lee SS (1988) Demonstration of cryptic incompatibility in distylous *Amsinckia douglasiana*. Evolution 42:248–253
Charlesworth D (1979) The evolution and breakdown of tristyly. Evolution 33:486–498
Charlesworth D (1989) Allocation to male and female function in hermaphrodites, in sexually polymorphic populations. J Theor Biol 139:327–342
Charnov EL (1975) Sex ratio selection in an age-structured population. Evolution 29:366–368
Charnov EL (1979) Simultaneous hermaphroditism and sexual selection. Proc Natl Acad Sci USA 76:2480–2484
Charnov EL (1982) The theory of sex allocation. Princeton Univ Press, Princeton
Charnov EL (1987) On sex allocation and selfing in higher plants. Evol Ecol 1:30–36
Charnov EL, Bull JJ (1986) Sex allocation, pollinator attraction and fruit dispersal in cosexual plants. J Theor Biol 118:321–325
Charnov EL, Maynard Smith J, Bull JJ (1976) Why be an hermaphrodite? Nature (Lond) 263:125–126
Clark AB (1978) Sex ratio and local resource competition in a prosimian primate. Science 201:163–165
Darwin C (1877) The different forms of flowers on plants of the same species. Univ Chicago Press (1986), Chicago
Dowrick VPJ (1956) Heterostyly and homostyly in *Primula obconica*. Heredity 10:219–236
Finney DJ (1952) The equilibrium of a self-incompatible polymorphic species. Genetica 26:33–64
Fisher RA (1930) The genetical theory of natural selection. Oxford Univ Press, Oxford
Fisher RA (1941) The theoretical consequences of polyploid inheritance for the mid style form of *Lythrum salicaria*. Ann Eugenics 11:31–38
Fisher RA (1944) Allowance for double reduction in the calculation of genotype frequencies with polysomic inheritance. Ann Eugenics 12:169–171
Ganders FR (1976) Pollen flow in distylous populations of *Amsinckia* (Boraginaceae). Can J Bot 54:2530–2535
Ganders FR (1979) The biology of heterostyly. NZ J Bot 217:607–635
Goldman DA, Willson MF (1986) Sex allocation in functionally hermaphroditic plants: a review and critique. Bot Rev 52:157–194
Haldane JBS (1936) Some natural populations of *Lythrum salicaria*. J Genet 32:393–399
Halkka O, Halkka L (1974) Polymorphic balance in small island populations of *Lythrum salicaria*. Ann Bot Fenn 11:267–270
Hamilton WD (1967) Extraordinary sex ratios. Science 156:477–488
Heuch I (1979a) Equilibrium populations of heterostylous plants. Theor Popul Biol 15:43–57

Heuch I (1979b) The effect of partial self-fertilization on type frequencies in heterostylous plants. Ann Bot 44:611–616

Heuch I, Lie RT (1985) Genotype frequencies associated with incompatibility systems in tristylous plants. Theor Popul Biol 27:318–336

Hicks DJ, Wyatt R, Meagher TR (1985) Reproductive biology of distylous partridgeberry, *Mitchella repens*. Am J Bot 72:1503–1514

Horovitz A (1978) Is the hermaphrodite flowering plant equisexual? Am J Bot 65:485–486

Kohn JR, Barrett SCH (1992) Experimental studies on the functional significance of heterostyly. Evolution (in press)

Lack AJ, Kevan PG (1987) The reproductive biology of a distylous tree, *Sarcotheca celebica* (Oxalidaceae) in Sulawesi, Indonesia. Bot J Linn Soc 95:1–8

Leigh EG (1970) Sex ratio and differential mortality between the sexes. Am Nat 104:205–210

Levin DA (1968) The breeding system of *Lithospermum caroliniense*: adaptation and counteradaptation. Am Nat 102:427–444

Levin DA (1972) Plant density, cleistogamy, and self-fertilization in natural populations of *Lithospermum caroliniense*. Am J Bot 59:71–77

Levin DA (1974) Spatial segregation of pins and thrums in populations of *Hedyotis nigricans*. Evolution 28:648–655

Lloyd DG (1977) Genetic and phenotypic models of natural selection. J Theor Biol 69:543–560

Lloyd DG (1979) Evolution towards dioecy in heterostylous populations. Plant Syst Evol 131:71–80

Mather K, De Winton D (1941) Adaptation and counter-adaptation of the breeding system in *Primula*. Ann Bot 18:297–311

Maynard Smith J (1978) The evolution of sex. Cambridge Univ Press, Cambridge

Maynard Smith J (1982) Evolution and the theory of games. Cambridge Univ Press, Cambridge

Morgan MT, Barrett SCH (1988) Historical factors and anisoplethic population structure in tristylous *Pontederia cordata*: a reassessment. Evolution 42:496–504

Muenchow GE, Grebus M (1989) The evolution of dioecy from distyly: reevaluation of the hypothesis of the loss of long-tongued pollinators. Am Nat 133:149–156

Mulcahy DL (1964) The reproductive biology of *Oxalis priceae*. Am J Bot 51:1045–1050

Nicholls MS (1986) Population composition, gender specialization, and the adaptive significance of distyly in *Linum perenne* (Linaceae). New Phytol 102:209–217

Nicholls MS (1987) Pollen flow, self-pollination and gender specialization: factors affecting seed-set in the tristylous species *Lythrum salicaria* (Lythraceae). Plant Syst Evol 156:151–157

Opler PA, Baker HG, Frankie GW (1975) Reproductive biology of some Costa Rican *Cordia* species (Boraginaceae). Biotropica 7:234–247

Ornduff R (1966) The breeding system of *Pontederia cordata* L. (Pontederiaceae). Bull Torrey Bot Club 93:407–416

Ornduff R (1980) Heterostyly, population composition, and pollen flow in *Hedyotis caerulea*. Am J Bot 67:95–103

Ornduff R (1982) Heterostyly and incompatibility in *Villarsia capitata* (Menyanthaceae). Taxon 31:495–497

Ornduff R (1986) Comparative fecundity and population composition of heterostylous and non-heterostylous species of *Villarsia* (Menyanthaceae) in western Australia. Am J Bot 73:282–286

Price SD, Barrett SCH (1982) Tristyly in *Pontederia cordata* L. (Pontederiaceae). Can J Bot 60:897–905

Ray PM, Chisaki HF (1957) Studies on *Amsinckia*. I. A synopsis of the genus, with a study of heterostyly in it. Am J Bot 44:529–536

Riveros M, Arroyo MTK, Humana AM (1987) An unusual kind of distyly in *Quinchamalium chilense* (Santalaceae) on Volcan Casablanca, Southern Chile. Am J Bot 74:313–320

Schoch-Bodmer H (1938) The proportion of long-, mid-, and short-styled plants in natural populations of *Lythrum salicaria* L. J Genet 36:39–43

Shaw RF, Mohler JD (1953) The selective advantage of the sex ratio. Am Nat 87:337–342

Snow AA (1986) Pollination dynamics in *Epilobium canum* (Onagraceae): consequences for gametophytic selection. Am J Bot 73:139–151

Sobrevila C, Ramirez N, de Enrech NX (1983) Reproductive biology of *Palicourea fendleri* and *P. petiolaris* (Rubiaceae), heterostylous shrubs of a tropical cloud forest in Venezuela. Biotropica 15:161–169

Spieth PT (1971) A necessary condition for equilibrium in systems exhibiting self-incompatible mating. Theor Popul Biol 2:404–418
Stanton ML, Preston RE, Snow AA, Handel SN (1986) Floral evolution: attractiveness to pollinators influences male fitness. Science 232:1625–1627
Taha HA (1971) Operations research. MacMillan, New York
Taylor PD (1984) Evolutionary stable reproductive allocations in heterostylous plants. Evolution 38:408–416
Weller SG (1981a) Pollination biology of heteromorphic populations of *Oxalis alpina* (Rose) Knuth (Oxalidaceae) in southeastern Arizona. Bot J Linn Soc 83:189–198
Weller SG (1981b) Fecundity in populations of *Oxalis alpina* in southeastern Arizona. Evolution 35:197–200
Weller SG (1986) Factors influencing frequency of the mid-styled morph in tristylous populations of *Oxalis alpina*. Evolution 40:279–289
Weller SG, Ornduff R (1977) Cryptic self-incompatibility in *Amsinckia grandiflora*. Evolution 31:47–51
Werren JH (1980) Sex ratio adaptations to local mate competition in a parasitic wasp. Science 208:1157–1159
Willson MF (1979) Sexual selection in plants. Am Nat 113:777–790
Wyatt R (1983) Pollinator-plant interactions and the evolution of breeding systems. In: Real L (ed) Pollination biology. Academic Press, Orlando

Chapter 9

Pollen Competition in Heterostylous Plants

M.A. McKenna[1]

1 Introduction

The term pollen competition has been used to describe a variety of phenomena in plant reproductive systems. In the context of sexual selection it has been treated synonymously with male competition, including factors such as anther and pollen production, pollinator attraction, and the efficiency of pollen pickup and deposition on conspecific stigmas (Stephenson and Bertin 1983; Willson and Burley 1983; Galen et al. 1986). In a more restricted sense, pollen competition can be limited to studies of the interactions among pollen grains after deposition on a stigma. Pollen competition in this sense refers strictly to competition between microgametophytes during germination and growth.

Plant characteristics such as pollen production values, preferential attractiveness to pollen vectors, and floral morphology may play an important role in determining the likelihood of parentage by individual pollen donors (Bertin 1988). Pollinator behavior, pollen carryover and pollinator grooming patterns, and wider aspects of plant community structure will also play a significant part in determining the pre-pollination aspects of male competition in plants (Thomson 1986; Bertin 1988). The post-pollination aspects of male competition relate to differences in pollen germination and growth within various stylar environments. Post-pollination aspects of "pollen competition" differ from "male competition" in the usual sense of the word, since microgametophytes from the same pollen donor are competing against each other as well as against microgametophytes from other pollen donors. Pre-pollination aspects of pollen competition in heterostylous plants will be briefly reviewed at the beginning of this chapter; the remainder of the chapter will focus on post-pollination competition among microgametophytes.

2 Pre-Pollination Pollen Competition

Heterostyly is not common in the plant kingdom, but ever since Darwin's (1877) studies of mating systems in heterostylous plants, interest in the evolutionary basis of heterostyly has generated much research. These studies have revealed differences between long-styled morphs and short-styled morphs in several traits related to

[1]Department of Botany, Howard University, Washington, D.C., USA

Monographs on Theoretical and Applied Genetics 15
Evolution and Function of Heterostyly (ed. by S.C.H. Barrett)

pre-pollination aspects of pollen competition such as pollen production and pollen removal patterns. Pollen production differences are usually correlated with differences in the size of pollen grains produced by floral morphs; pollen from long anthers (found in the short-style or thrum morph) is almost always larger and less abundant than pollen from short anthers (found in the long-style or pin morph) (Ganders 1979; Price and Barrett 1982; Murray 1990). In *Palicourea lasiorrachis*, Feinsinger and Busby (1987) report no difference between morphs in pollen size or pollen production. In *Linum perenne*, thrums produce more pollen than pins, although there is no difference in the size of pollen produced by the two morphs (Nicholls 1986). Variation in pollen size between related species in the Onagraceae is correlated with style length (Baker and Baker 1983) but such a correlation was not found by Cruden and Miller-Ward (1981) in 19 representatives of ten angiosperm families. Pollen size differences between pin and thrum plants are often ascribed to the necessity for larger energy reserves in thrum pollen growing in a longer style (Darwin 1877), but Ganders (1979) found no correlation between the ratio of pin:thrum pollen volume and the ratio of pin:thrum style length in 23 distylous species.

Differences between floral morphs in the relative amounts of pollen removed from anthers have been reported in a few species. In *Lythrum junceum* and *Pontederia cordata*, honeybees collect more pollen from long anthers (Ornduff 1975; Wolfe and Barrett 1987); this suggests a preference for larger pollen grains. However, solitary bees preferentially collect pollen from short anthers of *Pontederia cordata*, and bumblebees appear to exhibit a random foraging pattern on this species (Price and Barrett 1982). Bumblebees also exhibit random pollen foraging among floral morphs of *Oxalis alpina* (Weller 1981).

The relative amount of pollen reaching pin and thrum stigmas and the proportion of "legitimate" pollen deposited has been widely studied, largely as a result of interest in the relationship between herkogamy and pollen flow patterns in heterostylous species. Although as little as 1–2% of pollen produced generally reaches a stigma (Ganders 1975), in most species pollen flow is asymmetric, with pin stigmas receiving more pollen than thrums (Ganders 1979). Differences in the proportion of legitimate pollen reaching pin and thrum stigmas are seen in both directions. In *Turnera subulata*, pin stigmas receive a greater proportion of legitimate pollen (Swamy and Bahadur 1984), while in *Jepsonia parryi*, pin stigmas receive more pollen than thrums, but a greater proportion of illegitimate pollen (Ornduff 1970). In general, however, asymmetry in legitimate pollen flow tends to favor the pin morph, with a few exceptions (Ornduff 1980; Feinsinger and Busby 1987).

Stigma surface characteristics may also exhibit dimorphism in traits such as surface secretions and the shape of stigmatic papillae, and these differences appear to relate to incompatibility mechanisms. In *Anchusa officinalis* (Philipp and Schou 1981) and *Eichhornia paniculata* (M.B. Cruzan and S.C.H. Barrett, pers. commun.) thrum stigmas are relatively dry compared to pin stigmas, while in *Lucelia gratissima* (Murray 1990) and *Linum* spp. (Dulberger 1987) the opposite pattern is observed. In pins, the papillae are usually longer (Ganders 1979), but in *Lucelia gratissima* thrum papilla cells are significantly longer than pin papillae (Murray 1990). In *Eichhornia paniculata*, morph differences in pollen size and stigma surface characteristics influence relative germination rates, perhaps as a result of differences in the rate of hydration (M.B. Cruzan and S.C.H. Barrett pers. commun.). Differences in pollen grain color, shape and exine properties are sometimes found between the morphs of heterostylous plants, but the importance of this variation is not generally known (Ganders 1979).

3 Post-Pollination Pollen Competition: An Overview

In both heterostylous and nonheterostylous species, the angiosperm flower provides an opportunity for natural selection to occur among microgametophytes when a sufficient number of pollen grains are deposited on a stigma to create competition to fertilize a limited number of ovules. Under these conditions the most rapidly germinating and growing pollen tubes will achieve fertilization, and will therefore be selected for. Microgametophytic selection can be defined as selection based on differences in pollen tube growth rates, reflecting differences in gametophytic genotypes and leading to a different probability of fertilization among pollen genotypes. Selection among microgametophytes can occur at different points during this life stage. The germination ability of pollen grains reflects the innate viability of individual genotypes; selection at this stage may also be influenced by random factors such as the position of grains on the stigmatic surface and environmental factors such as relative humidity (Hoekstra 1983; Thomson 1989).

A second level of selection depends on the relative growth rates of the pollen genotypes in the style. This may reflect differences in general metabolic efficiency of pollen genotypes and differences in the efficiency of the uptake of stylar nutrients during pollen tube growth. Pollen germination and tube growth is influenced by substances present in stigmatic and stylar tissue (Ascher and Drewlow 1971; Asano 1980; Heslop-Harrison and Heslop-Harrison 1975; Knox 1984; van Herpen 1986), and labeling experiments have demonstrated that carbohydrates from the pistil are used for pollen tube wall formation (Kroh et al. 1971). Pollen growth is also strongly influenced by environmental factors such as temperature, floral age, and physiological status of the pistillate plant (Dane and Melton 1973; Gilissen 1977; van Herpen 1981, 1984, 1986; Schlichting 1986). Environmental effects on pollen tube growth rates may result from ambient conditions during pollen development and during pollen germination and growth in the style (van Herpen and Linskens 1981).

A final point of selection occurs in the ovary, based on the ability of pollen tubes to penetrate the ovule and achieve fertilization. In some species the stylar transmitting tissue ends in thickened "obturator" tissue adjacent to the micropyle; pollen tube growth in this region may be influenced by physical and chemical characteristics of the obturator tissue (Tilton and Horner 1980; Tilton et al. 1984; Arbeloa and Herrero 1987). Except in cases of specific "selective fertilization" mechanisms as in *Oenothera* (Harte 1975; Schwemmle 1968), most researchers have assumed that pollen is received by ovules roughly in order of arrival time at the micropyle. Synchronous receptivity of ovules and linear fertilization patterns have also generally been assumed, although this may not always be the case (Hill and Lord 1986).

Intergametophytic interactions can also affect pollen growth rates. Interactions between pollen grains can facilitate their growth as seen in the "pollen population effect" (Brewbaker and Majumder 1961), the "mentor pollen" method (Stettler

1968; Mulcahy and Mulcahy 1986) and the "pioneer pollen" effect (Visser and Verhaegh 1980). Negative intergametophytic effects caused by physical obstruction have also been reported (d'Eeckenbrugge 1986; d'Eeckenbrugge et al. 1986), and the morphology of the pistil can contribute to the relative strength of obstructive interactions (Cruzan 1986; Lord and Kohorn 1986).

The interaction between specific pollen and style genotypes may influence pollen growth rates in subtle and complex ways. The observation that crosses between certain individuals are more successful than others has long been made in agronomic literature, where it has been termed specific combining ability. In view of the potential sporophytic influences on pollen tube growth, and maternal effects on seed and seedling characteristics, experiments designed to relate the expression of genetic vigor in the gametophyte and sporophyte must be carefully designed (Baker 1975; Charlesworth 1988). A prediction concerning the "fitness" (i.e., survivorship and fertilization ability) of an individual haploid microgametophyte in a particular stylar environment involves consideration of its innate genetic capacity for growth, the efficiency of the stylar partner in providing nutrients, the typical environment and its effects on stylar and pollen genotypes, and the presence or absence of competing gametophytes.

The potential for microgametophytic selection to effect directed evolutionary change depends on three premises: (1) there is post-meiotic gene expression in haploid pollen, (2) there are genetically mediated differences in pollen tube growth rates, and (3) there is a correlation of genetic expression in the haploid gametophyte and the diploid sporophyte.

3.1 Post-Meiotic Gene Expression in Pollen

Post-meiotic gene expression in pollen has been extensively studied through the use of genetic endosperm markers such as waxy, shrunken, sugary, and opaque in *Zea mays* (Brink and MacGillivray 1924; Ericksson 1969; Pfahler 1975). Deficiency alleles such as alcohol dehydrogenase deficiency in *Zea mays* pollen (Freeling and Cheng 1978; Schwartz and Osterman 1976) and B-galactosidase deficiency in *Brassica campestris* pollen (Singh and Knox 1986) provide additional biochemical markers. The development of electrophoretic techniques (Makinen and Brewbaker 1967; Hamill and Brewbaker 1969) has allowed further refinement of these studies, and microelectrophoresis techniques for single pollen grains have also been developed (Mulcahy et al. 1979; Miller and Mulcahy 1983; Gay et al. 1986) that demonstrate the expression of segregating alleles (isozymes) in individual pollen grains from the same diploid source. Results of an extensive series of studies on *Tradescantia* (Mascarenhas et al. 1986; Willing et al. 1984) and other studies on *Malus* (Bagni et al. 1981) and *Nicotiana* (Suss and Tupy 1979) have demonstrated mRNA, rRNA, and protein synthesis during pollen tube germination and growth. Half of the protein synthesis within the first hour of pollen tube growth in *Tradescantia paludosa* is based on mRNA synthesized by the pollen grain prior to germination; the remaining protein synthesis is based on newly synthesized mRNA. The mRNAs in a mature pollen grain of *Tradescantia* are the products of approximately 20 000 genes, and electrophoretic analysis indicates that the same genes are active during pollen maturation and during pollen germination and tube growth (Mascarenhas et al. 1986).

3.2 Genetic Influences on Pollen Growth

One of the earliest descriptions of genetically mediated differences in pollen tube growth rate was made by Heribert-Nilsson (1920), who found that pollen carrying genes for red leaf veins tended to preferentially transmit this character in *Oenothera erythrosepala* (*O. lamarckiana*); he termed this process certation. Research since that time involving direct observation of pollen growth as well as differential fertilization studies has revealed that pollen tube growth is under genetic control.

Studies investigating variations in in vitro and in vivo pollen tube growth rates in inbred lines of *Zea mays* have demonstrated multifactorial genetic control of growth rate in the microgametophyte (Sari-Gorla et al. 1975). Genotypic growth rates of pollen sources are also influenced by the degree of genetic relatedness between the pollen and the style. Significant differences in growth rates between self- and outcross pollen have been widely reported, even in the absence of classical self-incompatibility (Barnes and Cleveland 1963; Baluch et al. 1973; Levin 1975; Currah 1981; Eenink 1982; Hessing 1986). The degree of genetic relatedness of pollen and style may theoretically result in a prezygotic expression of heterosis or of "outbreeding depression" (Waser and Price 1983); investigations in this area have been limited (Schemske and Pautler 1984). It appears likely that subtle maternal and paternal gene interactions influence microgametophytic competition. Ottaviano et al. (1983) found strong heritability in pollen competitive ability in *Zea mays*, but Snow and Mazer (1988) find little evidence for heritability of pollen growth rates in *Raphanus raphanistrum*.

3.3 Gene Overlap in the Gametophyte and Sporophyte

The evolutionary significance of gametophytic selection is largely dependent on the extent of overlap in genetic expression during the gametophytic and sporophytic life stage. An electrophoretic study of pollen and sporophytic tissue (leaves, roots, and seeds) in *Lycopersicon esculentum* (Tanksley et al. 1981) revealed that 60% (18/30) of the isozymes expressed in the sporophyte are also expressed postmeiotically in the pollen. Similar electrophoretic techniques indicate an 81% overlap (26/32) between isozymes expressed in pollen and leaf tissue of *Malus* (Weeden 1986). In *Zea mays* Sari-Gorla et al. (1986) reported a 72% overlap between isozymes expressed in the gametophyte and sporophyte; 22% of isozymes expressed are specific to the sporophyte and 6% are unique to pollen. Recombinant DNA techniques have been used to investigate the extent of overlap between gametophytic and sporophytic tissue in *Tradescantia paludosa* (Willing and Mascarenhas 1984; Willing et al. 1984). RNA and cloned DNA from pollen and shoot tissue were compared by heterologous hybridizations and a 60% overlap was found between shoot and pollen sequences. Since the complexity of pollen RNA (number of pollen RNA nucleotides) is only 66% of shoot RNA complexity, a 60% overlap is nearly the maximum possible.

3.4 Effects of Pollen Competition on Sporophytic Traits

The joint expression of genes in the gametophytic and sporophytic stages allows gametophytic selection to influence the character of the sporophytic generation. What are the consequences of this relationship? One approach to this question is to select for genes controlling specific traits such as salt tolerance and heavy metal tolerance during the gametophytic stage, and investigate the expression of these genes in the sporophytic stage. Gametophytic selection for tolerance to heavy metals and salinity has resulted in a correlated response in the sporophyte (Sacher et al. 1983; Searcy and Mulcahy 1986). After carrying out classical sporophytic selection for herbicide resistance in sugar beet, Smith and Moser (1985) found a correlation between sporophytic and gametophytic tolerance.

The effect of microgametophytic selection on the sporophytic generation has also been investigated by focusing on how general fitness-related sporophytic characters such as seedling vigor and time to flowering vary with the intensity of pollen competition. The intensity of pollen competition has been experimentally varied in two ways: (1) by varying the number of pollen grains on the stigma, and (2) by varying the placement of pollen proximal or distal to the ovary, thereby varying the stylar distance over which differences in pollen growth rates can be expressed. Distal pollinations allow for greater microgametophytic selection because fast growing pollen tubes have greater opportunity to surpass slower tubes over the longer stylar distance (Correns 1928).

Pollen competition experiments have been carried out in a variety of species, and most studies have reported greater offspring vigor in plants derived from pollination conditions that allow greater competition among microgametophytes. Plants derived from relatively intense pollen competition show significantly lower variation in characters such as plant height and yield (Ter-Avanesian 1978; Stephenson et al. 1988), greater seedling height, weight, faster time to flowering, and greater seedling competitive ability (Mulcahy and Mulcahy 1975; Mulcahy et al. 1975; Fingerett 1979; McKenna and Mulcahy 1983; Lee and Hartgerink 1986; Stephenson et al. 1986; McKenna 1987; Bertin 1990). *Cassia fasciculata* flowers pollinated with heavy pollen loads are more likely to produce mature fruits than those receiving light loads even when total seed set is equivalent (Lee and Bazzaz 1982). Snow (1990) did not find increased sporophytic vigor in response to increased pollen competition in *Raphanus raphanistrum*, however, and Mazer (1987) detected no significant paternal genetic influence on seed number or weight in this species. Snow (pers. commun.) also found no evidence for greater offspring vigor in response to increased pollen competition in two *Epilobium* species. In *Zea mays*, Mulcahy (1974) found a significant correlation between relative pollen growth rate and relative seedling weight. Recurrent selection for gametophytic competitive ability in *Zea mays* through the development of separate mating lines derived from intense and limited pollen competition has also demonstrated a response to selection in corn, and a correlated response in sporophytic vigor in selected lines (Ottaviano et al. 1982, 1986).

4 Post-Pollination Pollen Competition and Heterostyly

The extent of pollen competition and its potential influence on sporophytic vigor in natural populations of heterostylous plants is generally not known. Some studies of microgametophytic competition in heterostylous plants have been designed to address this question, including consideration of relevant features of the floral dimorphism. Other studies of microgametophytic competition in heterostylous plants have been prompted by an interest in the possible consequences of pollen growth differences on mating systems. Pollen competition studies in heterostylous plants fall into three general categories: (1) comparisons of relative germination and growth rates of intermorph, intramorph, and self-pollen in heterostylous species that do not exhibit the typical intramorph incompatibility system, (2) investigations of the relationship between style length and the strength of microgametophytic competition among pollen from single donors, (3) studies employing variation in the size of legitimate pollen loads on stigmas to study the effects of interference or "stigma clogging" by illegitimate (self- or intramorph) pollen grains or to study the effects of differences in the intensity of pollen competition.

4.1 Competition Between Intramorph and Intermorph Pollen

In most heterostylous species there is no opportunity for direct competition between pin and thrum pollen because of intramorph incompatibility. Rare examples of heterostylous species in which intramorph crosses are compatible do exist, however, particularly in the Boraginaceae: *Cryptantha flava* (Casper 1985) *Amsinckia* spp. (Ray and Chisaki 1957; Ganders 1979), *Anchusa officinalis* (Philipp and Schou 1981) and *Anchusa hybrida* (Dulberger 1970), and also in *Narcissus tazetta* (Amaryllidaceae) (Dulberger 1964), *Eichhornia paniculata* (Pontederiaceae) (Barrett et al. 1987), and *Palicourea petiolaris* (Rubiaceae) (Sobrevila et al. 1983). In *Cryptantha flava* and *Anchusa officinalis*, there is no apparent difference in germination and growth of pin and thrum pollen in intramorph or intermorph crosses. In *Anchusa officinalis* pollen from both morphs reach the ovary within 18 h of pollination, and can be traced into the ovules through the micropyle (Schou and Philipp 1983a). After hand pollinations with abundant pollen, a very small number of tubes (six or seven) enter the ovary, however, so it is possible that some selection of pollen tubes occurs at the base of the style. *Anchusa officinalis* is self-incompatible; self-pollen enters the ovary, but ovules do not swell, and seeds are not produced. In *Cryptantha flava* pollen from pin and thrum morphs reaches the base of the style within 2 days of pollination (Casper 1985), and there is no difference between self, intramorph or intermorph crosses in pollen tube number or length at intervals ranging from 2 to 24 h following pollination.

Competition experiments between pin and thrum pollen were carried out in *Amsinckia grandiflora* (Weller and Ornduff 1977) by applying equal mixtures of pin and thrum pollen to the stigmas of pins and thrums. Offspring morph ratios following mixed pollinations did not differ significantly from a 1:1 ratio, suggesting that intermorph fertilizations predominated. In addition, they found that pollen growth in intramorph crosses was slower than in intermorph crosses, so they concluded that

Amsinckia grandiflora possesses a cryptic self-incompatibility system similar to *Chieranthus cheiri* (Bateman 1956). Ganders (1975) did not find any evidence for differential pollen growth between pins and thrums in *Amsinckia spectibilis* and *Amsinckia vernicosa* var. *furcata*. Pin seed parents that received a mixture of pin and thrum pollen, produced predominantly pin offspring (3:1 ratio pins:thrums) and pin flowers that were self-pollinated and pollinated with thrum pollen after delays of 0.5 h, 1 h or 1.5 h produced only pin progeny.

In some species of *Amsinckia*, there seems to be selection operating against the thrum allele. Intramorph crosses between heterozygous thrums result in an excess of pin progeny and intermorph crosses with thrum seed parents also result in an excess of pins (Weller and Ornduff 1977; Ganders 1979). In *Amsinckia grandiflora* intermorph pollinations with pin seed parents and mixed pin and thrum pollinations with thrum seed parents yield the expected 1:1 ratio, indicating that the excess of pin progeny is not due to selection against *pollen* carrying the dominant allele (Weller and Ornduff 1977). Selection may be operating against the thrum allele in thrum seed parents during ovule formation or embryo development. Shore and Barrett (1985) also report an excess of pins obtained from selfing rare thrum plants of *Turnera ulmifolia* that exhibit a breakdown in self-incompatibility. In this case they suggest that the self-compatibility gene imparts a selective advantage to male gametophytes carrying recessive alleles, but this hypothesis has not been tested experimentally.

Ornduff (1964) reported evidence for a competitive disadvantage of pollen from the mid-style morph of tristylous *Oxalis suksdorfii* (Oxalidaceae). Lower seed set is found in legitimate and illegitimate crosses of long and short morph seed parents with pollen from the mid-style morph. Results from mixed pollinations of short morph seed parents with pollen from the long and mid-style morphs suggest that pollen from the long style morph is fertilizing a greater proportion of seed, although seed parentage was not conclusively determined in this study.

Eichhornia paniculata (Pontederiaceae) is a heterostylous plant that exhibits a breakdown of the typical genetic polymorphism and incompatibility system, resulting in self-compatible tristylous, dimorphic, and monomorphic populations. The breeding system of this species has been extensively studied through an elegant set of experiments carried out by Barrett and coworkers. Glover and Barrett (1986) carried out a series of mixed pollinations in a tristylous self-compatible population of *Eichhornia* to compare the competitive ability of self- and outcross pollen. An electrophoretic marker was used to evaluate the parentage of the progeny from these crosses, and a significant excess of outcross progeny was observed. Further electrophoretic studies of tristylous *Eichhornia* populations (Morgan and Barrett 1990) revealed that even though these plants are self-compatible, a high level of outcrossing persists in these populations. A recent examination of pollen germination and tube growth following legitimate and illegitimate crosses in *Eichhornia* indicates that within each style morph, legitimate pollen grows faster than illegitimate pollen (M.B. Cruzan and S.C.H. Barrett, pers. commun.), despite a faster germination rate in pollen from short anthers. Thus it appears that microgametophytic selection contributes towards disassortative mating and high outcrossing rates in *Eichhornia paniculata*. A comparison of morph frequencies in 110 populations of *Eichhornia* (Barrett et al. 1989) demonstrated anisoplethy resulting from a deficiency of the short morph in trimorphic populations, and a deficiency of the long morph in dimorphic populations, suggesting the possibility that gametophytic selection could be operating against the *S* and *m* alleles. Offspring morph ratios obtained from a series of controlled crosses revealed some evidence of selection against these alleles in self crosses, and no evidence of selection against these alleles in crosses with a homozygous recessive (long morph) maternal parent.

4.2 Pollen Competition and Style Length Polymorphisms

Heterostylous plants exhibit a style length polymorphism that provides a natural system for studying the effects of variation in the intensity of pollen competition due to relative distance of pollen from the ovary. The design of these studies is complicated by the intramorph incompatibility found in most species, because it is impossible to compare the same pollen source on all style lengths. Heterostylous species lacking intramorph incompatibility are ideal candidates for such studies. A set of experiments in *Anchusa officinalis* were designed to compare the quality of offspring derived from pollination conditions that allow for relatively intense or limited pollen competition (McKenna 1986). These experiments employed the stylar dimorphism to vary the *stylar distance* over which differences in pollen tube growth rates are expressed; one expects a greater opportunity for microgametophytic selection in longer styles.

In *Anchusa officinalis*, distyly is controlled by a single locus (Philipp and Schou 1981; Schou and Philipp 1983a,b), dominance is expressed in the thrums, and pins are homozygous recessive. Natural populations of *Anchusa officinalis* often show a strong deviation from a 1:1 ratio of pins to thrum, always with an excess of pins. In 11 populations in Northern Denmark, Schou and Philipp found deviations ranging from 2:1 to 28:1. This is particularly curious since the unusual compatibility of crosses between thrums carrying the dominant allele would lead one to expect a deviation from 1:1 ratio towards an excess of thrums. Another curious feature of this system is that nearly all thrums found in nature are heterozygous, although homozygotes are viable and can be obtained in artificial crosses.

A series of crosses between six seed parents (three pin and three thrum) with four pollen donors (two pin, two thrum) were carried out in a greenhouse population of *Anchusa* established from seeds collected in northern Denmark. Pin plants had considerably longer styles than thrums (Table 1). Seed parent flowers were emasculated by removing the corolla prior to anther dehiscence, and pollen quantity was controlled by brushing one newly dehisced anther across the stigma. There is no significant difference in the amount of pollen per anther or pollen viability in thrum and pin morphs of *Anchusa officinalis* (Philipp and Schou 1981). All crosses were carried out in a controlled temperature chamber at 23 °C.

Seeds were harvested at maturity, individually weighed and planted singly in 2-inch pots placed in random positions on a greenhouse bench. Germination was monitored daily. After 10 days seedling root lengths were measured and seedlings were transplanted to 45 cm × 15 cm cylindrical plastic pots to reduce restrictions on root growth. After 3 weeks the rosette and upper 15 cm of the root were harvested

Table 1. *Anchusa officinalis* seed parent style length (mm)

	Pin		
	P1	P3	P4
Y	9.08	8.52	7.56
s	0.30	0.22	0.39
n	30	30	30
		Thrum	
	T1	T2	T6
Y	5.19	4.84	4.96
s	0.24	0.20	0.20
n	30	30	30

in half of the plants; after 9 weeks a similar harvest was carried out in the remaining plants. Harvested root and shoot samples were oven dried at 80 °C and individually weighed.

Seed set of pins and thrums did not differ (38% in thrums and 39% in pins). *Anchusa officinalis* typically exhibits low seed set in natural and hand pollinations (Philipp and Schou 1981; Andersson 1988); Philipp and Schou reported a 38% seed set in plants from this population under natural conditions. A similar pattern of low seed set has been reported in other distylous members of the Boraginaceae such as *Lithospermum caroliniense* (Weller 1985) and *Amsinckia grandiflora* (Weller and Ornduff 1977). In *Cryptantha flava* low seed sets are due to routine embryo abortion unrelated to environmental factors (Casper and Wiens 1981).

Offspring quality was measured by comparing seed weight, germination percentage, germination time and root and shoot growth in progeny from pin and thrum parents (Tables 2 and 3). There was no difference in germination time between seeds produced by the two morphs, but seeds produced from the pin morph had a significantly higher germination percentage. Seeds produced by the pin morph were also significantly heavier than seeds produced by the thrum morph. Comparing the results of pin and thrum crosses within individual pollen donors, seeds produced by the pin morph were heavier than those produced by the thrum morph for each individual pollen donor (Table 2). There was no significant difference in the length of the root from 10-day-old seedlings produced by pin or thrum seed parents and there was no difference in the shoot biomass after 3 or 9 weeks. There is a trend for heavier root weight in the offspring produced from pin plants and at 9 weeks this difference is bordering significance at the 5% level. Root weight may be an important component of fitness in *Anchusa*, since studies of other biennial plants with a rosette growth form indicate that the probability of flowering and reproductive success is strongly influenced by root biomass (Gross 1981).

The results of this experiment indicate that within this set of *Anchusa* pollen donors, heavier seeds with a greater germination percentage are produced in pin seed parents than in thrum seed parents. Under natural conditions, if there is a similar pattern for more vigorous offspring in crosses involving the pin morph as seed parent, these offspring may be over represented in *Anchusa* populations. This could help

Table 2. Comparison of seed weights from *Anchusa* pin vs. thrum seed parents crossed with four pollen donors

Pollen Parent	Seed Pin	Parent Thrum
P5 – Seed weight (mg)	Y = 8.322	Y = 6.203
	s = 1.622	s = 1.563
	n = 25	n = 17
P7 – Seed weight (mg)	Y = 8.833	Y = 5.889
	s = 1.394	s = 0.979
	n = 25	n = 39
T5 – Seed weight (mg)	Y = 8.259	Y = 6.280
	s = 1.433	s = 1.578
	n = 47	n = 22
T7 – Seed weight (mg)	Y = 7.075	Y = 6.591
	s = 1.737	s = 1.378
	n = 16	n = 41

Table 3. Comparison of progeny from *Anchusa* pin vs. thrum seed parents crossed with an identical set of four pollen donors

	Pin	Thrum	
Mean seed weight (mg)	8.006	6.169	$p < 0.001$
Germination time (days)	5.301	5.293	n.s.
Germination percentage	93	72	$p < 0.001$
Mean 10-day root length (cm)	4.826	4.529	
Mean 3-week shoot weight (g)	1.131	1.149	
Mean 3-week root weight (g)	0.125	0.104	
Mean 9-week shoot weight (g)	4.410	4.605	
Mean 9-week root weight (g)	1.007	0.901	$p < 0.100$

explain the excess of the pin morph in all populations and it would also account for the apparent rarity of homozygous thrum individuals in natural populations, since the thrum offspring of a pin plant will necessarily be heterozygous. Of course the excess of pins in natural populations could be due to other factors such as direct sporophytic selection against the thrum phenotype or some factor relating to the natural pollination efficiency of pins vs thrums. Andersson (1988) reports that bumblebee pollinators of *Anchusa officinalis* tend to restrict their foraging activity to nearest neighbors; this could contribute to the maintenance of anisoplethy if pin and thrum plants differ in microsite requirements. No evidence exists for selection against the dominant allele in ovules or embryos as described earlier for *Amsinckia*; controlled intramorph crosses between thrums in *Anchusa officinalis* result in a 3:1 ratio of thrums:pins (Schou and Philipp 1983a).

In a recent study (S.W. Graham and S.C.H. Barrett, pers. commun.) the style length polymorphism in *Eichhornia paniculata* has been used to test the effect of style length on the strength of the competitive advantage exhibited by first pollinations ("pollen precedence", Waser and Fugate 1986). A series of double pollinations were carried out in the three style morphs of *Eichhornia*, with various time intervals between the application of first and second pollen loads. The proportion of offspring sired by pollen from the first and second loads was determined through the use of electrophoretic markers. In all style morphs and at all time treatments, most progeny were fertilized by pollen from the first pollen load. The pollen precedence advantage is more pronounced in long styles than in shorter styles. Assuming that there are no differences in seed survivorship among progeny from the first and second pollinations, the pollen precedence pattern in *Eichhornia* does not support Correns' (1928) hypothesis that longer style lengths will allow greater opportunity for growth differences among pollen tubes to be expressed. The differences observed between the style morphs in the strength of pollen precedence may relate to differences in pollen germination rates on the three style morphs. Pollen germination in *Eichhornia paniculata* is fastest for pollen placed on stigmas of the long morph, and slowest for pollen on stigmas of the short morph (M.B. Cruzan and S.C.H. Barrett, pers. commun.) Further studies of pollen germination and tube growth following single and double pollinations with these pollen donors might provide more insight regarding the mechanisms responsible for these intriguing patterns.

4.3 Pollen Competition and Variable Pollen Loads

A common way in which pollen competition can vary in natural populations is through variation in the size of compatible pollen loads on stigmas. Pollen loads in heterostylous plants are strongly influenced by floral morphology, and stigmatic pollen loads have been extensively studied as indicators of pollen flow patterns in heterostylous plants. Pin stigmas generally have greater surface area, and pin stigmas capture more pollen grains than thrum stigmas in virtually every species studied (Ganders 1979). Pin stigmas tend to capture more incompatible (pin) pollen than compatible (thrum) pollen, however. There does not appear to be much evidence that seed set in heterostylous species is limited by the amount of compatible pollen on stigmas (Casper 1983; Nicholls 1986), but pollen competition cannot occur without a sufficiently high ratio of compatible pollen to receptive ovules. The presence of incompatible pollen can also reduce the effective number of compatible grains if incompatible grains result in "clogging" of the stigma or stylar tissue (Yeo 1975; Lloyd and Yates 1982). Stigma clogging experiments involving pollinations with a mixture of compatible and incompatible pollen indicate no effect of incompatible pollen on seed set in *Pontederia cordata* (Barrett and Glover 1985) or on growth of compatible tubes in *Lucelia gratissima* (Murray 1990).

Shore and Barrett (1984) investigated the effect of stigma clogging on seed set in *Turnera ulmifolia* (Turneraceae), a neotropical weed. *Turnera* exhibits the typical morphological features associated with distylous plants, with reciprocal style and stamen lengths, pollen size dimorphism, and the classical self- and intramorph incompatibility system found in many distylous species (Barrett 1978; Shore and Barrett 1984, 1985). The stigma clogging experiments in *Turnera* involved varying the amount and sequence of compatible and incompatible pollen added to stigmas, including delayed pollen additions at 1.5 and 3.5 h. *Turnera* flowers are ephemeral; the time interval in delayed pollinations represents the approximate lifetime of a flower. Reduced seed set due to clogging occurred only in the pin morph under the most extreme treatments, when one anther of compatible pollen was applied 1.5 or 3.5 h after the application of five anthers of incompatible pollen. The thrum morph did not show any effect of stigma clogging treatments. Thrum flowers received less incompatible pollen than pin flowers since pollen production in thrum anthers is less than pin anthers, but the extreme treatments employed in this study would make this difference relatively minor, particularly since thrum stigmas have a smaller surface area. Incompatible tubes were arrested at the stigma/style interface in thrums, but incompatible tubes grew at least half the length of the style in pins. Thus the reduced seed set effects observed in the pins may be due to stylar obstruction or interference, rather than stigma clogging.

Shore and Barrett (1984) conclude that interference from incompatible pollen is unlikely to be an important effect in *Turnera ulmifolia* since the conditions required to demonstrate reduced seed set are not commonly expected in this species. *Turnera* tends to grow in dense populations of both floral morphs, is attractive to a wide range of pollinators, and observed pollinator foraging patterns tend to reduce the level of geitonogamy (Barrett 1978).

Shore and Barrett (1984) also studied the effects of variation in pollination intensity in *Turnera ulmifolia* by controlled intermorph pollinations with 0 to 95 pollen grains per flower. There was an increase in seed set per flower with increasing numbers of pollen grains, but at all pollination levels the number of seeds set was less than the number of pollen grains applied. Two to seven pollen grains are required per seed and seed set at the highest pollination intensity did not approach the maximum seed set observed in experiments with 2000 compatible grains. Maximum seed set is less than the number of ovules, suggesting that ovule sterility or ovule/embryo abortion may be present. There was no evidence that pin and thrum morphs respond differently to variation in pollination intensity.

A further study in *Turnera ulmifolia* was designed to compare the quality of offspring derived from intermorph (legitimate) pollinations of two intensities (McKenna 1986). The *amount* of pollen applied to the stigma was varied to create conditions that provide opportunity for relatively intense or limited pollen competition (McKenna 1986). An excess and limited pollination treatment was created by applying a 6:1 ratio of pollen. The limited treatment consisted of rubbing one newly dehisced anther across the stigma; the excess treatment used six anthers per stigma. Fifty random crosses of each treatment were made in a greenhouse population of 20 thrum and 20 pin plants. Seeds were harvested at maturity and a random subsample of seeds without the elaisome were weighed. Seeds were planted individually in 2-inch pots placed in random positions on a greenhouse bench. Individual seedling heights were measured at weekly intervals. At 6 weeks seedlings were weighed, transplanted to 6-inch pots and grown to flowering to determine the morph ratio in the offspring.

The mean number of seeds per fruit did not differ significantly between the excess and limited treatments (Table 4). In both treatments mean seed set was less than mean ovule number per capsule, suggesting that some ovules are nonfunctional or that some ovules or embryos are aborted. Mean seed weight was significantly greater in offspring from excess pollinations. Mean 6-week seedling height and weight was also significantly greater in offspring from excess pollinations.

Table 4. Comparison of *Turnera* progeny derived from excess and limited pollination treatments

	Pollination treament Excess	Limited	
No. of seeds/fruit	Y = 22.78 s = 5.97 n = 49	Y = 21.78 s = 6.98 n = 54	
Seed weight (mg)	Y = 1.54 s = 0.32 n = 51	Y = 1.13 s = 0.18 n = 33	$p < 0.001$
6-Week seedling height (mm)	Y = 79.60 s = 18.00 n = 52	Y = 68.62 s = 16.51 n = 52	$p < 0.001$
6-Week seedling weight (g)	Y = 1.66 s = 0.50 n = 48	Y = 1.39 s = 0.46 n = 51	$p < 0.01$

A comparison between the results of excess and limited crosses in pin and thrum seed parents shows the same general trend for increased seedling weight and height in offspring from the excess pollination treatment, although this effect is stronger in offspring from thrum seed parents (Table 5). When crosses in pin and thrum seed parents are considered separately, the excess treatment resulted in greater seed set in pin parents, and lower seed set in thrum parents. Despite these differences in seed set, the excess treatment results in greater seed weight in both pin and thrum seed parents. A 1:1 ratio of pins to thrums was found in offspring from both excess and limited pollination treatments (Table 6). There was no difference in the 6-week seedling height of offspring that later produced pin or thrum flowers, so the excess pollination treatment did not seem to be preferentially selecting for either morph.

Natural pollen loads have not been quantified in *Turnera ulmifolia*, but the number of pollen grains used in the excess pollination treatment (12 000) most likely far exceeds the pollination intensity found in natural populations. In order to maximize seed set the limited treatment also contained a large number of grains (2000); it is likely that significant pollen competition occurs in pollen loads this large. This is over six times the number of pollen grains Shore and Barrett (1984) estimate to be necessary for full fertilization. Thus the results of this experiment do not indicate whether pollen competition will occur under natural pollination intensities, or the strength of the phenotypic effects of pollen competition under natural pollination intensities.

Table 5. Comparison of progeny from *Turnera* pin vs. thrum seed parents after excess and limited pollinations

A. Progeny from thrum seed parents

	Pollination treatment		
	Excess	Limited	
No. of seeds/fruit	Y = 23.39	Y = 26.86	$p < 0.05$
	s = 6.00	s = 5.13	
	n = 28	n = 22	
Seed weight (mg)	Y = 1.45	Y = 1.16	$p < 0.001$
	s = 0.16	s = 0.13	
	n = 36	n = 18	
6-Week seedling height (mm)	Y = 82.33	Y = 68.27	$p < 0.001$
	s = 16.64	s = 16.98	
	n = 33	n = 33	
6-Week seedling weight (g)	Y = 1.68	Y = 1.39	$p < 0.05$
	s = 0.47	s = 0.45	
	n = 31	n = 33	

B. Progeny from pin seed parents

	Pollination treatment		
	Excess	Limited	
No. of seeds/fruit	Y = 21.95	Y = 18.28	$p < 0.05$
	s = 6.12	s = 6.01	
	n = 21	n = 32	
Seed weight (mg)	Y = 1.77	Y = 1.11	$p < 0.001$
	s = 0.47	s = 0.22	
	n = 15	n = 15	
6-Week seedling height (mm)	Y = 74.84	Y = 69.21	
	s = 16.71	s = 16.08	
	n = 19	n = 19	
6-Week seedling weight (g)	Y = 1.64	Y = 1.39	
	s = 0.56	s = 0.51	
	n = 17	n = 18	

Table 6. Morph ratio in *Turnera ulmifolia* progeny derived from excess and limited pollinations

Pollination treatment	Pins	Thrums
Excess – Thrum seed parent	29	18
– Pin seed parent	17	17
	46	35
Limited – Thrum seed parent	30	25
– Pin seed parent	17	26
	47	51

5 Conclusion

The ecological and evolutionary implications of microgametophytic selection and its consequences in plant mating systems have attracted significant attention (Janzen 1977; Bertin 1982; Mulcahy 1983; Mulcahy et al. 1983; Stephenson and Bertin 1983; Willson and Burley 1983; Bawa and Webb 1984; Ellstrand 1984; Schemske and Pautler 1984; Marshall and Ellstrand 1985, 1986; Galen et al. 1986; Snow 1986, 1990; Charlesworth 1988; Stephenson et al. 1988). The overlap in gametophytic and sporophytic expression in angiosperms also presents a valuable opportunity for the use of pollen selection in applied breeding programs (Negrutiu et al. 1986; Pallais et al. 1986; Walden and Greyson 1986).

Many questions remain to be explored concerning pollen competition in heterostylous species. Heterostylous plants provide certain advantages for the study of pollen competition. The pollen polymorphism of heterostylous plants can be used to distinguish compatible from incompatible grains. This reduces the complications inherent in pollen competition studies with species that exhibit sporophytic or gametophytic incompatibility, where it may be difficult to determine the "effective" pollen load on naturally pollinated stigmas. Style length polymorphisms can also be utilized in pollen competition studies; species that do not exhibit intramorph incompatibility are ideal for these studies. Much more research is needed to evaluate the evolutionary importance of pre-pollination and post-pollination aspects of pollen competition in natural populations of heterostylous plants.

References

Andersson S (1988) Size-dependent pollination efficiency in *Anchusa officinalis* (Boraginaceae): causes and consequences. Oecologia 76:125–130

Arbeloa A, Herrero M (1987) The significance of the obturator in the control of pollen tube entry into the ovary in peach (*Prunus persica*). Ann Bot 60:681–685

Asano Y (1980) Studies on crosses between distantly related species of lilies. VI. Pollen tube growth in interspecific crosses of *Lilium longiflorum* (I.) J Jpn Soc Hortic Soc 49:392–396

Ascher PD, Drewlow LW (1971) Effect of stigmatic exudate injected into the stylar canal on compatible and incompatible pollen tube growth in *Lilium longiflorum* Thunb. In: Heslop-Harrison J (ed) Pollen: development and physiology. Appleton Century Crofts, New York, pp 267–272

Bagni N, Adamo P, Serafini-Fracassini D (1981) RNA, proteins and polyamines during tube growth in germinating apple pollen. Plant Physiol 68:727–730

Baker HG (1975) Sporophyte-gametophyte interactions in *Linum* and other genera with heteromorphic self-incompatibility. In: Mulcahy DL (ed) Gamete competition in plants and animals. North Holland, Amsterdam, pp 191–200

Baker H, Baker I (1983) Some evolutionary and taxonomic implications of variation in the chemical reserves of pollen. In: Mulcahy DL, Ottaviano E (eds) Pollen: biology and implications for plant breeding. Elsevier Biomedical, New York, pp 43–52

Baluch SJ, Risius ML, Cleveland RW (1973) Pollen germination and tube growth after selfing and crossing *Coronilla varia*. Crop Sci 13:303–306

Barnes DK, Cleveland RW (1963) Genetic evidence for nonrandom fertilization in alfalfa as influenced by differential pollen growth. Crop Sci 3:295–297

Barrett SCH (1978) Heterostyly in a tropical weed: the reproductive biology of the *Turnera ulmifolia* complex (Turneraceae) Can J Bot 56:1713–1725

Barrett SCH, Glover D (1985) On the Darwinian hypothesis of the adaptive significance of tristyly. Evolution 39:766–774

Barrett SCH, Brown AHD, Shore JS (1987) Disassortative mating in tristylous *Eichhornia paniculata* (Pontederiaceae). Heredity 58:49–55

Barrett SCH, Morgan MT, Husband BC (1989) The dissolution of a complex genetic polymorphism: the evolution of self-fertilization in tristylous *Eichhornia paniculata* (Pontederiaceae). Evolution 43:1398–1416

Bateman AJ (1956) Cryptic self-incompatibility in the wallflower *Cheiranthus cheiri* L. Heredity 10:257–261

Bawa KS, Webb CJ (1984) Flower, fruit and seed abortion in tropical forest trees: implications for the evolution of paternal and maternal reproductive patterns. Am J Bot 71:736–751

Bertin RI (1982) Floral biology, hummingbird pollination and fruit production of trumpet creeper (*Campsis radicans*, Bignoniaceae). Am J Bot 69:122–134

Bertin RI (1988) Paternity in plants. pp 30–59 in J. Lovett Doust and L. Lovett Doust (eds) Plant reproductive ecology. Oxford Univ Press, New York

Bertin RI (1990) Effects of pollination intensity in *Campsis radicans*. Am J Bot 77:178–187

Brewbaker JL, Majumder SK (1961) Cultural studies on the pollen population effect and the self-incompatibility inhibition. Am J Bot 48:457–464

Brink RA, MacGillivray JH (1924) Segregation for the waxy character in maize pollen and differential development of the male gametophyte. Am J Bot 11:465–469

Casper BB (1983) The efficiency of pollen transfer and rates of embryo initiation in *Cryptantha* (Boraginaceae). Oecologia 59:262–268

Casper BB (1985) Self-compatibility in distylous *Cryptantha flava* (Boraginaceae). New Phytol 99:149–154

Casper BB, Wiens D (1981) Fixed rates of random ovule abortion in *Cryptantha flava* (Boraginaceae) and its possible relation to seed dispersal. Ecology 62:866–869

Charlesworth D (1988) Evidence for pollen competition in plants and its relationship to progeny fitness: a comment. Am Nat 132:298–302

Correns C (1928) Bestimmung, Vererbung und Verteilung des Geschlechtes bei den höheren Pflanzen. Handb Vererbungswiss II:1–138

Cruden RW, Miller-Ward S (1981) Pollen-ovule ratio, pollen size, and the ratio of stigmatic area to the pollen-bearing area of the pollinator: an hypothesis. Evolution 35:964–974

Cruzan MB (1986) Pollen tube distributions in *Nicotiana glauca*: evidence for density-dependent growth. Am J Bot 73:902–907

Currah L (1981) Pollen competition in onion, *Allium cepa* L. Euphytica 39:687–696

Dane F, Melton B (1973) Effect of temperature on self and cross compatibility and in vitro pollen growth characteristics in alfalfa. Crop Sci 13:587–591

Darwin C (1877) The different forms of flowers on plants of the same species. Murray, Lond

d'Eeckenbrugge GC (1986) Incompatibility reaction and gametophytic competition in *Cichorium intybus* L. In: Mulcahy DL, Mulcahy GB, Ottaviano E (eds) Biotechnology and ecology of pollen. Springer, Berlin Heidelberg New York, pp 473–476

d'Eeckenbrugge GC, Ngendahayo M, Louant BP (1986) Intra- and interspecific incompatibility in *Brachiaria ruziziensis* Germain et Evrard (Panicoideae). In: Mulcahy DL, Mulcahy GB, Ottaviano E (eds) Biotechnology and ecology of pollen. Springer, Berlin Heidelberg New York, pp 257–264

Dulberger R (1964) Flower dimorphism and self-incompatibility in *Narcissus tazetta* L. Evolution 18:361–363

Dulberger R (1970) Floral dimorphism in *Anchusa hybrida* Ten. Isr J Bot 19:37–41

Dulberger R (1987) Fine structure and cytochemistry of the stigma surface and incompatibility in some distylous *Linum* species. Ann Bot 59:203–217

Eenink AH (1982) Compatibility and incompatibility in witloof-chicory (*Cichorium intybus* L.). 3. Gametic competition after mixed pollinations and double pollinations. Euphytica 31:773–786

Ellstrand NC (1984) Multiple paternity within the fruits of the wild radish, *Raphanus sativus*. Am Nat 123:819–828

Ericksson G (1969) The waxy character. Heredity 63:180–204

Feinsinger P, Busby WH (1987) Pollen carryover: experimental comparisons between morphs of *Palicourea lasiorrachis* (Rubiaceae), a distylous, bird-pollinated, tropical treelet. Oecologia 73:231–235

Fingerett ER (1979) Pollen competition in a species of evening primrose, *Oenothera organensis* Munz. MS Thesis, Washington State Univ, Washington

Freeling M, Cheng DSK (1978) Radiation induced alcohol dehydrogenase mutants in maize following allyl alcohol selection of pollen. Genet Res 31:107–131

Galen C, Shykoff JA, Plowright RC (1986) Consequences of stigma receptivity schedules for sexual selection in flowering plants. Am Nat 127:462–476

Ganders FR (1975) Mating patterns in self compatible distylous populations of *Amsinckia* (Boraginaceae). Can J Bot 53:773–779

Ganders FR (1979) The biology of heterostyly. NZ J Bot 17:607–635

Gay G, Kerhoas C, Dumas C (1986) Micro-isoelectric focusing of pollen grain proteins in *Cucurbita pepo* L. In: Mulcahy DL, Mulcahy GB, Ottaviano E (eds) Biotechnology and ecology of pollen. Springer, Berlin Heidelberg New York, pp 496–497

Gilissen LJW (1977) Style-controlled wilting of the flower. Planta 133:275–280

Glover DE, Barrett SCH (1986) Variation in the mating system of *Eichhornia paniculata* (Spreng.) Solms. (Pontederiaceae). Evolution 40:1122–1131

Gross KL (1981) Predictions of fate from rosette size in four biennial plant species: *Verbascum thapsus, Oenothera biennis, Daucus carota* and *Tragopogon dubius*. Oecologia 48:209–213

Hamill E, Brewbaker R (1969) Peroxidase isoenzymes of maize. Physiol Plant 22:943–958

Harte C (1975) Competition in the haploid generation in *Oenothera*. In: Mulcahy DL (ed) Gamete competition in plants and animals. North Holland, Amsterdam, pp 31–42

Heribert-Nilsson N (1920) Zuwachsgeschwindigkeit der Pollensclauche und gestorte Mendelzahlen bie *Oenothera lamarckiana*. Hereditas 1:41–67

Heslop-Harrison J, Heslop-Harrison Y (1975) Enzymic removal of the proteinaceous pellicle of the stigma papilla prevents pollen tube entry in the Caryophyllaceae. Ann Bot 39:163–165

Hessing MB (1986) Pollen growth following self and cross pollination in *Geranium caespitosum* James. In: Mulcahy DL, Mulcahy GB, Ottaviano E (eds) Biotechnology and ecology of pollen. Springer, Berlin Heidelberg New York, pp 467–472

Hill JP, Lord EM (1986) Dynamics of pollen tube growth in the wild radish, *Raphanus raphanistrum* (Brassicacae) I. Order of fertilization. Evolution 40:1328–1333

Hoekstra FA (1983) Physiological evolution in angiosperm pollen: possible role of pollen vigor. In: Mulcahy DL, Ottaviano E (eds) Pollen: biology and implications for plant breeding. Elsevier Biomedical, New York, pp 35–41

Janzen DH (1977) A note on optimal mate selection by plants. Am Nat 111:365–371

Knox BR (1984) Pollen-pistil interactions. In: Linskens HF, Heslop-Harrison J (eds) Encyclopedia of plant physiology. Springer, Berlin Heidelberg New York

Kroh M, Labarca C, Loewus F (1971) Use of pistil exudate for pollen tube wall biosynthesis in *Lilium longiflorum*. In: Heslop-Harrison J (ed) Pollen: development and physiology. Appleton Century Crofts, New York, pp 273–278

Lee TD, Bazzaz FA (1982) Regulation of fruit maturation pattern in an annual legume, *Cassia fasciculata*. Ecology 63:1374–1388

Lee TD, Hartgerink AP (1986) Pollination intensity, fruit maturation pattern, and offspring quality in *Cassia fasciculata* (Leguminosae). In: Mulcahy DL, Mulcahy GB, Ottaviano E (eds) Biotechnology and ecology of pollen. Springer, Berlin Heidelberg New York, pp 417–422

Levin DA (1975) Gametophytic selection in *Phlox*. In: Mulcahy DL (ed) Gamete competition in plants and animals. North Holland, Amsterdam, pp 207–218

Lloyd DG, Yates JMA (1982) Intrasexual selection and the segregation of pollen and stigmas in hermaphrodite plants exemplified by *Wahlenbergia albomarginata* (Campanulaceae). Evolution 36:903–913

Lord EM, Kohorn LU (1986) Gynoecial development, pollination, and the path of pollen tube growth in the tepary bean, *Phaseolus acutifolius*. Am J Bot 73:70–78

Makinen Y, Brewbaker JL (1967) Isoenzyme polymorphism in flowering plants I. Diffusion of enzymes out of intact pollen grains. Physiol Plant 20:477–482

Marshall DL, Ellstrand NC (1985) Proximal causes of multiple paternity in wild radish. Am Nat 126:596–605

Marshall DL, Ellstrand NC (1986) Sexual selection in *Raphanus sativus*: experimental data on nonrandom fertilization, maternal choice, and consequences of multiple paternity. Am Nat 127:446–461

Mascarenhas JP, Stinson JS, Willing RP, Pe ME (1986) Genes and their expression in the male gametophyte of flowering plants. In: Mulcahy DL, Mulcahy GB, Ottaviano E (eds) Biotechnology and ecology of pollen. Springer, Berlin Heidelberg New York, pp 39–44

Mazer SJ (1987) Parental effects on seed development and seed yield in *Raphanus raphanistrum*: implications for natural and sexual selection. Evolution 41:355–371

McKenna MA (1987) The effect of microgametophytic competition on sporophytic vigor in *Dianthus chinensis* L. Caryophyllaceae. PhD Thesis, State Univ New York, Stony Brook, New York

McKenna MA (1986) Heterostyly and microgametophytic selection: The effect of pollen competition on sporophytic vigor in two distylous species. In: Mulcahy DL, Mulcahy GB, Ottaviano E (eds) Biotechnology and ecology of pollen, Springer-Verlag, NY, pp 443–448

McKenna MA, Mulcahy DL (1983) Ecological aspects of pollen competition in *Dianthus chinensis*. In: Mulcahy DL, Ottaviano E (eds) Pollen: biology and implications for plant breeding. Elsevier Biomedical, New York, pp 419–424

Miller J, Mulcahy DL (1983) Microelectrophoresis and the study of genetic overlap. In: Mulcahy DL, Ottaviano E (eds) Pollen: biology and implications for plant breeding. Elsevier Biomedical, New York, pp 317–322

Morgan MT, Barrett SCH (1990) Outcrossing rates and correlated mating within a population of *Eichhornia paniculata* (Pontederiaceae). Heredity 64:271–280

Mulcahy DL (1974) Correlation between speed of pollen tube growth and seedling height in *Zea mays* L. Nature (Lond) 249:491–493

Mulcahy DL (1983) Models of pollen tube competition in *Geranium maculatum*. In: Real L (ed) Pollination Biology. Academic Press, Lond New York, pp 152–162

Mulcahy DL, Mulcahy GB (1975) The influence of gametophytic competition on sporophytic quality in *Dianthus chinensis*. Theor Appl Genet 46:277–280

Mulcahy DL, Mulcahy GB, Ottaviano E (1975) Sporophytic expression of gametophytic competition in *Petunia hybrida*. In: Mulcahy DL (ed) Gamete competition in plants and animals. North Holland, Amsterdam, pp 227–232

Mulcahy DL, Mulcahy GB, Robinson RW (1979) Evidence for postmeiotic genetic activity in pollen of *Cucurbita* species. J Hered 70:365–368

Mulcahy DL, Curtis PS, Snow AA (1983) Pollen competition in a natural population. In: Jones CE, Little RL (eds) Handbook of experimental pollination biology. Van Nostrand Reinhold, New York, pp 330–338

Mulcahy GB, Mulcahy DL (1986) Pollen-pistil interaction. In: Mulcahy DL, Mulcahy GB, Ottaviano E (eds) Biotechnology and ecology of pollen. Springer, Berlin Heidelberg New York, pp 173–178

Murray BG (1990) Heterostyly and pollen-tube interactions in *Luculia gratissima* (Rubiaceae). Ann Bot 65:691–698

Negrutiu I, Heberle-Bors E, Potrykus I (1986) Attempts to transform for kanamycin-resistance in mature pollen of tobacco. In: Mulcahy DL, Mulcahy GB, Ottaviano E (eds) Biotechnology and ecology of pollen. Springer, Berlin Heidelberg New York, pp 65–70

Nicholls MS (1986) Population composition, gender specialization, and the adaptive significance of distyly in *Linum perenne* (Linaceae). New Phytol 102:209–217

Ornduff R (1964) The breeding system of *Oxalis suksdorfii*. Am J Bot 51:307–314

Ornduff R (1970) Incompatibility and the pollen economy of *Jepsonia parryi*. Am J Bot 57:1036–1041

Ornduff R (1975) Pollen flow in *Lythrum junceum*, a tristylous species. New Phytol 75:161–166

Ornduff R (1980) Heterostyly, population composition and pollen flow in *Hedyotis caerulea*. Am J Bot 67:95–103

Ottaviano E, Sari-Gorla M, Pe E (1982) Male gametophytic selection in maize. Theor Appl Genet 63:249–254

Ottaviano E, Sari-Gorla M, Arenari I (1983) Male gametophyte competitive ability in maize: selection and implications with regard to the breeding system. In: Mulcahy DL, Ottaviano E (eds) Pollen: biology and implications for plant breeding. Elsevier, New York

Ottaviano E, Sidotti P, Villa M (1986) Pollen competitive ability in maize selection and single gene analysis. In: Mulcahy DL, Mulcahy GB, Ottaviano E (eds) Biotechnology and ecology of pollen. Springer, Berlin Heidelberg New York, pp 21–26

Pallais N, Malagamba P, Fong N, Garcia R, Schmiediche P (1986) Pollen selection through storage: a tool for improving true potato seed quality? In: Mulcahy DL, Mulcahy GB, Ottaviano E (eds) Biotechnology and ecology of pollen. Springer, Berlin Heidelberg New York, pp 153–158

Pfahler P (1975) Factors affecting male transmission in maize (*Zea mays* L.). In: Mulcahy DL (ed) Gamete competition in plants and animals. North Holland, Amsterdam, pp 115–124

Philipp M, Schou O (1981) An unusual heteromorphic incompatibility system. Distyly, self-incompatibility, pollen load and fecundity in *Anchusa officinalis* (Boraginaceae). New Phytol 89:693–703

Price SD, Barrett SCH (1982) Tristyly in *Pontederia cordata* (Pontederiaceae). Can J Bot 60:897–905

Ray PM, Chisaki HF (1957) Studies of *Amsinckia*. I. A synopsis of the genus with a study of heterostyly in it. Am J Bot 44:529–536

Sacher R, Mulcahy DL, Staples R (1983) Developmental selection for salt tolerance during self pollination of *Lycopersicon* × *Solanum* F1 for salt tolerance of F2. In: Mulcahy DL, Ottaviano E (eds) Pollen: biology and implications for plant breeding. Elsevier Biomedical, New York, pp 329–334

Sari-Gorla M, Ottaviano E, Faini D (1975) Genetic variability of gametophytic growth rate in maize. Theor Appl Genet 46:289–294

Sari-Gorla M, Frova C, Redaelli R (1986) Extent of gene expression at the gametophyte phase in maize. In: Mulcahy DL, Mulcahy GB, Ottaviano E (eds) Biotechnology and ecology of pollen. Springer, Berlin Heidelberg New York, pp 27–32

Schemske DW, Pautler LP (1984) The effects of pollen composition on fitness components in a neotropical herb. Oecologia 62:31–36

Schlichting CD (1986) Environmental stress reduces pollen quality in *Phlox*: compounding the fitness deficit. In: Mulcahy DL, Mulcahy GB, Ottaviano E (eds) Biotechnology and ecology of pollen. Springer, Berlin Heidelberg New York, pp 483–488

Schou O, Philipp M (1983a) An unusual heteromorphic incompatibility system. II. Pollen tube growth and seed sets following compatible and incompatible crossings within *Anchusa officinalis* L. (Boraginaceae) In: Mulcahy DL, Ottaviano E (eds) Pollen: biology and implications for plant breeding. Elsevier Biomedical, New York, pp 219–227

Schou O, Philipp M (1983b) An unusual heteromorphic incompatibility system. III. On the genetic control of distyly and self-incompatibility in *Anchusa officinalis* L. (Boraginaceae). Theor Appl Genet 68:139–244

Schwartz D, Osterman J (1976) A pollen selection system for alcohol dehydrogenase negative mutants in plants. Genetics 83:63–65

Schwemmle J (1968) Selective fertilization in *Oenothera*. Adv Genet 14:225–324

Searcy KB, Mulcahy DL (1986) Gametophytic expression of heavy metal tolerance. In: Mulcahy DL, Mulcahy GB, Ottaviano E (eds) Biotechnology and ecology of pollen. Springer, Berlin Heidelberg New York, pp 159–165

Shore JS, Barrett SCH (1984) The effect of pollination intensity and incompatible pollen on seed set in *Turnera ulmifolia* (Turneraceae). Can J Bot 62:1298–1303

Shore JS, Barrett SCH (1985) The genetics of distyly and homostyly in *Turnera ulmifolia* L. (Turneraceae). Heredity 55:167–174

Singh MB, Knox RB (1986) Expression of B-galactosidase gene in pollen of *Brassica campestris*. In: Mulcahy DL, Mulcahy GB, Ottaviano E (eds) Biotechnology and ecology of pollen. Springer, Berlin Heidelberg New York, pp 3–8

Smith GA, Moser HS (1985) Sporophytic-gametophytic herbicide tolerance in sugar beet. Theor Appl Genet 71:231–237

Snow AA (1986) Pollination dynamics in *Epilobium canum* (Onagraceae): consequences for gametophytic selection. Am J Bot 73:139–151

Snow AA (1990) Effects of pollen load size and number of donors on sporophyte fitness in wild radish (*Raphanus raphanistrum*). Am Nat 136:742–758

Snow AA, Mazer SJ (1988) Gametophytic selection in *Raphanus raphanistrum*: a test for heritable variation in pollen competitive ability. Evolution 42:1065–1075

Sobrevila C, Ramirez N, Xena de Enrech N (1983) Reproductive biology of *Palicourea fendleri* and *P. petiolaris* (Rubiaceae), heterostylous shrubs of a tropical cloud forest in Venezuela. Biotropica 15:161–169

Stephenson AG, Bertin RI (1983) Male competition, female choice, and sexual selection in plants. In: Real L (ed) Pollination biology. Academic Press, Lond New York, pp 109–149

Stephenson AG, Winsor JA, Davis LE (1986) Effects of pollen load size on fruit maturation and sporophyte quality in zucchini. In: Mulcahy DL, Mulcahy GB, Ottaviano E (eds) Biotechnology and ecology of pollen. Springer, Berlin Heidelberg New York, pp 429–434

Stephenson AG, Winsor JA, Schlichting CD, Davis LE (1988) Pollen competition, nonrandom fertilization, and progeny fitness: a reply to Charlesworth. Am Nat 132:303–308

Stettler RF (1968) Irradiated mentor pollen: its use in remote hybridization of black cottonwood. Nature (Lond) 219:746–747

Suss J, Tupy J (1979) Poly(A) RNA synthesis in germinating pollen of *Nicotiana tabacum* L. Biol Plant 21:365–371

Swamy NR, Bahadur B (1984) Pollen flow in dimorphic *Turnera subulata* (Turneraceae). New Phytol 98:205–209

Tanksley SD, Zamir D, Rick CM (1981) Evidence for extensive overlap of sporophytic and gametophytic gene expression in *Lycopersicon esculentum*. Science 213:453–455

Ter-Avanesian DV (1978) The effect of varying the number of pollen grains used in fertilization. Theor Appl Genet 52:77–79

Thomson JD (1986) Pollen transport and deposition by bumblebees in *Erythronium*: influences of floral nectar and bee grooming. Oecologia 69:561–566

Thomson JD (1989) Germination schedules of pollen grains: implications for pollen selection. Evolution 43:220–223

Tilton VR, Horner HT (1980) Stigma, style and obturator of *Ornithogalum caudatum* (Liliaceae) and their function in the reproductive process. Am J Bot 67:1113–1131

Tilton VR, Wilcox LW, Palmer RG, Albertsen MC (1984) Stigma, style and obturator of soybean, *Glycine max* (L.) Herr. (Leguminosae) and their function in the reproductive process. Am J Bot 71:676–686

van Herpen MMA (1981) Effect of season, age and temperature on the protein pattern of pollen and styles in *Petunia hybrida*. Acta Bot Neerl 30:277–287

van Herpen MMA (1984) Extracts from styles, developed at different temperatures, and their effect on compatibility of *Petunia hybrida* in excised style culture. Acta Bot Neerl 33:195–203

van Herpen MMA (1986) Biochemical alterations in the sexual partners resulting from environmental conditions before pollination regulate processes after pollination. In: Mulcahy DL, Mulcahy GB, Ottaviano E (eds) Biotechnology and ecology of pollen. Springer, Berlin Heidelberg New York, pp 131–134

van Herpen MMA, Linskens HF (1981) Effect of season, plant age and temperature during plant growth on compatible and incompatible pollen tube growth in *Petunia hybrida*. Acta Bot Neerl 30:209–218

Visser T, Verhaegh JJ (1980) Pollen and pollination experiments. II. The influence of the first pollination on the effectiveness of the second one in apple. Euphytica 29:385–390

Walden DB, Greyson RI (1986) Maize pollen research: Preliminary reports from two projects investigating gamete selection. In: Mulcahy DL, Mulcahy GB, Ottaviano E (eds) Biotechnology and ecology of pollen. Springer, Berlin Heidelberg New York, pp 139–146

Waser NM, Fugate ML (1986) Pollen precedence and stigma closure: a mechanism of competition for pollination between *Delphinium nelsonii* and *Ipomopsis aggregata*. Oecologia 70:573–577

Waser NM, Price MV (1983) Optimal and actual outcrossing in plants and the nature of plant/pollinator interactions. In: Jones CE, Little RJ (eds) Handbook of experimental pollination biology. Van Nostrand Reinhold, New York pp 341–359

Weeden NF (1986) Identification of duplicate loci and evidence for post-meiotic gene expression in pollen. In: Mulcahy DL, Mulcahy GB, Ottaviano E (eds) Biotechnology and ecology of pollen. Springer, Berlin Heidelberg New York, pp 9–14

Weller SG (1981) Pollination biology of heteromorphic populations of *Oxalis alpina* (Rose) Knuth (Oxalidaceae) in southeastern Arizona. Bot J Linn Soc 83:189–198

Weller SG (1985) The life history of *Lithospermum caroliniense*, a long-lived herbaceous sand dune species. Ecol Monogr 55:49–67

Weller SG, Ornduff R (1977) Cryptic self incompatibility in *Amsinckia grandiflora*. Evolution 31:47–51

Willing RP, Mascarenhas JP (1984) Analysis of the complexity and diversity of mRNAs from pollen and shoots of *Tradescantia*. Plant Physiol 75:865–868

Willing RP, Eisenberg A, Mascarenhas JP (1984) Genes active during pollen development and the construction of cloned cDNA libraries to messenger RNAs from pollen. Plant Cell Incompatibility Newslett 16:11–12

Willson MF, Burley N (1983) Mate choice in plants. Princeton Univ Press, Princeton, New York

Wolfe LM, Barrett SCH (1987) Pollinator foraging behavior and pollen collection on the floral morphs of tristylous *Pontederia cordata* L. Oecologia 74:347–351

Yeo PF (1975) Some aspects of heterostyly. New Phytol 75:147–153

Chapter 10

Evolutionary Modifications of Tristylous Breeding Systems

S.G. WELLER[1]

1 Introduction

Tristyly is perhaps the most complex breeding system found in the flowering plants. Three floral morphs are found in populations; in the short-styled morph, stigmas occur in the low position, and the two sets of anthers occur in the mid and high levels of the flower (Fig. 1). In the mid-styled morph, anthers occur in the low and high positions, and the stigmas occupy the mid position. In the long-styled morph, the stigmas occur in the high position, and anthers occupy the mid and low positions. Floral trimorphism is accompanied by strong self-incompatibility. Pollinations occurring between anthers and stigmas at the same level, by necessity those between different floral morphs, were termed legitimate by Darwin (1877); these are normally the only pollinations that lead to fertilization and seed production (Fig. 1). Illegitimate crosses include self- and cross-pollinations between anthers and stigmas occurring at different levels. These pollinations usually fail to produce seed, although in some modified tristylous breeding systems illegitimate crosses may be compatible, and result in seed production.

The level of complexity in tristylous reproductive systems is considerably greater than in distylous systems; for example, in species with typical trimorphic incompatibility there are six categories of compatible crosses and 18 categories of intermorph crosses and self-pollinations that fail to produce seeds. Tristyly is unique among flowering plant breeding systems in that two different incompatibility reactions are found within a single individual (Fig. 1). For each floral form, pollen grains produced by only one of the two whorls will germinate and lead to seed production of a second floral form. It is this differentiation in incompatibility reaction within a floral form that leads to the complexity in number of compatible crosses in typical trimorphic species. Given the complexity of tristyly, it is not surprising that the breeding system is found in only three families of flowering plants: the Pontederiaceae, Oxalidaceae, and Lythraceae.[2] Despite the rarity of the breeding system, the scattered distribution of tristyly among both monocotyledons and dicotyledons shows that it has evolved independently in each of the three families.

Because of the rarity of tristyly, the steps involved in its evolution will probably never be understood fully. Theoretical considerations indicate that the initial stages in the evolution of tristyly may have involved the appearance of a self-sterile

[1] Department of Ecology and Evolutionary Biology, University of California, Irvine, CA 92717, USA
[2] Reports of tristyly in the Amaryllidaceae, Connaraceae and Thymelaeaceae are discussed in Chapter 1 and 6

Monographs on Theoretical and Applied Genetics 15
Evolution and Function of Heterostyly (ed. by S.C.H. Barrett)

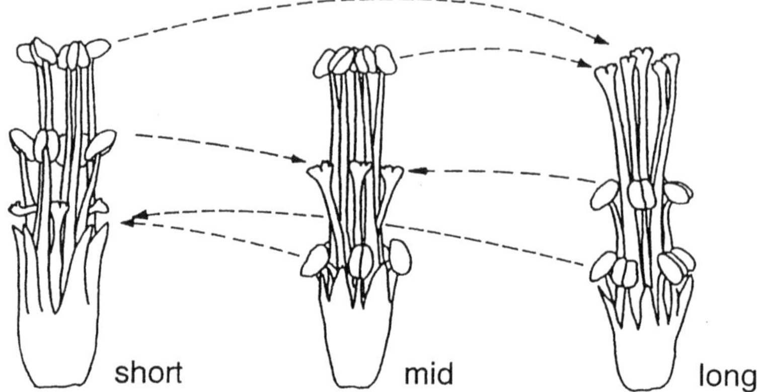

Fig. 1. The reproductive system of a trimorphic species. Pollinations between anthers and stigmas occurring at the same level, shown by *arrows*, are termed legitimate; normally, these are the only pollinations leading to seed production

genotype, as in the case of many plant breeding systems involving unisexuality or self-incompatibility (Charlesworth 1979). This genotype is able to spread if the product of the selfing rate and inbreeding depression is greater than one-half. Alternatively, Charlesworth (1979) considered the possibility that morphological complementarity may evolve first, followed by modifications of the incompatibility reaction. The second pathway has the advantage of requiring only that inbreeding depression be greater than one-half. A problem with any model for the evolution of tristyly is the number of assumptions required before any progress can be made. Since the assumptions are difficult to test, and there are virtually no data on the genetic basis for modifications of incompatibility in tristylous species, further progress in understanding the evolution of tristyly is likely to be slow.

Tristyly appears to be an unstable breeding system. In each of the families where it occurs there are numerous examples of modifications that lead to loss of incompatibility, modification of morphology, or both. A great deal of information is available on the modifications leading to the breakdown in tristyly, and various mechanisms have been modeled and tested empirically. The purpose of this chapter is to compare patterns in the breakdown of tristyly in families where the breeding system occurs. Preliminary observations indicate that modifications of tristyly differ substantially in the Lythraceae, Oxalidaceae, and Pontederiaceae. A more detailed comparison of these pathways should point the way to future research designed to elucidate factors underlying the loss of tristyly.

Modifications of tristyly will be discussed in turn for the Lythraceae, Oxalidaceae, and Pontederiaceae. The number of species differs greatly among the families; consequently, the range of variability in modes of breakdown of tristyly differs considerably for the three families. The Lythraceae is broadly distributed in tropical and temperate regions of the Old and New World, with a total of approximately 22 genera and 450 species. Tristyly is known in a relatively small number of species distributed in three of the genera. In the Oxalidaceae three of the five genera are known to exhibit tristyly. The largest genus in the family is *Oxalis*, with close to 900 species. Two centers of diversity occur for *Oxalis*, one in South Africa and the other in South America. Because of the very large size of the genus and the occurrence of two centers of distribution, strikingly different patterns in the breakdown of tristyly should be expected. The Pontederiaceae is a relatively small family with 9 genera and 34 species, although many of these have wide distributions because of their weedy characteristics.

2 Lythraceae

Tristyly occurs in the Old World species of *Lythrum* and *Nesaea*, and the New World monotypic *Decodon*. Distyly has evolved in several New World species of *Lythrum* as well as *Nesaea* (Ornduff 1979). Two genera, *Rotala* and *Pemphis*, exhibit distyly but not tristyly. Without a better understanding of the phylogeny of the Lythraceae, it is difficult to determine whether tristyly in *Lythrum, Nesaea,* and *Decodon* represents a primitive breeding system retained in these genera, or a breeding system that has evolved on several occasions in the family. More information on the evolution of distyly in *Rotala* and *Pemphis* would be helpful; in the case of *Pemphis* it appears that distyly has evolved from tristyly (Lewis and Rao 1971).

Tristyly in *Lythrum* has been studied most thoroughly in *Lythrum salicaria* (Haldane 1936; Schoch-Bodmer 1938; Fisher and Mather 1943; Mulcahy and Caporello 1970; Halkka and Halkka 1974; Heuch 1980) and *Lythrum junceum* (Dulberger 1970; Ornduff 1975). Early genetic studies of *L. salicaria* (Fisher and Mather 1943), combined with later theoretical studies (Spieth and Novitski 1969; Heuch 1979), provided the basis for predicting isoplethy (equal numbers of style forms) in tristylous populations, a prediction that is not immediately obvious based on the genetic mechanism controlling expression of style forms (Charlesworth 1979). Surveys of style forms in populations of varying sizes were carried out by Haldane (1936) in Great Britain and Halkka and Halkka (1974) on islands in the Baltic Sea, and even in the smallest populations (n=5) all three style forms were represented. Halkka and Halkka (1974) believed that extensive migration among the islands maintained all style forms, even in those populations with small numbers of individuals (Halkka and Halkka 1974). Modeling studies by Heuch (1980) showed that in populations of 20 or more individuals maintenance of three floral morphs was likely; in Halkka and Halkka's study, only one population had fewer than 20 individuals, indicating that gene flow among isolated populations may not have been as prevalent as they suggested, or necessary for the maintenance of floral trimorphism. Neither *L. salicaria* nor *L. virgatum*, both members of section *Mesolythrum*, show any indication of disruption of tristyly, and there is no indication of the evolution of distyly in populations of either species.

Species placed in section *Euhyssopifolia* show the greatest variation in breeding systems (Dulberger 1970; Ornduff 1975). Old World species (*L. junceum*) placed in this section are tristylous, while those found in the New World (*L. curtissii, L. californicum*, and *L. lineare*) are distylous. Several species in section *Euhyssopifolia* are homostylous (Koehne 1903), and presumably derived from heterostylous ancestors, although the mechanism for the change has not been studied. Dulberger (1970) sampled seven populations of *L. junceum* in Israel and found isoplethy in the majority of cases; a single population sampled by Ornduff (1975) in Morocco also proved to be isoplethic. Studies of the morphology and incompatibility system of this species indicate that *L. junceum* possesses typical trimorphic incompatibility, with the highest seed production occurring in the mid-styled form (Dulberger 1970). Pollen production is equivalent for the long- and mid-styled forms, while the short-styled morph produces fewer pollen grains, in keeping with many heterostylous species (Ganders 1979). Dulberger compared the breeding systems of *L. junceum* and *L. salicaria* and found few differences, although in *L. salicaria* additional differences, such as pollen color, distinguish the floral morphs. Self-incompatibility is less strongly developed in *L. salicaria*, which Dulberger believed might be the result of the higher ploidal level. Alternatively, repeated bouts of colonization of *L. salicaria*, a widespread weed, may have led to selection for partially self-compatible races.

Ornduff's (1975) study of pollen flow in *L. junceum* in Morocco showed that long stigmas captured more pollen from long anthers than expected on the basis of random pollen flow. Stigmas of short-styled forms had less pollen from anthers of long stamens than expected, and stigmas of mid-styled plants had approximately the number of pollen grains from long-level anthers expected on the basis of random pollen flow. Pollen from mid-level and short-level anther whorls could not be distinguished reliably for *L. junceum*. Ornduff used plants whose flowers had not been emasculated, indicating that legitimate pollen flow (pollen movement between anthers and stigmas occurring at the same level) is likely to be considerable (Ganders 1974). Results obtained by Ornduff for pollen flow in *L. junceum* mirror the earlier results of Mulcahy and Caporello (1970), who studied pollen flow in an introduced population of *L. salicaria*. In the latter study, pollen loads on stigmas of emasculated and intact flowers were compared, although interpretation of the results is complicated by the very high levels of contamination resulting from the process of emasculation (Ganders 1979).

Studies of closely related species showing breeding system variation are most likely to provide indications of the evolutionary pathway for observed modifications. In the Lythraceae, Ornduff's (1979) comparisons of tristylous and distylous species in section of *Euhyssopifolia* reveal a potential pathway for the evolution of distyly from tristyly (Fig. 2). Distylous species in section *Euhyssopifolia* have a single set of six anthers, while tristylous species have two sets of six anthers each. Stamen insertion for the short and long morphs in distylous populations is alternate with the petals. In *Lythrum junceum* anther insertion for the short set of anthers is opposite the petals, while the stamens of the mid and long anther sets are inserted alternate with the petals. Stamens and styles of the distylous species are found in the approximate positions of stamens and styles in the mid and long forms of trimorphic *L. junceum*. Ratios calculated using measurements of style length, stamen length, pollen diameter, and several other floral characters showed that long:mid ratios of tristylous species were very similar to long to short ratios of the dimorphic species, and generally quite different from long:short ratios of the tristylous species (Ornduff 1979). The simplest interpretation for these results is that distyly has evolved in section *Euhyssopifolia* as a result of loss of the short-styled morph of a tristylous ancestor or ancestors (Fig. 2; Ornduff 1979). Accompanying loss of the short-styled

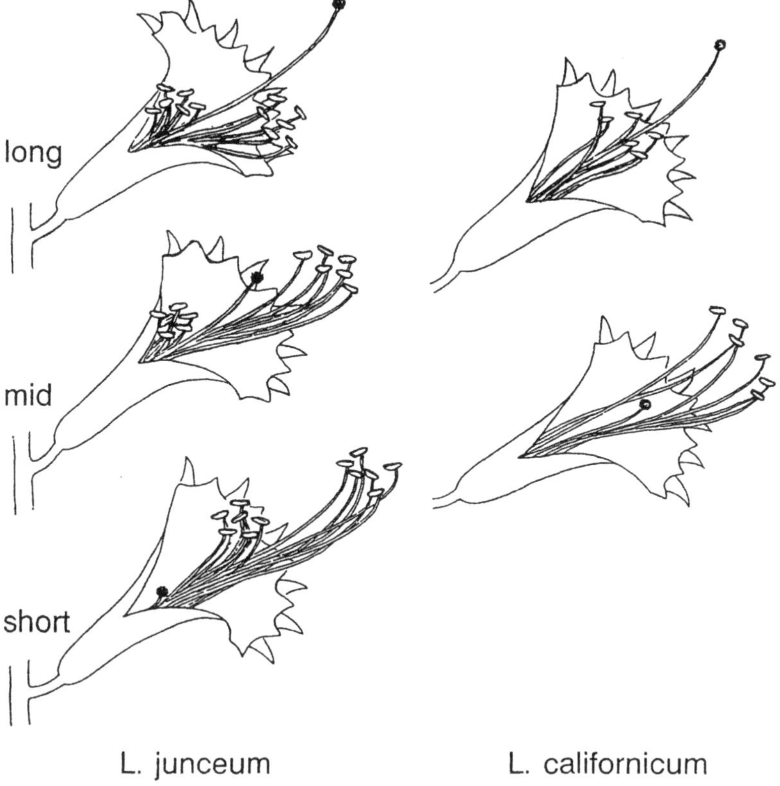

Fig. 2. Reproductive systems of tristylous *Lythrum junceum* and distylous *L. californicum*, both section *Euhyssopifolia*. Loss of the short-styled morph and the low set of stamens of the mid- and long-styled morphs have led to the evolution of distyly in section *Euhyssopifolia*. (After Ornduff 1979)

morph was loss of one set of stamens (those in the lowest position, with insertion opposite the petals) from each of the remaining morphs, converting the mid-styled morph to the present short-styled morph of distylous species (Fig. 2). The remaining instance of distyly in *Lythrum* occurs in *L. rotundifolia* (section *Hochstetteria*), a species with two sets of stamens (Koehne 1903; Ornduff 1979). It seems unlikely that the selective forces resulting in the evolution of distyly are the same in the two sections, since stamen loss occurs only in section *Euhyssopifolia*. Retention of two sets of stamens in *L. rotundifolia* may mean that loss of the mid-styled morph, rather than the short-styled morph, led to the evolution of distyly in this species, an interpretation complicated by the lack of tristyly in section *Hochstetteria*. Ornduff (1978, 1979) concluded that distyly has evolved on at least two occasions within *Lythrum*.

Similar variation in factors leading to the evolution of distyly may occur in *Nesaea*, where tristylous and distylous species occur in two sections (Ornduff 1978). In section *Salicariastrum* trimorphic species have two sets of stamens, and the distylous species have one set of stamens; in section *Heimiastrum* trimorphic and dimorphic species have two sets of stamens. These observations indicate that distyly is likely to have at least two separate origins in *Nesaea* (Ornduff 1978), as in the case of *Lythrum*. Further comparative studies of *Nesaea* species would be of great interest.

The mechanism for the evolution of distyly in *Pemphis acidula* may be similar to the mechanism found in *Lythrum* section *Hochstetteria* and *Nesaea* section *Heimiastrum*. Populations of this species appear to be uniformly dimorphic, with 1:1 style morph ratios occurring throughout the range (Lewis and Rao 1971; Gill and Kyauka 1977). In *P. acidula*, two whorls of stamens are found in the long and short morphs. The whorl of stamens that would be equivalent to the mid whorl in a fully trimorphic species converges on either the long stamen whorl of the short-styled morph or the short stamen whorl of the long-styled morph. The degree of convergence in stamen length as well as pollen size varies among populations of *P. acidula*, which led Lewis and Rao to conclude that this species had been trimorphic in the past, and still showed signs of "adjusting to a fully dimorphic state by means of mutation and selection." Despite the lack of trimorphism in the other species of *Pemphis*, or the occurrence of closely related genera showing trimorphism, the argument that the recent ancestor of *P. acidula* was trimorphic seems reasonable in view of the presence of the mid whorl of stamens, and the variability of stamen length and pollen size of this whorl relative to the upper and lower whorls, in which these measures are constant (Lewis and Rao 1971). The assumption, however, that the variability is under genetic control and selection is favoring convergence in stamen length and pollen size is difficult to evaluate, since there are no populations retaining the mid morph, and there is no way of knowing how recently the mid morph was present in any of the extant populations. Incompatibility reactions provide no further insights: pollen tubes from upper and lower anther whorls of the long-styled morph show similar patterns of growth or inhibition when used in compatible or incompatible pollinations. Patterns of pollen tube growth for pollen from the two whorls of the short-styled morph were similarly indistinguishable. Because of the absence of mid-styled morphs in any populations of *P. acidula*, the potential role in the evolution of distyly of loss of incompatibility differentiation between the stamen whorls of the short- and long-styled morphs cannot be evaluated.

The distribution of distyly in the Lythraceae, and the occurrence of distylous species without close tristylous relatives, indicate that distyly appeared in the distant evolutionary past of this family (Ornduff 1979). Because many distylous species lack close tristylous relatives, or show well-developed distylous syndromes, with little evidence of intermediacy between typical trimorphism versus dimorphism, the selective basis for the evolution of distyly is unclear. Ornduff (1978) found that pollen flow among dimorphic species appeared to show a greater random component than in the related tristylous *L. junceum*; thus greater levels of legitimate pollen flow in a distylous breeding system are unlikely to explain the evolution of distyly in this genus. In any case, such a view would entail a form of group selection, since loss of the mid morph would be the proximate factor increasing the level of legitimate pollen flow. Since distylous species of *Lythrum* have one or two whorls of stamens, it is clear that different selective pressures may occur among closely related species.

A possible explanation for the evolution of distyly involving loss of a stamen whorl is suggested by Baker (1959) for distylous *Mussaenda chippii* (Rubiaceae).

Lack of pollen transfer from the anthers of the long-styled morph to stigmas of the short-styled morph was hypothesized to explain the evolution of dioecy in this species. The long-styled form, which receives adequate pollen loads, evolves into the female form, while the short-styled form serves primarily as a pollen donor and evolves toward complete maleness. A modification of Baker's argument could apply to distylous species of *Lythrum* and *Nesaea*, where loss of the short set of stamens may have occurred in long- and mid-styled floral morphs. If transfer of pollen from anthers of the short stamen sets of the mid- and long-styled forms was insufficient to lead to full seed set of the short-styled morph, eventual loss of the latter morph from populations would be predicted. Depending on the nature of the developmental control of the short set of stamens in the remaining long- and mid-styled morphs, either loss of these sets or modification of stamen length would be predicted. Lengthening of the short level of stamens of the mid morph, however, would lead to a form where anthers were closely juxtaposed against the stigmas. Given the disadvantages of this morphological arrangement, loss of the short stamen set may be favored, and subsequent loss of the short stamen set of the long-styled morph would be expected due to pleiotropic effects. Alternatively, natural selection would favor equal investment by the two remaining sexual morphs, leading to loss or reduction in function of the short set of anthers of the long-styled morph.

In *Lythrum rotundifolia* and *Pemphis acidula*, distylous species with two sets of stamens in each floral morph, it seems probable that modifications of trimorphic incompatibility similar to those that have occurred in *Oxalis* (Ornduff 1964; Weller 1976a) may explain loss of the mid-styled morph. A detailed discussion of these incompatibility modifications will be discussed in the next section on the Oxalidaceae.

The remaining genera of the Lythraceae exhibiting heterostyly include *Decodon*, which is uniformly tristylous, and *Rotala*, which is distylous (Ganders 1979). Neither genus has been studied in detail, so that few inferences can be made about the function of tristyly in *Decodon* and the evolution of distyly in *Rotala*.

3 Oxalidaceae

Tristyly and distyly are known in *Averrhoa, Biophytum*, and *Oxalis* in the Oxalidaceae. *Dapania* and *Sarcotheca*, which are sometimes placed with *Averrhoa* in the Averrhoaceae, have distylous breeding systems (Lack and Kevan 1987). Except for *Oxalis* and *Biophytum*, the genera contain only a few species. In the latter genus, the incompatibility relationships of a single species, *B. sensitivum*, have been analyzed (Mayura Devi 1964). The six categories of legitimate crosses yielded seeds, and self-pollinations failed to produce seeds. The remaining categories of illegitimate crosses were not completed, making a full analysis of incompatibility impossible for this species. Homostylous forms of *B. sensitivum* are also known (Mayura Devi and Hashim 1966); capsules were produced by 50% of the flowers included in a bagging experiment, indicating that the homostyles are self-compatible but not necessarily autogamous. Reports of distyly in *Biophytum* (Reiche 1890, cited in East 1940) need confirmation.

Oxalis is the largest genus in the family. The approximately 900 species are distributed fairly evenly between the New World and Old World, primarily in the southern hemisphere. Raven and Axelrod (1974) suggested that the family Oxalidaceae probably evolved before the separation of Africa and South America, which would explain the centers of diversity in these areas. Following fusion of the South American and North American continents there appears to have been a northward migration of several of the South American sections of *Oxalis*, although far fewer species occur on the North American continent compared to South America. A northward migration has occurred in the Old World as well. The reproductive systems of only a few sections of *Oxalis* have been studied in detail. In particular, virtually no South American species have received attention, despite the diversity of species occurring there. The most detailed studies have been of the primarily North American section *Corniculatae* (Eiten 1963; Mulcahy 1964; Ornduff 1964, 1972), the North American species of section *Ionoxalis* (Denton 1973; Ornduff 1983; Weller and Denton 1976; Weller 1976a,b, 1978, 1979, 1980, 1981a,b, 1986). Additional studies of the breeding systems of southern hemisphere species include those of Ornduff on South African species (Ornduff 1973, 1974), and Gibbs' (1976) study of the South American food plant *O. tuberosa*. Thus, the most detailed information is available for species of *Oxalis* occurring on the periphery of the major centers of distribution.

A wide range of modifications, including the evolution of distyly as well as the evolution of self-compatibility, are known to occur in *Oxalis* section *Corniculatae* (Ornduff 1972). Trimorphic species of section *Corniculatae* inhabit stable woodlands, while autogamous species exhibit much broader distributions and frequently become weedy. These considerations, as well as the details of the reproductive systems of several tristylous and distylous species indicate that tristyly is the ancestral condition in this section (Mulcahy 1964; Ornduff 1964, 1972). No species in section *Corniculatae* shows retention of all features associated with trimorphism. The species most closely resembling a fully trimorphic species possessing typical trimorphism and the expected incompatibility reactions is *O. suksdorfii*, a species that has been studied in detail by Ornduff (1964, 1972). Populations studied by Ornduff in Oregon were significantly heterogeneous in style form representation ($\chi^2 = 39.2$, df = 12, calculated from surveys in Table 3, Ornduff, 1964; one population was eliminated to minimize cells in the chi-square analysis with less than five individuals). When data from all populations were combined, the floral morphs were equally represented ($\chi^2 = 3.42$, df = 2). Despite the appearance of morphological tristyly in *O. suksdorfii* the incompatibility relationships of this species are substantially modified, and the mid-styled form is virtually seed sterile (Ornduff 1964). In the short-styled morph pollen from the mid and long stamen whorls is equally capable of leading to fertilization of ovules of the long-styled morph. Conversely, pollen from the mid and short stamen whorls of the long-styled morph is equally capable of producing seed when used in pollinations of the short-styled morph (Fig. 3; Table 5 in Ornduff 1964). In the long- and short-styled morphs the mid level stamens have diverged away from the level of mid stigmas; presumably this increases the likelihood that pollen from these whorls will participate in compatible (though illegitimate) pollinations (Ornduff 1972). Because of seed sterility of the

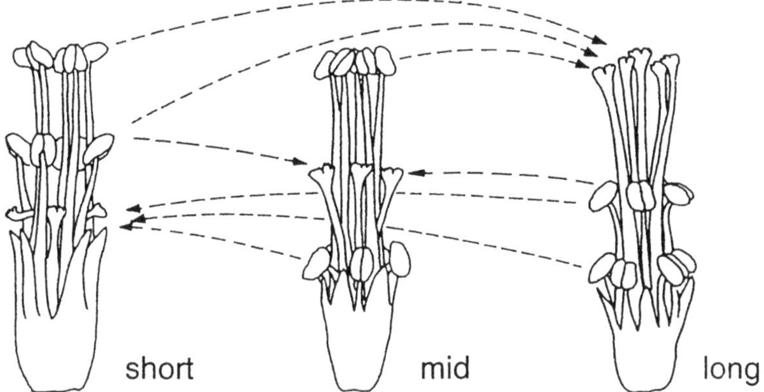

Fig. 3. The reproductive system of *Oxalis alpina* (section *Ionoxalis*), showing incompatibility modifications. *Arrows* indicate directions of pollinations leading to seed production. Note the loss of differentiation in incompatibility reaction between the stamen whorls of the long and short morphs. As a result of incompatibility modifications, two illegitimate crosses yield seed. The reproductive system of *O. suksdorfii* (section *Corniculatae*) is similar, except for nearly complete seed sterility of the mid-styled morph. (After Weller 1976a)

mid-styled morph and divergence of the mid level stamens of the remaining floral forms, the breeding system of *O. suksdorfii* resembles a typical distylous species in several important respects.

The dissimilarity of the breeding system of *Oxalis suksdorfii* to a typical tristylous breeding system is further heightened by the seed sterility of the mid-styled form. Low seed production was found for mids used in artificial crossing programs, and for plants growing under field conditions (Ornduff 1964). Very low mid seed production following legitimate crosses was attributed by Ornduff (1972) to failure of pollen tube growth in styles and failure of embryo sac development. These results indicate that once the incompatibility reactions of pollen grains of the mid stamen levels are modified, pollen grains from mid anther whorls lose the ability to grow in mid styles. Failure of embryo sac development in mid-styled plants may result if the absence of pollen capable of leading to seed production of the mid morph leads to diminished selection for viability of ovules.

These observations present several interesting problems. First, application of sexual selection theory suggests that pollen grains of the mid anther level, regardless of their ability to fertilize ovules of the short- and long-styled morphs, would be selected to fertilize ovules of the mid-styled morph as well. This argument assumes that incompatibility modification results from a change in the ability of stigmas of the short- and long-styled morphs to discriminate among pollen produced by the anther whorls of the opposite form. Loss of incompatibility differentiation is not complete in *O. suksdorfii*, since illegitimate crosses occurring over a distance of two reproductive whorls fail to produce seed (S × l/M, L × s/M, Table 5 in Ornduff 1964). If modified incompatibility results from a change in the incompatibility reaction occurring in the pollen grains, the inability of mid level pollen to grow on mid stigmas is explained more readily.

The second difficulty posed by the reproductive system of *Oxalis suksdorfii* is retention of the mid-styled form. The breeding system of this species is functionally androdioecious, and as Charlesworth (1984) has demonstrated, male fitness of the mid morph would have to be very high in order for the morph to be maintained. Strong self-incompatibility of the short and long morphs eases the conditions for maintenance of a female sterile form, but even in this case it is difficult to understand why the mid morph is retained in populations of *O. suksdorfii*. Clonal growth and infrequent episodes of sexual reproduction may slow loss of the mid morph (Ornduff 1972), but seem unlikely to explain the occurrence of the mid morph in all populations surveyed by Ornduff (1972), even those with very small numbers of individuals. Androdioecy in this species may represent one of the few bone fide cases known (Charlesworth 1984).

Despite the difficulties in interpreting several aspects of the reproductive biology of *Oxalis suksdorfii*, the evidence available indicates that the breeding system of this species is intermediate between tristyly and distyly, and the evolution of distyly will be the eventual outcome of selection favoring the short- and long-styled forms in populations.

Observations of other species in section *Corniculatae* clearly indicate that there may be several pathways by which tristyly may be modified. Morphologically tristylous breeding systems are known for several other species in section *Corniculatae* (Ornduff 1972), although none of these species possesses more than remnant self-incompatibility. In tristylous *Oxalis dillenii* subsp. *filipes* floral morphs show no evidence of convergence of stamen whorls associated with evolution of distyly. Trimorphic pollen is retained to a certain degree; in the populations Ornduff studied loss of pollen size differentiation was found in either the short or long morphs in each case. Flower size in *O. dillenii* subsp. *filipes* varies throughout the season; in the latter part of the season natural selfing rates may be high due to a reduction in flower size and the consequent increases in the level of self-pollination (Ornduff 1972). Some races of this subspecies are consistently autogamous, when one stamen whorl occurs at the same level as the stigmas (semi-homostyly), or when the flower is reduced in size and the reproductive whorls are juxtaposed against one another, even though the spatial relationships of tristyly are maintained (quasi-homostyly, Ornduff 1972). The spatial relationships of the styles and stamens of the autogamous species of *Oxalis* section *Corniculatae* argue against the concept of a crossover in a supergene explaining the evolution of homostyly. Accumulation of small mutations affecting positioning of the reproductive whorls seems more likely.

Crosses of two of the three forms of tristylous *Oxalis grandis* indicated a slight degree of self-incompatibility in the long-styled morph (Ornduff 1972). Mulcahy (1964) carried out legitimate and illegitimate crosses of trimorphic *O. priceae* subsp. *colorea*, a species lacking strong self-incompatibility. Self-pollinations of the mid form of this subspecies were found to produce fewer surviving offspring than the self-pollinations of the remaining forms. From the data presented, it is not clear whether residual self-incompatibility or inbreeding depression contributed to the reduction in offspring number. In either case, Mulcahy (1964) suggested that evolution of distyly in the related *O. priceae* subsp. *priceae* may have resulted from the increased tendency toward clonal growth. If clonal growth led to greater geitonogamous pollen flow, higher levels of inbreeding depression or greater self-incompatibility could result in selection against the mid form relative to the short and long forms. As in the case of *O. suksdorfii*, loss of the mid form in *O. priceae* subsp. *priceae* has been accompanied by an adjustment in the positions of the mid stamen whorls of the short and long forms, such that a single level of anthers is approximated in each of the forms (Mulcahy 1964; Ornduff 1972). Presumably, the adjustments result in a higher proportion of pollen from the mid anther whorls reaching compatible stigmas. Mid level stamens show less convergence toward the upper and lower stamen whorls in distylous *O. priceae* subsp. *texana*, perhaps indicating the more recent evolution of distyly in this subspecies (Ornduff 1972). Much less is known of the details of heterostyly in the remaining taxa of section *Corniculatae*, although several species are autogamous and weedy in nature. South American species may show the same degree of variation in breeding systems as the North American species (Ornduff 1972). Further study of the South American species of section *Corniculatae* is warranted.

Information presently available for section *Corniculatae* indicates a diversity of pathways resulting in modification of tristyly, including the evolution of distyly by at least two different mechanisms, and the evolution of autogamy. A comparison of *Oxalis suksdorfii*, with subspecies of *O. priceae* possessing distyly and tristyly, indicates that loss of incompatibility differentiation between the mid and long stamen whorls within the short morph, and between the short and mid stamen whorls within the long-styled morph, is not essential for the evolution of distyly. Apparently, complete or near complete loss of self-incompatibility does not preclude evolution of distyly, as in the case of *O. priceae* subsp. *priceae* and *O. priceae* subsp. *texana*. While there is no direct evidence that the ancestor of the two distylous subspecies of *O. priceae* was self-compatible, the occurrence of self-compatibility in trimorphic *O. priceae* subspecies *colorea* makes this likely. If self-incompatibility was lost before the evolution of distyly, the adjustments in stamen lengths found in *O. priceae* subsp. *priceae* and *O. priceae* subsp. *texana* are difficult to understand, since loss of self-incompatibility would be expected to diminish selection for the modifications of stamen length noted in both distylous subspecies.

Available population surveys of tristylous, self-compatible species in section *Corniculatae* provide few clues for factors underlying the evolution of distyly. *Oxalis dillenii* subsp. *filipes* showed excesses of the long-styled morph, but no indication of reduced mid-styled morph frequency relative to the remaining morphs (Ornduff 1972). Faberge's (1959) extensive survey of a population of *O. grandis* also showed a significant excess of the long-styled morph, but no shortage of the mid-styled morph relative to the short-styled morph, and Ornduff's survey of two small populations of the same species failed to reveal any bias against the mid-styled morph.

A more complete analysis of potential residual self-incompatibility, as well as an analysis of inbreeding depression and outcrossing rates among the tristylous, self-compatible species of *Oxalis* section *Corniculatae* would be of interest, and help determine whether selection occurs against the mid form due to inbreeding, as suggested by Mulcahy (1964). A problem with Mulcahy's argument for the evolution of distyly is that the level of inbreeding depression resulting from self-fertilization is likely to change, as selection purges deleterious alleles from the population (Lande and Schemske 1985). Such a process would be especially likely in the mid morph, where self-pollination is likely to be higher relative to the short and long morphs, since the stigmas are located between two whorls of anthers. The dynamics of this situation are clearly complex, and a modeling approach would be useful for understanding the interactions of selfing rates and expression of inbreeding depression among the morphs. Until the occurrence of inbreeding depression rather than residual self-incompatibility can be verified for *O. priceae* subsp. *priceae*, these arguments are probably premature. The occurrence of significant interspecific variation in the strength of self-incompatibility is apparent, however, when comparisons are made between species such as *O. suksdorfii* and the eastern North American species of *Oxalis* section *Corniculatae*.

Modifications of heterostylous reproductive systems have been studied in detail in the North American *Oxalis* section *Ionoxalis*. Denton (1973) recognized 25 North American taxa of section *Ionoxalis*, and based on floral morphology, documented the occurrence of heterostyly among the North American taxa. Approximately half of the taxa were noted as tristylous; the remaining species were described as distylous, with the exception of three apparently homostylous species. Despite the occurrence of morphological homostyly, these species are likely to be self-incompatible or completely sterile (Denton 1973). Two of the three morphologically homostylous species produce no seed, based on studies of herbarium specimens, and may represent sterile chromosomal races perpetuated by clonal reproduction in a manner similar to sterile pentaploid races of *O. pes-caprae*, originally native to South Africa (Ornduff 1987). The remaining morphologically homostylous species of *Oxalis* section *Ionoxalis*, based on observations of herbarium specimens, has about the same level of capsule production as fully heterostylous species, indicating that homostyly is not necessarily associated with the acquisition of self-compatibility (Denton 1973).

Artificial crossing programs have been used to analyze the reproductive systems of 13 of the 20 heterostylous species in *Oxalis* section *Ionoxalis* (Weller 1976a, 1980, unpubl.; Ornduff 1983). Studies of six tristylous species from southern Mexico without close distylous relatives showed conventional incompatibility relationships: crosses between anthers and stigmas at the same level in the flower produced abundant seed, while self-pollinations and any other crosses resulted in complete, or nearly complete failure of seed production (Weller 1980). In populations of three of these species equal style morph representation was found (Table 1), and there was no evidence for heterogeneity in morph frequencies when populations were subdivided. The incompatibility relationships of *O. nelsonii*, studied by Ornduff (1983), were different from those of the remaining tristylous species from southern Mexico in that two classes of self-pollinations and one class of illegitimate cross-pollination produced considerable seed. Most tristylous species in section *Ionoxalis* are diploid and restricted to southern Mexico. In contrast, distylous species usually have higher chromosome numbers, and often have very wide-ranging distributions. The association of higher ploidy levels with distyly indicates that distyly is a derived condition in section *Ionoxalis* (Weller and Denton 1976).

The incompatibility relationships of distylous populations of six species were studied in detail to ascertain whether residual traces of trimorphic self-incompatibility could be detected (Weller 1976a, unpubl.). Based on field observations

Table 1. Style morph representation in trimorphic species of *Oxalis* section *Ionoxalis* from southern Mexico

Species	Style Morph			χ^2	Prob.
	S	M	L		
Oxalis lasiandra					
Site 1	11	17	13		
Site 2	9	7	6		
Total	20	27	19	1.73	ns
Oxalis lunulata					
Site 1	24	20	21		
Site 2	23	18	32		
Total	47	38	53	2.48	ns
Oxalis magnifica					
Site 1	9	11	6	1.46	ns

(*O. alpina*) or herbarium specimens (*O. gregaria*), two of the species possessed tristylous as well as distylous populations. Despite the presumably recent derivation from tristyly, and the intraspecific nature of the variation, no evidence of residual trimorphic incompatibility was found. Crosses involving the mid stamen whorl of the long and short floral morphs were as fecund as legitimate crosses involving the uppermost and lowermost stamen whorls (Weller 1976a, unpubl.)

A potential difficulty in comparisons across species is uncertainty over the phylogenetic relationships. A similar problem exists among the Mexican populations of *O. alpina*, which show substantial variability in ploidy level (Weller and Denton 1976). To avoid comparisons among populations of potentially polyphyletic origin, the incompatibility systems of uniformly tetraploid tristylous and distylous populations of *O. alpina* from southeastern Arizona were studied in detail (Weller 1976a, 1978, 1979, 1981a,b, 1986). An inter-populational crossing program indicated that these populations share a common, tristylous ancestor (Weller 1978). Heterostylous breeding systems show great variability among the populations, which occur at elevations above about 2000 m, and are isolated by stretches of lower elevation desert. Tristyly is known in the Chiricahua, Huachuca, and White Mts.; in the remaining ranges only short and long morphs are encountered. In areas with tristylous populations, the mid form varies greatly in frequency, ranging from a high of 46% in the Huachuca Mts., to values as low as 1–4% in the Chiricahua Mts. The frequency of the mid morph shows clinal variation in the Chiricahua Mts.; at the south end of the range mids occur at a frequency of 22%, but drop steadily to the north, and finally disappear from populations (Weller 1979). The variability encountered in the expression of tristyly in this region presents an unparalleled opportunity for investigating the factors responsible for the evolution of distyly in section *Ionoxalis*.

Results from crosses demonstrated that incompatibility relationships in trimorphic populations are modified and resemble the incompatibility relationships characteristic of distylous species, even before loss of the mid morph occurs (Weller

1976a). Loss of incompatibility differentiation in the short and long morphs is the major change that occurs (Fig. 3); in two of the three populations of *O. alpina* investigated, spread of incompatibility modifiers appeared to be complete (Weller 1976a). In the third population from the Huachuca Mts., where mid morphs occurred in the highest frequency, a few short and long individuals appeared to retain incompatibility differentiation.

The breeding system of *O. alpina* is similar to that of *O. suksdorfii*, except that mid morphs of *O. alpina* are as fecund as the short and long morphs (Weller 1981a). Pollen from the mid level anthers of the short and long morphs is capable of effecting fertilization when used in illegitimate pollinations of stigmas of the long and short morphs, as well as stigmas of the mid morph. Interpretation of compatibility relationships of the mid morph is complicated by uniform seed sterility of some individuals, regardless of the nature of the cross (Weller 1976a). Inadvertent use of these individuals in the crossing program resulted in low seed set and the appearance of incompatibility among some legitimate crosses (M × m/S, Table 2; M × m/S and M × m/L, Table 4 in Weller 1976). Additional controlled crosses have revealed that there is no association of seed sterility with any morph (Weller, unpubl.). Reduced seed production could not be detected for of any of the morphs when natural fecundity was investigated (Weller 1981a). Because pollen from mid level anthers can effect fertilizations of the short and long morphs in reciprocal crosses, but retains function in pollinations of the mid morph, it seems likely that stigmas of shorts and longs have been modified (Weller 1976a). This would mean that stigmas of the long morph, for example, no longer differentiate between pollen grains produced by the mid and long anther whorls of the short-styled morph, but retain the ability to discriminate against pollen produced by the anthers of the short stamen whorl of the mid-styled form. Charlesworth (1979) considered this modification of stigmas unlikely, and suggested that modification of the incompatibility reaction of pollen grains was more likely, but this interpretation may have been complicated by her difficulty in assessing incompatibility reactions of the mid-styled morph, due to seed sterility of morphs and the sampling problem described above. A change in the incompatibility reaction of the pollen grains of *O. suksdorfii* does seem very likely, since there is no question that all mid individuals of this species produce very few seeds (Ornduff 1964, 1972; Charlesworth 1979).

Regardless of the mechanism for incompatibility modification in *Oxalis alpina* from southeastern Arizona, the modifications that have occurred should result in selection against the mid morph. Loss of incompatibility differentiation in the short and long morphs gives these forms the potential to outcompete the mid morph in outcrosses, leading to the prediction that the mid morph should disappear eventually from populations with modified incompatibility (Weller 1976a). Loss of the mid morph could also occur, especially under conditions of pollinator limitation, because more pollen in populations can effect fertilizations of the short and long morphs. Charlesworth (1979) used computer simulations of the breeding system of *O. alpina* with modified incompatibility, and also predicted that loss of the mid morph should occur. Once the incompatibility modifier was fixed in populations, factors leading to retention of the mid morph included abundant pollination of all forms, highly specific pollen transfer among the whorls, and highly specific incompatibility relationships (Charlesworth 1979).

Empirical tests of models for loss of the mid morph from populations of *Oxalis alpina* have produced mixed results. Some potential factors are easily eliminated: all morphs have equivalent seed production under field conditions (Weller 1981a), ruling out the possibility that reduced levels of pollination result in selection against the mid morph. Differential visitation by pollinators was also ruled out, since the solitary bees that visit *O. alpina* flowers show no preference for any of the forms (Weller 1981b).

Progeny testing was carried out in several populations to test the possibility that the mid morph was under-represented in the progeny of long and short morphs (Weller 1986). Different patterns of pollen flow were assumed, since the differences in size of pollen produced by stamen whorls occurring at different levels are insufficient for analysis of pollen flow. In the first case, expected progeny distributions were calculated on the basis of completely legitimate pollen flow (Fig. 4, legitimate model); in the second case, pollination was assumed to be random, such that all pollen grains capable of effecting fertilization had an equal probability of being deposited on a stigma (Fig. 4, compatible model). Completely legitimate pollen flow would favor retention of the mid morph (Charlesworth 1979), but seems unlikely based on observations of pollinators of *O. alpina*, which show few of the precise movements expected to contribute to legitimate pollen flow in a heterostylous species (Ganders 1979; Weller 1981b). Models for pollen flow do not affect the mid morph, because only a single whorl of pollen is capable of effecting fertilization of this form, and the only compatible pollinations are also legitimate. In addition to the pattern of pollen flow, segregations expected based on the genetic system underlying expression of tristyly in *Oxalis* section *Ionoxalis* were used to calculate progeny distributions (Weller 1976b).

Three tristylous populations were progeny-tested for morph frequencies over a 3-year period, and two tristylous populations were tested over 2 years, yielding 13 data sets. In eight out of 13 cases, morph frequencies among the progeny differed from the parental distributions; in seven out of eight cases the mid morph occurred

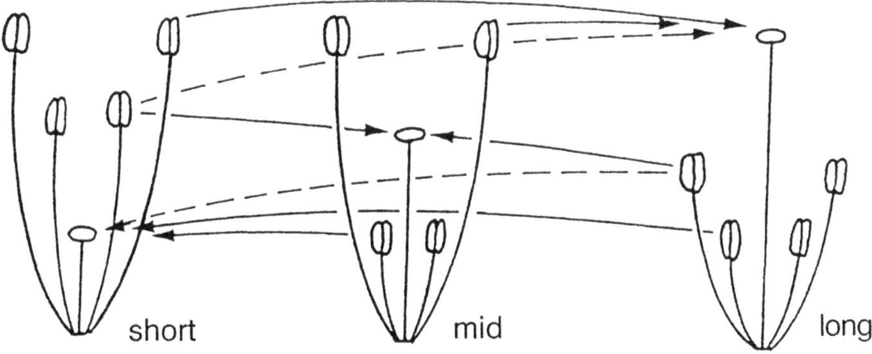

Fig. 4. Diagrammatic representation of the breeding system of *O. alpina* showing the legitimate (*solid lines*) and compatible (*dashed* and *solid lines*) models of pollen flow. (After Weller 1986)

in greater than expected numbers (Weller 1986). Hence although models based on loss of incompatibility modification predicted reduced frequency of the mid morph, the reverse result was obtained. A more detailed analysis of three tristylous populations revealed the cause of greater than expected frequencies of the mid morph. In two of the three populations, where the mid morph occurred in reduced numbers (11–22%), mid morphs produced significant excesses of mid progeny among their offspring (Weller 1986). In the remaining population, from the Huachuca Mts., where the frequency of the mid morph is much higher (24–46%), mid excess did not occur among the progeny.

Interpretation of segregation patterns among progeny from short and long morphs depends on whether pollen flow follows the legitimate versus compatible models. If the compatible model (all pollen grains capable of effecting fertilization have an equal probability of being deposited on a stigma) is accepted as more realistic, mid excess is typical in the progeny of short and long morphs in tristylous populations with reduced representation of the mid morph (Weller 1986). With this model of pollen flow, segregations from the Huachuca Mts. population did not deviate from those expected in most cases. These results can be categorized as follows: in populations where the mid morph is relatively uncommon, mid excess was frequent among the progeny. When the mid morph is more common, and loss of incompatibility differentiation in shorts and longs may not be complete, mid excess was rare. One potential explanation for mid excess is gametophytic selection favoring pollen tubes from mid individuals, or pollen tubes carrying the mid allele. Preferential abortion of ovules carrying non-mid alleles could also contribute to mid excess. Selection might favor such mechanisms if modified incompatibility resulted in selection against the mid morph. The frequency of the mid morph in populations would depend on the balance of these opposing forces, and loss of the mid morph from populations with modified incompatibility may not be inevitable (Weller 1986). Further information on pollen tube growth rates and selective ovule abortion in *O. alpina* would be useful in resolving these issues.

Measurements of pollen size and floral dimensions of the floral morphs of *Oxalis alpina* in populations with varying frequencies of the mid morph provide evidence that modifications of floral morphology occur in concert with changes in frequency of the mid morph. Indices of convergence in stamen length and pollen size for the long and short morphs were negatively correlated with the frequency of the mid morph (Weller 1979), suggesting that floral morphology is a sensitive indicator of morph representation. For this reason, it would be difficult to argue that populations of *O. alpina* have retained the mid morph through the inefficient action of selection, perhaps as a result of clonal growth or low rates of establishment from seed. The present distribution of tristyly and distyly in southeastern Arizona seems more likely to result from a dynamic equilibrium incorporating differing selective forces.

The model for loss of tristyly described for *Oxalis alpina* may well explain the origin of distyly in other species in section *Ionoxalis*. Apparently, loss of incompatibility differentiation is common in the section; the instability of this feature is perhaps not surprising, given the restriction of incompatibility differentiation to tristylous species. The generality of the model could be tested using *O. gregaria*, a species from central Mexico with apparently monophyletic populations showing similar variation in reproductive systems. Nothing is known of the South American species of section *Ionoxalis*, and it would be of great interest to determine the distribution of heterostylous reproductive systems among these species. In the North American species of section *Ionoxalis*, self-incompatibility appears to be very strong in nearly all of the species that have been tested. There is no evidence for the evolution of distyly following an intermediate self-compatible stage, as in the case of several distylous species in section *Corniculatae*. The strength and pervasiveness of the incompatibility system is a major distinguishing feature between section *Ionoxalis* and section *Corniculatae*, despite the remarkable similarities in the reproductive systems of *O. alpina* and *O. suksdorfii*.

Among the South American species of *Oxalis* only *O. tuberosa* (section *Ortgieseae*) has been studied in detail (Gibbs 1976). Although this species has been cultivated asexually for hundreds and possibly thousands of years, typical trimorphic incompatibility was present in plants studied by Gibbs (1976). The morphological expression of tristyly was typical of tristylous species showing strong self-incompatibility. Without more information on the relatives of *O. tuberosa* and other South American species of *Oxalis* further generalizations are impossible.

As a result of Ornduff's (1974) studies of South African *Oxalis*, and Salter's (1944) taxonomic treatment of *Oxalis*, more is known of the reproductive systems of these species. Apparently, South African *Oxalis* are uniformly trimorphic (Marloth 1925; Ornduff 1974; Salter 1944), although trimorphism does not necessarily imply that incompatibility occurs in all species. In surveys for potential incompatibility, Ornduff found that eight species showed strong self-incompatibility, one species showed some seed production following selfing, and one species appeared to be autogamous, despite the occurrence of structural trimorphism (Ornduff 1974). In the latter case, the mechanism of autogamy was not investigated, although the breeding system of this species might be similar to the quasi-homostylous species of section *Corniculatae*. Salter (1944) found that three additional species showed strong self-incompatibility. In surveys of 15 populations of nonweedy South African *Oxalis* species, Ornduff (1974) found that isoplethy occurred in slightly over half the cases. In a few cases, the long morph predominated very strongly. If Ornduff's and Salter's sample of South African *Oxalis* species is representative, trimorphism accompanied by strong self-incompatibility appears to be widespread, although more information on the remaining 300 to 400 species would be useful.

The contrast between the North American sections of *Oxalis* that have been studied in detail, and the South African species is striking. In North America, and perhaps in the related South American species, evolution of distyly has occurred on several occasions in unrelated sections. In South Africa, where an equally diverse array of *Oxalis* species is found, there is no evidence for distyly. In both regions, modifications of heterostyly have occurred involving loss of incompatibility, but in South Africa there is no evidence for the occurrence of incompatibility modifiers that appear to be responsible for the evolution of distyly in sections *Corniculatae* and *Ionoxalis*.

4 Pontederiaceae

Evolutionary modifications of tristyly in the Pontederiaceae, which have been analyzed thoroughly by Barrett and his associates, involve acquisition of self-compatibility and loss of floral trimorphism, but there is no evidence for the evolution of distyly in the family. Of the nine genera in the family, tristyly is found in *Eichhornia* and *Pontederia*. Four species of *Pontederia* retain varying levels of self-incompatibility (Barrett and Anderson 1985); a fifth species, *P. parviflora*, is homostylous and presumably fully self-compatible (Lowden 1973). Species of *Eichhornia* are moderately to strongly self-compatible (Barrett 1977a, 1978, 1985).

The reproductive system of *Pontederia cordata* has been studied in great detail (Ornduff 1966; Price and Barrett 1982; Barrett et al. 1983; Price and Barrett 1984; Barrett and Glover 1985; Barrett and Anderson 1985; Morgan and Barrett 1988). Self-incompatibility is most strongly developed in the short and long morphs, although self-pollinations using pollen from the anthers closest to the stigmas yielded moderate levels of seed production for both of the forms (Ornduff 1966; Barrett and Anderson 1985). Corresponding illegitimate cross-pollinations resulted in similar patterns of seed production. Self-pollinations of the mid morph using pollen from the long anthers, and corresponding illegitimate cross-pollinations of the mid form yielded large numbers of seeds relative to the other forms (Ornduff 1966; Barrett and Anderson 1985). Substantial variability was detected in the strength of self-incompatibility; among 12 clones of long-styled plants for example, self-fertility ranged from 4 to 94% (Barrett and Anderson 1985). Several lines of evidence indicate that variation in self-incompatibility in *P. cordata* is genetically based (Barrett and Anderson 1985). All three style forms were represented in most populations of *P. cordata* studied by Barrett et al. (1983). Significant heterogeneity in morph frequency characterized the populations, and overall, the short-styled morph occurred in significant excess and the long-styled morph was under-represented, a result attributed to the greater pollen production of the mid anther level of the short morph relative to the mid anther level of the long morph (Barrett et al. 1983), as well as historical factors (Morgan and Barrett 1988).

The breeding systems of *Pontederia sagittata* and *P. cordata* are very similar. The mid morph showed high seed production following self-pollinations using pollen from the long anthers (Glover and Barrett 1983). Intermorph illegitimate crosses of mids using pollen from long anthers of the short and mid morphs also produced abundant seed. Self-incompatibility in the short and long morphs was very strong. Floral morph frequencies were sampled in eight populations, and in six of the eight populations the morphs were equally frequent. There was a tendency for under-representation of the long morph in all populations, and when results were pooled the deficiency of longs was significant (Glover and Barrett 1983). As in the case of *P. cordata*, the deficiency of long morphs in the populations may result from unequal pollen production of mid level anthers from the short and long morphs (Glover and Barrett 1983). An analysis of pollen loads by Glover and Barrett (1983) indicated that considerable illegitimate pollination occurred in populations. Stigmas of long-styled individuals received a greater proportion of legitimate pollen than the short and mid morphs. Self-incompatibility and perhaps strong inbreeding depression among progeny produced via selfing are likely to explain the near equality of morphs typical of populations of *P. sagittata*.

In *P. rotundifolia*, strong floral trimorphism is present, although the incompatibility system is weaker than in *P. cordata* (Barrett 1977b). As in the case of the preceding two species, self-pollinations and illegitimate cross-pollinations of mids using pollen from long anthers produced far more seed than other illegitimate combinations. Morphs were equally represented in two disturbed agricultural environments in the Lower Amazon; in contrast, significant anisoplethy was found in less disturbed marsh and riverbank populations from Costa Rica (Barrett 1977b). The role of differential pollen production in producing anisoplethy was not investigated in this species. No consistent bias toward the short morph could be detected in the population surveys, although the number of populations surveyed was small. Fruit set was highest in the isoplethic populations, although some fruit set occurred in a population composed of only the short morph. The degree of self-compatibility and levels of self-pollination were obviously sufficient to allow some fruit production in the absence of one or more of the floral morphs (Barrett 1977b). The occurrence of populations monomorphic for a floral morph attests to a potentially limited role for establishment from seed under some conditions. Given the degree of variation in self-incompatibility, and the likely occurrence of considerable illegitimate pollen flow, the maintenance of isoplethy in any populations of *P. rotundifolia* is surprising.

Barrett and Anderson (1985) attribute the greater seed production of the mid morph of *Pontederia* species following self- or illegitimate pollinations to the different positional relationships and developmental patterns of the mid anther levels of the short and long morphs. Flowers of *Pontederia* are zygomorphic and the shorter set of stamens is inserted on the upper side of the perianth. The longer set of stamens is inserted on the lower side of the perianth. Two of the three stamens inserted on the upper side of the perianth are attached to narrow tepals; the remaining stamen is attached to a broad tepal. The situation is reversed for stamens attached to the lower side of perianth. Because of these spatial relationships, the mid level set of stamens differs in tepal attachment for the short and long floral morphs. Barrett and Anderson (1985) hypothesize that such differences as the greater pollen production of the mid anthers of the short floral morph can be attributed to the influence of stamen insertion on developmental patterns. They further hypothesize that mid-level pollen may be biochemically heterogeneous, which has selected for mid morphs with a broader range of compatibility responses. If this is true, the ability of the mid morph to produce seed after self-pollinations and several categories of illegitimate cross-pollinations would be explained. Pollen from mid-level anthers of the short and long morphs showed a differentiated response when used in own-form illegitimate crosses and intermorph illegitimate crosses (Table 4 in Barrett and Anderson 1985), providing support for the idea that developmental pathways may influence the nature of the self-incompatibility response. As Barrett and Anderson suggest, it would be of interest to investigate the effect of insertion patterns on the expression of incompatibility in other tristylous species.

Tristyly is found in three species of *Eichhornia* (Barrett 1988). The remaining five species are largely monomorphic. The tristylous species are characterized by a high degree of self-compatibility in most cases, and the frequent loss of floral morphs from populations (Barrett 1977a, 1978, 1985; Barrett and Forno 1982). The occurrence of populations containing one or two of the style morphs is associated with the well-developed tendency towards weediness in *Eichhornia*; *E. crassipes* is a notorious pest occurring in artificial and natural freshwater habitats (Barrett and Forno 1982). The occurrence of semi-homostylous, highly self-compatible variants of the mid morph is associated with weediness in each of the species.

The reproductive system of *Eichhornia azurea* maintains the greatest similarity to trimorphism in *Pontederia* species. Surveys of five populations in Brazil showed that three floral forms were present in each case, although there was no indication of isoplethy in any of the populations, and style form representation was heterogeneous among populations (Barrett 1978). Populations monomorphic for style morphs were also found in the same area. *Eichhornia azurea* lacks the extreme specializations for vegetative spread found in *E. crassipes*, which probably explains the retention of three style forms in most populations (Barrett 1978). Pollinations of a single clone of a long-styled individual showed a high level of self-incompatibility when pollen from short stamens was placed on the stigmas. Abundant seed production occurred when pollen from mid level anthers was used in self-pollinations of the long morph (Barrett 1978). In view of these data, and the occurrence of strong pollen trimorphism in *E. azurea*, Barrett (1978) characterized the reproductive system of this species, as compared to the other tristylous *Eichhornia* species, as most similar to *Pontederia*. It would be of interest to determine the strength of self-incompatibility in the remaining floral morphs of *E. azurea*, and compare values for the mid morph with *Pontederia* species. Semi-homostylous forms of *E. azurea* were found in a single population from Costa Rica. The derivation of this form is not clear, though pollen size measurements of typical tristylous morphs and the semi-homostyle provided by Barrett (1978) indicate that the homostyle may be derived from a short-styled ancestor.

The breakdown of tristyly is more complete in *Eichhornia crassipes* and *E. paniculata*. In both species populations with three reproductive morphs are found only in Brazil, presumably within the native range of the species (Barrett 1977a, 1985). Because of its weedy nature, *E. crassipes* has a very broad distribution, and in the adventive areas only mid and long forms, or more commonly the mid form alone, are found in populations (Barrett 1977a). Extensive population surveys by Barrett and Forno (1982) showed that even within geographic areas where all three forms occurred populations were never isoplethic; three style forms were found in seven of the 123 populations sampled in Brazil. Pollen of *E. crassipes* exhibited weak trimorphism, and there was no evidence for self-incompatibility in any of the floral morphs. The mid morph showed a greater tendency toward autogamy than the short and long morphs (Barrett 1979). Monomorphic populations of *E. crassipes* produced large numbers of seeds, although values were well below those obtained in an artificial crossing program (Barrett 1977a). *Eichhornia crassipes* reproduces from seed in habitats where water levels fluctuate, a characteristic feature of the Amazonian region where this species is presumed to be native. Reproduction by seed is probably a rare event in the adventive regions of the distribution of *E. crassipes*, where stable water levels promote clonal growth and prevent seedling establishment (Barrett 1980).

Eichhornia paniculata is an emergent aquatic with an annual or short-lived perennial life cycle, but in several respects shows similarities to the reproductive system of *E. crassipes*. Three style morphs are found in many parts of Brazil, although at the southern part of the distribution dimorphic and monomorphic populations are encountered commonly (Barrett et al. 1989). Disjunct populations in Jamaica are usually monomorphic for self-pollinating semi-homostylous variants of the mid morph, and show reduced electrophoretic variation compared to mainland populations of *E. paniculata* (Glover and Barrett 1987, Husband and Barrett 1991). Controlled crosses in the greenhouse demonstrated a high degree of self-compatibility (Barrett 1985). Morphological modifications of the mid morph of *E. crassipes* and *E. paniculata* leading to semi-homostyly are likely to account for the predominance of this form in monomorphic populations (Barrett 1979; Barrett et al. 1989). Semi-homostylous mids show a high degree of autogamy. In three dimorphic populations of *E. paniculata*, the long morph had significantly higher outcrossing rates than the mid morph, using a multilocus procedure to estimate outcrossing rates (Glover and Barrett 1986; Barrett et al. 1989). In a Brazilian population mids were composed of a mixture of typical forms and semi-homostylous variants. The differences in outcrossing rates were even greater when mids and longs from Jamaica were compared (Barrett et al. 1989). In these populations, only semi-homostylous mids were found.

Stochastic events during colonization, automatic selection of selfing alleles, and fertility assurance (the assurance of seed production through autogamy in the absence of pollinators) are likely explanations for the breakdown of tristyly and occurrence of populations of *E. paniculata* monomorphic for semi-homostylous mids (Glover and Barrett 1986). Dimorphic populations consisting of longs and self-pollinating variants of the mid morph constitute the intermediate stage between trimorphism and monomorphism (Fig. 5). Computer simulations by Heuch (1980) and Barrett et al. (1989) indicate that reduced population size and periodic founder events would lead to preferential loss of the short allele from populations. This could

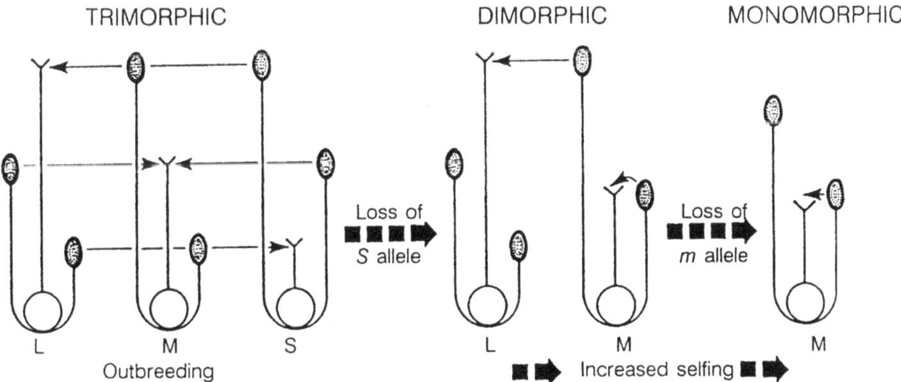

Fig. 5. Schematic representation of the evolutionary breakdown of tristyly in *Eichhornia paniculata*, with *arrows* showing the predominant matings in the population. Note stamen modification of mid-styled morph in the dimorphic and monomorphic populations. (After Glover and Barrett 1986)

explain the absence of the short morph of *E. paniculata* from Jamaica, and the absence of this morph throughout the adventive range of *E. crassipes* (Barrett et al. 1989). In dimorphic populations, the mid morph is favored as a result of morphological modifications leading to high rates of self-fertilization and high levels of seed production under conditions where pollinators are unreliable. This condition appears to apply outside the native ranges of *E. crassipes* and *E. paniculata*, where the long-tongued pollinators capable of depositing pollen on stigmas of all floral forms are absent (Barrett et al. 1989). Automatic selection for selfing variants of the mid morph is likely to occur because only a single set of anthers occurs at the same level as the stigma. The remaining set of anthers is likely to participate fully in pollinations of the long morph (Fig. 5; Barrett et al. 1989), leading to the spread of the selfing morph. This model for the evolution of monomorphism predicts that in dimorphic populations, the frequency of the long morph should be inversely related to the frequencies of self-pollinating variants of the mid morph. This prediction was borne out in a survey of 19 populations containing mids and longs (Barrett et al. 1989). The model for loss of trimorphism and the evolution of monomorphism in *E. paniculata* may well account for the occurrence of monomorphism in the five remaining species of *Eichhornia*, as well as *E. azurea* and *E. crassipes* (Barrett 1988).

A common factor underlying modification of tristylous breeding systems in the Pontederiaceae is the acquisition of self-compatibility, a process which has occurred to a moderate extent in *Pontederia*, and to a great extent in *Eichhornia*. If modifications that have occurred in *Pontederia* are viewed as potentially indicative of the stages that may have taken place in *Eichhornia*, the important role of preferential loss of self-incompatibility in the mid morph becomes obvious. Whatever the causes for greater self-compatibility of the mid morph (Barrett and Anderson 1985), it seems likely that once self-compatibility has arisen, semi-homostylous mutants are favored by a combination of fertility assurance and/or automatic selection. Developmental constraints associated with floral morphology may be of great significance in determining overall evolutionary patterns in the Pontederiaceae (Barrett and Anderson 1985).

5 Comparison of Evolutionary Modifications of Tristyly in Lythraceae, Oxalidaceae, and Pontederiaceae

In comparing the evolutionary modifications that have occurred in the Lythraceae, Oxalidaceae, and Pontederiaceae, the clearest pattern that emerges is the tendency toward loss of the mid morph and evolution of distyly in the first two families, and the complete absence of distyly in the Pontederiaceae. In the latter family, the evolution of self-compatibility and autogamy is a more pronounced trend, with selection favoring the mid morph. The root of these differences probably lies with the nature of self-incompatibility in each of the families. The Lythraceae and Oxalidaceae appear to be similar in that loss of incompatibility differentiation between the stamen whorls of both the short and long morphs may lead to selection against the mid morph. This is clearly not the only pathway to distyly: the occurrence of distylous *Lythrum* species lacking a stamen whorl, and the occurrence of self-compatible, distylous taxa of *Oxalis* section *Corniculatae* that are likely to have evolved from self-compatible, tristylous ancestors indicate different mechanisms for the evolution of distyly. Although generalizations regarding selective factors underlying the evolution of distyly from tristyly are difficult, the contrast with modifications that have occurred in the Pontederiaceae is striking. If Barrett and Anderson (1985) are correct, insertion patterns of stamens may be an important factor in leading to the preferential breakdown of self-incompatibility in the mid morph. Because of the arrangement of stamens and styles in the mid morph, where two sets of anthers are equally close to the stigmas, mutations leading to greater proximity of anthers and stigmas may be common. If so, autogamy, and automatic selection of self-compatible mid morphs would be likely, as appears to have occurred in all heterostylous species of *Eichhornia* to varying extents.

A comparison of *Eichhornia* with *Pontederia* reveals a gap in the sequence of evolutionary events associated with modifications leading to spread of autogamous mid individuals. In *Pontederia*, the mid morph shows the greatest self-compatibility, while the short and long morphs retain substantial self-incompatibility. In *Eichhornia*, all floral morphs are largely self-compatible. The sequence of evolutionary events associated with acquisition of self-compatibility in the short and long morphs is unclear. In *E. crassipes* these changes seem likely to have taken place before the complete disruption of isoplethy and spread of a single style form, since sexual reproduction in most parts of the adventive range of this species appears to be rare (Barrett and Forno 1982). Self-compatibility in the short and long morphs would not necessarily provide an advantage to these forms, since floral morphology is likely to restrict autogamous seed production. Despite this observation, and the fact that monomorphic populations of *E. crassipes* and *E. paniculata* are usually composed of mids, self-compatibility is found in populations most nearly representing the ancestral condition, by virtue of the presence of all three floral morphs.

6 Conclusions

A full understanding of the patterns of evolutionary modification of tristyly clearly awaits more information on the biochemistry and molecular genetics of tristyly. With this information, the significance of incompatibility differentiation in the Lythraceae and Oxalidaceae, and the developmental peculiarities of the mid level stamens in the Pontederiaceae may become clear. At present, biochemical and molecular studies have been restricted to species with homomorphic, multi-allelic self-incompatibility (Nasrallah et al. 1985). The extension of such studies to heteromorphic self-incompatibility would lead to a far greater understanding of these complex, often poorly understood reproductive systems.

Acknowledgments. I thank William R. Anderson, Spencer C. H. Barrett, Melinda Denton, Fred R. Ganders, and Robert Ornduff for many stimulating discussions of heterostylous reproductive systems. I am grateful to Ann K. Sakai, Robert Ornduff, and Spencer C. H. Barrett for reading the manuscript; I thank S. C. H. Barrett for permission to cite unpublished data.

References

Baker HG (1959) Reproductive methods as factors in speciation of flowering plants. Cold Spring Harbor Symp Quant Biol 24:177–191

Barrett SCH (1977a) Tristyly in *Eichhornia crassipes* (Mart.) Solms (water hyacinth). Biotropica 9:230–238

Barrett SCH (1977b) The breeding system of *Pontederia rotundifolia* L., a tristylous species. New Phytol 78:209–220

Barrett SCH (1978) Floral biology of *Eichhornia azurea* (Swartz) Kunth (Pontederiaceae). Aquat Bot 5:217–228

Barrett SCH (1979) The evolutionary breakdown of tristyly in *Eichhornia crassipes* (Mart.) Solms (water hyacinth). Evolution 33:499–510

Barrett SCH (1980) Sexual reproduction in *Eichhornia crassipes* (water hyacinth). II. Seed production in natural populations. J Appl Ecol 17:113–124

Barrett SCH (1985) Floral trimorphism and monomorphism in continental and island populations of *Eichhornia paniculata* (Spreng.) Solms. (Pontederiaceae). Biol J Linn Soc 25:41–60

Barrett SCH (1988) Evolution of breeding systems in *Eichhornia* (Pontederiaceae): a review. Ann MO Bot Gard 75:741–760

Barrett SCH, Anderson JM (1985) Variation in expression of trimorphic incompatibility in *Pontederia cordata* L. (Pontederiaceae). Theor Appl Genet 70:355–362

Barrett SCH, Forno IW (1982) Style morph distribution in New World populations of *Eichhornia crassipes* (Mart.) Solms-Laubach (water hyacinth). Aquat Bot 13:299–306

Barrett SCH, Glover DE (1985) On the Darwinian hypothesis of the adaptive significance of tristyly. Evolution 39:766–774

Barrett SCH, Price SD, Shore JS (1983) Male fertility and anisoplethic population structure in tristylous *Pontederia cordata* (Pontederiaceae). Evolution 37:745–759

Barrett SCH, Morgan MT, Husband BC (1989) The dissolution of a complex genetic polymorphism: the evolution of self-fertilization in tristylous *Eichhornia paniculata* (Pontederiaceae). Evolution 43:1398–1416

Charlesworth D (1979) The evolution and breakdown of tristyly. Evolution 33:486–498

Charlesworth D (1984) Androdioecy and the evolution of dioecy. Biol J Linn Soc 23:333–348

Darwin C (1877) The different forms of flowers on plants of the same species. Murray, Lond

Denton MF (1973) A monograph of *Oxalis*, section *Ionoxalis* (Oxalidaceae) in North America. Publ Mus Mich State Univ Biol Ser 4(10):455–615

Dulberger R (1970) Tristyly in *Lythrum junceum*. New Phytol 69:751–759

East EM (1940) The distribution of self-sterility in flowering plants. Proc Am Philos Soc 82:449–518

Eiten G (1963) Taxonomy and regional variation of *Oxalis* section *Corniculatae*. I. Introduction, keys and synopsis of the species. Am Midl Nat 69:257–309

Faberge AC (1959) Populations of *Oxalis* with floral trimorphism. (Abstr.) Genetics 44:509

Fisher RA, Mather K (1943) The inheritance of style length in *Lythrum salicaria*. Ann Eugenics 12:1–23

Ganders FR (1974) Disassortative pollination in the distylous plant *Jepsonia heterandra*. Can J Bot 52:2401–2406

Ganders FR (1979) The biology of heterostyly. NZ J Bot 17:607–635

Gibbs PE (1976) Studies on the breeding system of *Oxalis tuberosa* Mol. Flora 165:129–138

Gill LS, Kyauka PS (1977) Heterostyly in *Pemphis acidula* Forst. (Lythraceae). Adansonia 17:139–146

Glover DE, Barrett SCH (1983) Trimorphic incompatibility in Mexican populations of *Pontederia sagittata* Presl. (Pontederiaceae). New Phytol 95:439–455

Glover DE, Barrett SCH (1986) Variation in the mating system of *Eichhornia paniculata* (Spreng.) Solms. (Pontederiaceae). Evolution 40:1122–1131

Glover DE, Barrett SCH (1987) Genetic variation in continental and island populations of *Eichhornia paniculata* (Pontederiaceae). Heredity 59:7–17

Haldane JBS (1936) Some natural populations of *Lythrum salicaria*. J Genet 32:393–397

Halkka O, Halkka L (1974) Polymorphic balance in small island populations of *Lythrum salicaria*. Ann Bot Fenn 11:267–270

Heuch I (1979) Equilibrium populations of heterostylous plants. Theor Popul Biol 15:43–57
Heuch I (1980) Loss of incompatibility types in finite populations of the heterostylous plant *Lythrum salicaria*. Hereditas 92:53–57
Husband BC, Barrett SCH (1991) Colonization history and population genetic structure of *Eichhornia paniculata* in Jamaica. Heredity 66:287–296
Koehne E (1903) Lythraceae. In: Engler A (ed) Das Pflanzenreich IV. 216:1–326 W Engelmann, Leipzig
Lack A, Kevan PG (1987) The reproductive biology of a distylous tree, *Sarcotheca celebica* (Oxalidaceae) in Sulawesi, Indonesia. Biol J Linn Soc 95:1–8
Lande R, Schemske DW (1985) The evolution of self-fertilization and inbreeding depression in plants. I Genetic models. Evolution 39:24–40
Lewis D, Rao AN (1971) Evolution of dimorphism and population polymorphism in *Pemphis acidula* Forst. Proc R Soc Lond Ser B 178:79–94
Lowden RM (1973) Revision of the genus *Pontederia* L. Rhodora 75:426–487
Marloth R (1925) Oxalidaceae. In: Flora of South Africa, vol 2. Darter Bros and Co, Capetown, South Africa, pp 92–96
Mayura Devi P (1964) Heterostyly in *Biophytum sensitivum* DC. J Genet 59:41–48
Mayura Devi P, Hashim M (1966) Homostyly in heterostyled *Biophytum sensitivum* DC. J Genet 59:245–249
Morgan MT, Barrett SCH (1988) Historical factors and anisoplethic population structure in tristylous *Pontederia cordata*: A reassessment. Evolution 42:496–504
Mulcahy DL (1964) The reproductive biology of *Oxalis priceae*. Am J Bot 51:1045–1050
Mulcahy DL, Caporello D (1970) Pollen flow within a tristylous species: *Lythrum salicaria*. Am J Bot 57:1027–1030
Nasrallah JB, Kao T-H, Goldberg ML, Nasrallah ME (1985) A cDNA clone encoding an *S*-locus-specific glycoprotein from *Brassica oleracea*. Nature (Lond) 318:263–267
Ornduff R (1964) The breeding system of *Oxalis suksdorfii*. Am J Bot 51:307–314
Ornduff R (1966) The breeding system of *Pontederia cordata*. Bull Torrey Bot Club 93:407–416
Ornduff R (1972) The breakdown of trimorphic incompatibility in *Oxalis* section *Corniculatae*. Evolution 26:52–65
Ornduff R (1973) *Oxalis dines*, a new species from the western cape. J S Afr Bot 39:201–203
Ornduff R (1974) Heterostyly in South African flowering plants: a conspectus. J S Afr Bot 40:169–187
Ornduff R (1975) Pollen flow in *Lythrum junceum*, a tristylous species. New Phytol 75:161–166
Ornduff R (1978) Features of pollen flow in dimorphic species of *Lythrum* section *Euhyssopifolia*. Am J Bot 65:1077–1083
Ornduff R (1979) The morphological nature of distyly in *Lythrum* section *Euhyssopifolia*. Bull Torrey Bot Club 106:4–8
Ornduff R (1983) Heteromorphic incompatibility in *Oxalis nelsonii*. Bull Torrey Bot Club 110:214–216
Ornduff R (1987) Reproductive systems and chromosome races of *Oxalis pes-caprae* L. and their bearing on the genesis of a noxious weed. Ann MO Bot Gard 74:79–84
Price SD, Barrett SCH (1982) Tristyly in *Pontederia cordata*. Can J Bot 60:897–905
Price SC, Barrett SCH (1984) The function and adaptive significance of tristyly in *Pontederia cordata* L. (Pontederiaceae). Biol J Linn Soc 21:315–329
Raven PH, Axelrod DI (1974) Angiosperm biogeography and past continental movements. Ann MO Bot Gard 61:539–673
Reiche K (1890) Geraniaceae Engler and Prantl. Abs Just Bot Jahresber Leipzig, XVIII, 1892, pp 505–506
Salter TM (1944) The genus *Oxalis* in South Africa. J S Afr Bot (Suppl) 1:1–355
Schoch-Bodmer H (1938) The proportions of long-, mid- and short -styled plants in natural populations of *Lythrum salicaria* L. J Genet 36:39–42
Spieth PT, Novitski E (1969) Remarks on equilibrium conditions in certain trimorphic, self-incompatible systems. In: Morton NE (ed) Computer applications in genetics. Univ Hawaii Press, Honolulu, pp 161–167
Weller SG (1976a) Breeding system polymorphism in a heterostylous species. Evolution 30:442–454

Weller SG (1976b) The genetic control of tristyly in *Oxalis* section *Ionoxalis*. Heredity 37:387–393
Weller SG (1978) Dispersal patterns and the evolution of distyly in *Oxalis alpina*. Syst Bot 3:115–126
Weller SG (1979) Variation in heterostylous reproductive systems among populations of *Oxalis alpina* in southeastern Arizona. Syst Bot 4:57–71
Weller SG (1980) The incompatibility relationships of tristylous species of *Oxalis* section *Ionoxalis* of southern Mexico. Can J Bot 58:1908–1911
Weller SG (1981a) Fecundity in populations of *Oxalis alpina* in southeastern Arizona. Evolution 35:197–200
Weller SG (1981b) Pollination biology of heteromorphic populations of *Oxalis alpina* (Rose) Knuth (Oxalidaceae) in southeastern Arizona. Bot J Linn Soc 83:189–198
Weller SG (1986) Factors influencing frequency of the mid-styled morph in tristylous populations of *Oxalis alpina*. Evolution 40:279–289
Weller SG, Denton MF (1976) Cytogeographic evidence for the evolution of distyly from tristyly in the North American species of *Oxalis* section *Ionoxalis*. Am J Bot 63:120–125

Subject Index

www.ingramcontent.com/pod-product-compliance
Ingram Content Group UK Ltd.
Pitfield, Milton Keynes, MK11 3LW, UK
UKHW061701190726
13853UKWH00008B/2346

* 9 7 8 3 6 4 2 8 6 6 5 7 9 *